Microbes and Enzymes for Water Treatment and Remediation

The introduction of emerging contaminants through anthropogenic activities and industrial discharges has raised significant public health concerns worldwide. Various techniques, including bioremediation, have been explored for their effectiveness in removing pollutants from water bodies and effluents. Microorganisms, particularly bacteria and fungi, have emerged as promising candidates for bioremediation due to their abundance, diversity, and ability to thrive under various conditions. *Microbes and Enzymes for Water Treatment and Remediation* covers a range of topics, from the role of microorganisms and enzymes in efficient pollutant removal to recent advances in microbial immobilization and enzymatic systems for enhanced wastewater treatment. The book provides up-to-date insights into the potential of microbial and enzyme-based processes for wastewater treatment, addressing challenges and limitations while offering alternative methods for effluent treatment and water reclamation. It serves as a valuable resource for understanding the interplay between microbial, biological, and chemical components in the remediation of toxic aqueous pollutants, aiding both researchers and industrialists in advancing environmental stewardship efforts.

- Offers comprehensive coverage of emerging contaminants and their impact on public health.
- Provides in-depth exploration of bioremediation techniques utilizing microbial and enzymatic pathways.
- Addresses the limitations and challenges of the existing microbial and enzyme-based processes for wastewater treatment.

Microbial Biotechnology for Food, Health, and the Environment Series

Series Editor: Ashok Kumar Nadda

Plant-Microbial Interactions and Smart Agricultural Biotechnology
Edited by Swati Tyagi, Robin Kumar, Baljeet Saharan, Ashok Kumar Nadda

Microbial Products
Applications and Translational Trends
Edited by Mamtesh Singh, Gajendra Pratap Singh, Shivani Tyagi

Advances in Nanotechnology for Smart Agriculture
Techniques and Applications
Edited by Parul Chaudhary, Anuj Chaudhary, Ashok Kumar Nadda, and Priyanka Khati

Biohydrometallurgical Processes
Metal Recovery and Remediation
Edited by Satarupa Dey, Abhijit Dey, and Ashok Kumar Nadda

Microbes and Enzymes for Water Treatment and Remediation
Edited by Ashok Kumar Nadda, Priya Banerjee, and Swati Sharma

For more information about this series, please visit: www.routledge.com
https://www.routledge.com/Microbial-Biotechnology-for-Food-Health-and-the-Environment/book-series/CRCMBFHE

Microbes and Enzymes for Water Treatment and Remediation

Edited by
Ashok Kumar Nadda,
Priya Banerjee, and Swati Sharma

CRC Press is an imprint of the
Taylor & Francis Group, an informa business

Designed cover image: Shutterstock

First edition published 2025
by CRC Press
2385 NW Executive Center Drive, Suite 320, Boca Raton FL 33431

and by CRC Press
4 Park Square, Milton Park, Abingdon, Oxon, OX14 4RN

CRC Press is an imprint of Taylor & Francis Group, LLC

ISBN: 978-1-032-85075-7 (hbk)
ISBN: 978-1-032-85237-9 (pbk)
ISBN: 978-1-003-51723-8 (ebk)

DOI: 10.1201/9781003517238

Typeset in Times
by codeMantra

Contents

Preface

Anthropogenic activities and industrial discharges have resulted in the introduction of emerging contaminants into the natural environment in alarming quantities. Emerging contaminants in water, air, and land pose severe public health concerns. These pollutants have been widely studied on a global scale and researchers have attempted to eradicate these from the environment using various strategies. However, a complete solution has not been found to deal with continuously arising pollutants as of yet. In order to comply with the stringent environmental regulations and standards, several techniques have been investigated and implemented for removal of such pollutants from effluents and water bodies. In comparison to all other conventional methods of water treatment, bioremediation has gained significant attention in recent decades owing to its minimal chemical requirement, cost-effectiveness, eco-friendly implementation, and non-formation of solid sludge by-products. Bioremediation processes depend on utilization of biological mechanisms (live or dead biomass) to transform/degrade/detoxify pollutants to innoxious compounds, and subsequent mineralization of the transformed products to carbon dioxide, nitrogen, water, etc. Plants, fungi, bacteria, algae, and cyanobacteria and enzymes derived from the same have been widely investigated and applied for bioremediation of effluents. Microorganisms have been found to be most suitable for this purpose due to their convenient handling, culturing, and implementation. In recent years, microorganisms have gained recognition for their ability to adsorb, solubilize, and precipitate pollutants via processes like bioadsorption (biosorption), bioaccumulation, bioleaching, and bioprecipitation. Ubiquitous presence, abundance, diversity, small size, and unique ability to grow and propagate under both controlled and uncontrolled conditions with similar environmental resilience render bacteria and fungi as the best candidates for bioremediation. Chapters included in this book aim to compile recent investigations on the role of microorganisms and their derived enzymes in the efficient removal of emerging contaminants from different streams of effluents.

Chapter 1 introduces the readers to the different types of enzymes derived from widely investigated microorganisms and their respective efficiency for wastewater treatment. Chapter 2 discusses the diverse applications of microbial species for efficient removal of persistent organic pollutants and pharmaceutical compounds from effluents. Chapter 3 provides information on recent advances of microbe/enzyme immobilization on nanocarriers for enhanced pollutant uptake from effluents. Chapter 4 discusses the pollutant accumulation and subsequent detoxification of the same by different microbes and enzymes. Chapter 5 discusses different strategies of biotreatment of effluent in terms of their sustainability. Chapter 6 provides a detailed description of different enzymatic systems reported in the recent literature for efficient effluent treatment. Chapter 7 emphasizes on microbial remediation of effluents and subsequent use of treated water for agricultural purposes. Chapter 8 includes information on novel nanocatalysts and nanozymes reported in recent investigations for effluent treatment. Chapter 9 discusses the potential of extremophilic microorganisms for efficient removal of xenobiotic compounds from different types of effluents.

Chapter 10 provides a detailed description of different endocrine-disrupting compounds detected in effluents and efficient removal of the same through reported bacterial and fungal pathways.

This book is expected to serve as a comprehensive, accessible, up-to-date catalog of the potential role of microbes and enzymes in the treatment of wastewater and mechanisms of their action on various recalcitrant pollutants. This book compiles emerging and burning issues that challenge existing wastewater treatment strategies daily. Chapters included herein elucidate the microbial and enzymatic pathways for metabolism/degradation of hazardous aqueous pollutants, scale up, and optimize these processes and their application for the treatment of real industrial effluents to provide a holistic scenario of the concerned field. It also provides a comprehensive view of recent applications of microbial and enzyme systems for wastewater treatment and addresses the limitations and challenges of the existing microbial and enzyme-based processes for wastewater treatment. Each chapter provides up-to-date information on a specific topic. All the chapters in this book are written by potential authors worldwide. This reliable information from various active researchers and groups will be helpful to find out the alternative methods for effluent treatment and subsequent water reclamation and reuse. This book will help researchers and industrialists to better understand the correlation between microbial, biological, and chemical products and their roles in the mineralization of toxic aqueous pollutants.

About the Editors

Dr. Ashok Kumar Nadda is Assistant Professor in the Department of Biotechnology and Bioinformatics, Jaypee University of Information Technology, India. He completed his Ph.D. in Biotechnology from the Department of Biotechnology, Himachal Pradesh University, India. He worked as a postdoctoral fellow in the State Key Laboratory of Agricultural Microbiology, Huazhong Agricultural University, China. He also worked as a postdoctoral research associate at Konkuk University, South Korea. Dr. Nadda has a keen interest in microbial enzymes, immobilization of enzymes, nanobiotechnology, biocatalysis, wastewater management, biomass degradation, biofuel synthesis, and biotransformation. Dr. Nadda has published 58 research articles, 23 book chapters, and five books. He is also a member of the editorial board and reviewer committee of various journals of international repute.

Dr. Priya Banerjee is presently serving as Assistant Professor, Environmental Studies, Centre for Distance and Online Education, Rabindra Bharati University, Kolkata. She has completed her master's in Environmental Science from the University of Calcutta, Kolkata. After clearing UGC NET in Environmental Science, she completed her Ph.D. research from the Department of Environmental Science, University of Calcutta, Kolkata. She has 31 research articles, 36 book chapters, and two books to her credit. She has also participated in various national and international seminars to date and has received awards for best platform and poster presentations including the best poster presentation award at the 100th Indian Science Congress held at Kolkata. Her current research interests include integrated wastewater treatment, synthesis and application of nanomaterials, and toxicology. She is also an active reviewer of many reputed national and international journals.

Dr. Swati Sharma obtained her Ph.D. from Universiti Malaysia Pahang, Malaysia. She worked as a visiting researcher in the College of Life and Environmental Sciences at Konkuk University, South Korea. Dr. Sharma completed her M.Sc. from Dr. Yashwant Singh Parmar University of Horticulture and Forestry, Nauni Solan, Himachal Pradesh, India. She has also worked as a program co-coordinator at the Himalayan Action Research Center Dehradun and senior research fellow at the Indian Agricultural Research Institute. Dr. Sharma's research is in the field of bioplastics, hydrogels, keratin nanofibers and nanoparticles, biodegradable polymers and polymers with antioxidant and anticancerous activities and sponges.

Contributors

Roman Kumar Aneshwari
School of Pharmacy, MATS University, Raipur, Chhattisgarh, India

Priya Banerjee
Department of Environmental Studies, Centre for Distance and Online Education
Rabindra Bharati University
Kolkata, West Bengal, India

Rameshwari A. Banjara
Department of Chemistry
Rajiv Gandhi Government Post Graduate College
Ambikapur, Chhattisgarh, India

Debraj Biswal
Department of Zoology
Government General Degree College at Mangalkote
Burdwan, West Bengal, India

Nagendra Kumar Chandrawanshi
School of Studies in Biotechnology
Pt. Ravishankar Shukla University
Raipur, Chhattisgarh, India

Saba Ghattavi
Fisheries Department
Faculty of Marine Sciences, University of Hormozgan
Bandar Abbas, Iran

Ahmad Homaei
Department of Marine Biology
Faculty of Marine Science and Technology, University of Hormozgan
Bandar Abbas, Iran

Ashish Kumar
Department of Biotechnology, Guru Ghasidas Vishwavidyalaya (A Central University), Bilaspur, Chhattisgarh, India

Arindam Mitra
Department of Microbiology
School of Life Science and Biotechnology, Adamas University Kolkata
Kolkata, India

Aniruddha Mukhopadhayay
Department of Environmental Science, University of Calcutta
Kolkata, India

Ashok Kumar Nadda
Department of Biotechnology and Bioinformatics
Jaypee University of Information Technology
Waknaghat, Solan, India

Arkapriya Nandi
Department of Environmental Science
Rani Rashmoni Green University
Singur, Hooghly, West Bengal, India

Rusha Pal
Department of Pathology
Mass General Brigham
Harvard Medical School, Massachusetts, United States of America

Saumik Panja
Environmental Health and Safety
University of California San Francisco, California, United States of America

Arindam Rakshit
Department of Environmental Science
Rani Rashmoni Green University
Singur, Hooghly, West Bengal, India

Abhishek RoyChowdhury
School of Science
Navajo Technical University
Crownpoint, New Mexico, United States of America

Komal Sharma
Department of Environmental Science
University of Calcutta
Kolkata, India

Swati Sharma
University Institute of Biotechnology
Chandigarh University
Mohali, Punjab, India

Amit Kumar Verma
Department of Microbiology
National Institute of Medical Science & Research
Jaipur, Rajasthan India

Nikhi Verma
Department of Medical Microbiology
Postgraduate Institute of Medical Education and Research
Chandigarh, India

1 Microbial Enzymes and their Application in Wastewater Treatment

Amit Kumar Verma and Swati Sharma

1.1 INTRODUCTION

Water is the most fundamental biological molecule and an active matrix of life (Ball, 2017). Water-borne diseases result in a lot of clinically challenging cases. According to UNICEF, the WHO reported in 2019 there are one in three people in the world do not have safe drinking water. Modernization of the world leads to more and more industrial water and climate change, which decreases the freshwater supply and increases the population's need for clean water to increase the standard of living. The water scarcity problem is being spread in developing countries where the wastewater treatment is not well established (Zarei et al., 2020). In India, 256 of 700 districts have reported critical or exploitation of groundwater levels according to the 2017 Central Ground Water Board data. Most water exploitation is due to severe drought climate, overuse of water, and extensive groundwater harvesting (Qadir et al. 2010; Roson & Sartori 2015). Figure 1.1 shows the countries which are utilizing most of the water per year globally. Wastewater not only reduces the drinking or ground level

Countries	Water use yearly m^3/1000 liters	Population
India (2010)	761,000,000,000	1,234,281,170
China (2015)	598,100,000,000	1,406,847,870
United States (2016)	444,300,000,000	320,878,310
Indonesia (2016)	222,600,000,000	261,556,381
Pakistan (2008)	183,500,000,000	171,648,986
Iran (2004)	93,300,000,000	68,951,281
Mexico (2016)	86,580,000,000	123,333,376
Philippines (2005)	85,140,000,000	103,663,816
Vietnam (2016)	82,030,000,000	83,832,661
Japan (2009)	81,450,000,000	128,555,189

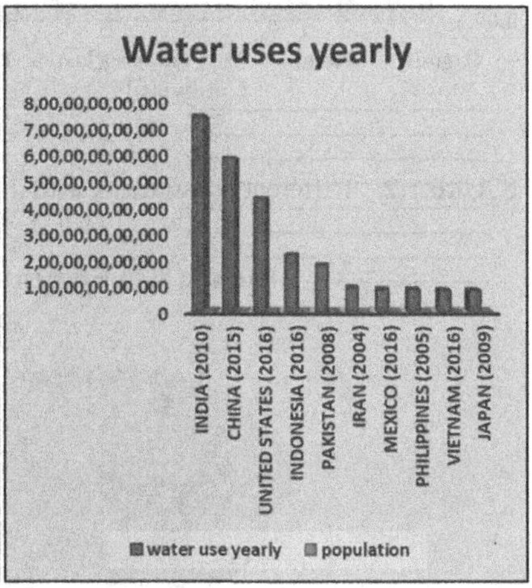

Water uses yearly
8,00,00,00,00,000
7,00,00,00,00,000
6,00,00,00,00,000
5,00,00,00,00,000
4,00,00,00,00,000
3,00,00,00,00,000
2,00,00,00,00,000
1,00,00,00,00,000
0
INDIA (2010)
CHINA (2015)
UNITED STATES (2016)
INDONESIA (2016)
PAKISTAN (2008)
IRAN (2004)
MEXICO (2016)
PHILIPPINES (2005)
VIETNAM (2016)
JAPAN (2009)
water use yearly
population

FIGURE 1.1 Data showing the countries utilizing most of the water per year globally. (World meters.)

DOI: 10.1201/9781003517238-1

water but also involves health issues, imbalance of ecosystem, loss of speciation, desertification, and agricultural deprivation (Meena et al., 2015).

Characteristics of wastewater are not different from the industrial and municipal wastewater discharge; both add nutrient-rich and organic and inorganic constituents to the habitation. Chemically, wastewater is composed of organic and inorganic compounds as well as various gases. Organic components may consist of carbohydrates, proteins, fats, greases, surfactants, oils, pesticides, phenols, etc. Inorganic components may consist of heavy metals, nitrogen, phosphorus, sulfur, chlorides, etc. The water pollutants are furthermore categorized as biological oxygen demands (BODs, organic waste), total suspended solid waste (TSSW, which includes organic fine particles of food or feces or inorganics like silt and clay), pathogen, nutrients, and contaminants of emerging concern (CECs, includes medical and personal care product wastes). The wastewater treatment by the biological process is responsible for the biodegradable constituents of BOD/chemical_oxygen_demand (COD) ratio greater than 0.5 (Metcalf and Eddy 2003). The industrial waste constituents with their percentage are shown in Figure 1.2.

Enzymes are biocatalysts and active protein (except RNase) molecules produced by living organisms, which mediate various biochemical reactions to produce metabolites and end products. The industrial enzyme market was worth more than US$6000 in 2021. According to the compound annual growth rate (CAGR) the market is projected to increase by 6% by 2027 (Figure 1.3) and, according to

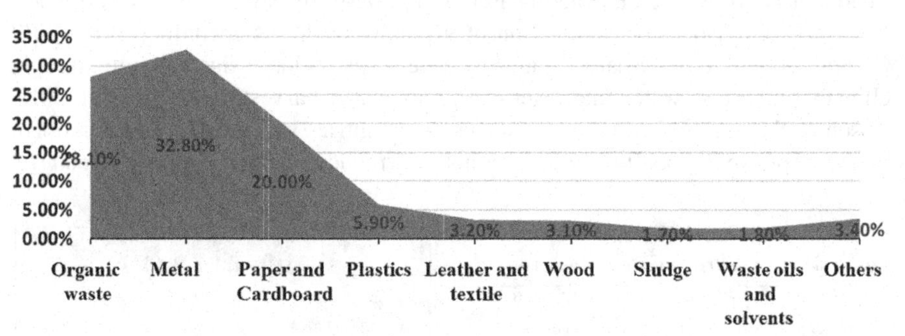

FIGURE 1.2 Different constituents with their percentage imparted to industrial waste.

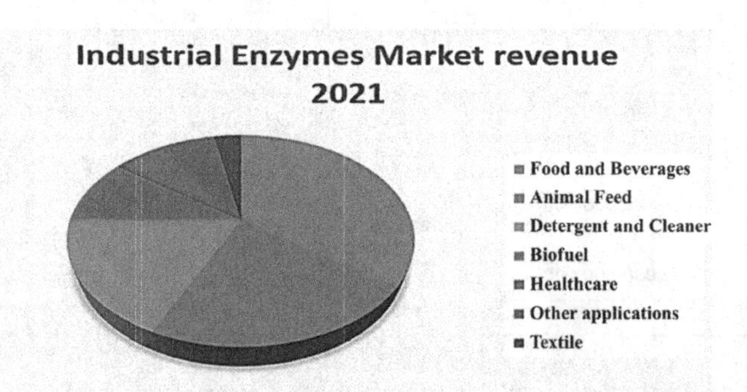

FIGURE 1.3 Industrial enzymes market revenue (Percentage) 2021. (Mordor intelligence.)

the CAGR India registered 3.8% growth during the period 2020–2025. Enzymes are highly specific in action and based on their activities they can be divided into oxidoreductases, transferases, hydrolases, lyases, isomerases, ligases, and translocases (Tao et al., 2020). Microbial sources of enzymes are being used extensively in various industrial and commercial applications (Singh et al., 2019). More than 4000 biological reactions are catalyzed by enzymes in which 150–170 enzymes are being commercially produced for around 500 scale-up industrial products. Today enzymes are not just bread and butter applied science, they are further advanced to be used in the biotechnological process, bioprocesses, medicine and therapy, textile, paper, and effluent and wastewater treatment technologies (Liu and Kokare, 2017) (Table 1.1).

The use of enzymes in wastewater treatment is an economical and eco-friendly process which is an improved technique over the physicochemical (flocculation, membrane filtration, desalination, percolation) and biological processes. The advantage of enzyme treatment is that it can be used at an extended spectrum of pH, temperature, saline condition; variable contaminate concentrations; no delay in acclimatization time; reduction in waste volume; and is easy to control (Nicell et al., 1993; Kaila et al., 2018). Enzymes act on the different waste components and remove the harmful chemicals by the process of precipitation and by changing them into valuable by-products. Development of commercial enzymes usually follows the screening of new or improved enzymes, selection of microorganisms (strains), quality and quantity improvement, fermentation to increase enzyme concentration, large-scale purification, and utilization (Patel et al., 2017). Before the scale-up utilization of enzymes in wastewater treatment, some issues need to be addressed for future research, including identification and characterization of enzymes and recalcitrant compounds, increasing their efficiency, and reducing their production and treatment costs.

1.2 ENZYMATIC TREATMENT OF WASTEWATER

Enzymes are specialized proteins that catalyze biological reactions. These enzymes are responsible for the transformation of the organic compounds produced largely by microbes and plants. The enzymes function by hydrolyzing, oxidizing, or reducing to metabolize the organic matter. One of the key end uses of enzymes is for industrial use, which has enormous commercial applications and the primary source is microbial extraction. The enzymes are more versatile in their action in mild reaction conditions, do not need the protection of the functional group, and have a longer life span. Furthermore, there has been more molecular editing and chemical modifications to improve the stability, specificity, and durability of the enzyme activity. The initial discharge of wastewaters from the industries loaded with a high level of inorganic effluents increases the TSSW, BOD, and COD load to the ecosystem. These issues have been strictly regulated by the government and private environmental protection agencies so as to reduce primary waste generation and treatment of effluent before releasing it into the environment. The enzymes which are currently used in industrial processes are classified into six groups: hydrolases, isomerases, lyasas, ligases, oxidoreductases, and transferases. There have been industries like breweries, dairy industry, pulp and paper industry, food industry, and complex organic chemical industry involved in generating by-products which are loaded with organic effluents.

TABLE 1.1
Mechanism of Action and Applications of Different Enzyme in Wastewater Treatment

Sources	Waste Contaminants	Enzymes	Mechanism of Action and Applications	References
Coal, petroleum, plastics, wood, metal, and dyes industries	Aerometric compounds, phenolic compounds	Horseradish peroxidase, chloroperoxidase	Oxidation of aromatic and phenolic compounds	Saini et al. (2021), Yu et al. (2019)
		Polyphenol oxidases	Oxidation of phenolic compound	Unuofin et al. (2019)
		Tyrosinase	Hydroxylation of monophenols and dehydrogenation	Osuoha et al. (2019)
Pulp and paper waste	Wood pulp, plant effluents, sugars and alcohol	Peroxidase	Kraft bleaching effluents	Sonkar et al. (2021)
		Laccase lactase	Bleaching plant effluent treatment, dairy waste processing	Dixit et al. (2021)
		Cellulolytic enzymes	Treatment of sludge	Sun and Cheng (2002), Runajak et al. (2020)
Pesticides	Cyanide	Cyanidase	Cyanide in industrial wastewaters to ammonia	Luque-Almagro et al. (2018); Mahendran et al. (2020)
		Cyanide hydratase	Hydrolyze cyanide to formamide	Luque-Almagro et al. (2018)
Food industries, fish and meat industries	Food waste, fish and meat waste, starch hydrolysis	Protease	Group of hydrolases which are widely used in the food industry	Kumar et al. (2021)
		Amylase	Amylases are polysaccharide hydrolases	Kumar et al. (2021)
Municipal solid waste and industrial solid waste	Heavy metals (cadmium, mercury, lead, uranium, strontium)	Cell-bound phosphatase	Heavy metal accumulation	Macaskie et al. (1994)
		Bromelain (EC: 3.4.22.32) and papain (EC: 3.4.22.2)	Adsorption of heavy metals	Chatterjee et al. (2013), Chatterjee et al. (2014)
Municipal wastewater	Sludge	Lipase	Removal of fats	Choudhury and Bhunia (2015)
		Lysozyme	Hydrolyzing loosely bound extracellular polymeric substance (LB-EPS)	Jafari et al. (2020)

1.3 HYDROLASES

The microbial enzyme hydrolases (e.g., lipases, esterases, peroxidase, cellulose, protease, etc.) are involved in the process of hydrolysis of the toxic organic chemicals into less toxic compounds. This method is preferred as it is a cheaper, safer, and eco-friendly way to remove environmental pollutants. It facilitates the cleavage of the carbon–carbon bond, sulfate–sulfate bond, carbon–nitrogen bond, carbon–oxygen bond, and other bonds with the addition of water. Pollutants such as carbofurans, carbaryl, diazinons, and coumaphos are successfully converted to less toxic by-products by the use of carbamate and parathion hydrolase from *Bacillus cereus, Pseudomonas, Flavobacterium, Nocardia,* and *Achromobacter* (Sutherland et al., 2002). Organophosphates (OP) are compounds that have been widely used as pesticides in agriculture having toxic by-products pyrethroids and malathion which are detoxified by the hydrolysis of phosphodiester bonds. The enzyme for the degradation of malathion is found in *Brevibacillus* sp., *Bacillus licheniformis*, and *Alicyclobacillus tengchongensis* (Littlechild 2015). Phthalates are chemicals that are widely used in the plastic industries as plasticizers and are ubiquitous environmental pollutants. The enzyme dialkyl phthalate ester (PE) hydrolases have been found in the *Acinetobacter* spp. *M63* and *Micrococcus* sp. *YGJI* catalyzed the phthalates to monoalkyl PEs (Aung et al., 2013). Parathion hydrolase hydrolyzes the OP pesticides (Zhao et al., 2021).

Lipases have shown a greater spectrum of catalyzing reactions in both aqueous and non-aqueous mediums. The demand for global microbial lipases has increased due to various applications in industries of food, meat, oil, paint, cosmetics, and pharmaceutics. Lipases are also potentially used in lipid-rich wastewater treatment. The oily layer formation on the surface of the water and the deposition of the lipid in the sewage is due to the discharge of treated and partially treated lipid molecules into the environment. Due to the omnipresent nature of lipases, it has vast industrial applications to hydrolyze the triacylglycerol into glycerol and fatty acids. The research and the study have been conducted to explore new sources, recombinant production, and developing application area to improve efficiency and recyclability. The fats, oil, and grease (FOG) are removed from the wastewater by chemical, mechanical, and biological methods. The chemical and mechanical methods are conventional, time consuming, costly, and ineffective. The microbial enzymes are being tested in so many studies to find the best suitable organisms for large-scale production and utilization in the industrial wastewater treatment (Wu et al., 2004; Cunha et al., 2008; Song et al., 2011; Mahdi et al., 2012; Dumore and Mukhopadhyay, 2012; Tsuji et al., 2013; Sarac and Ugur, 2016). *Pseudomonas aeruginosa* UKHL1 is more susceptible to the lipase enzyme oil-based wastewater treatment in which it degrades 37% of oil after 72 hours of experimental condition (Patel et al., 2020).

1.4 PEROXIDASES

These are groups of enzymes that catalyze the oxidation of substrate by peroxides. Most of the peroxidases are present as heme protein except the glutathione peroxidases which have the selenium-containing protein. They are widely distributed to all living organisms. The oxidoreductase includes reductases, oxidases, peroxidases, oxygenases, dehydrogenases, etc. The peroxidases mainly help in forming free radicals of phenolic compounds and these free radicals get precipitated by polymerization.

Lignin peroxidase (LiP), manganese-dependent peroxidases (MnP), and versatile peroxidases (VP) are used for many toxic compound phenols, antibiotics, pesticides, polycyclic aromatic hydrocarbons, and polychlorinated biphenyls (Agarwal et al., 2018). Immobilized horseradish peroxidase is used for the removal of P-chlorophenol from wastewater (Dalal and Gupta, 2007). The chlorinated phenol removal can be done by chloroperoxidase on magnetic nanoparticles in industrial wastewater (La Rotta et al., 2007). Laccases have also shown the potential for remediation which belong to the family of multi-copper oxidase. This class of enzymes oxidizes a large number of phenolic and non-phenolic compounds by the removal of enzymes from substrates. The production of the laccase enzyme uses many fungal species such as *Coprinus cinereus*, *Trametes versicolor*, *Phanerochaete chrysosporium*, etc. (Viswanath et al., 2014).

1.5 CELLULOLYTIC ENZYMES

Cellulose is the most abundant organic material in the biosphere. Due to being widely available most microbes possess cellulase enzyme for their hydrolysis. But few bacteria are identified which produce extracellular enzymes to hydrolyze cellulose extensively. Industrial to municipal solid waste (MSW) with a high content of organic waste (as cellulose) is seen to be degraded by a large number of microorganisms. Many studies have shown cellulase as a potential trigger for MSW and sludge treatment (Wood and Bhat, 1988; Gowthaman et al., 2001; Wei and Liu, 2006; Kumar et al., 2009; Gautam et al., 2012).

1.6 PROTEASE

Proteases perform proteolysis on protein molecules, which leads to the formation of smaller polypeptides or amino acids. Naturally, this enzyme is found everywhere as in plants, animals, and microbes. The evaluation has been made to show that 60% of effluent in wastewater is comprised of organic nitrogen (Westgate and Park, 2010). Alkaline protease is used for the production of short-chain fatty acids in the pre-treatment process of wastewater sludge activation. In a study, the sludge with 70.78% of protein is placed for protease hydrolysis to form the peptide chain and amino acid utilized by the microbes (Pei et al., 2010).

1.7 AMYLASE

Amylase has been isolated from plants and animals and predominantly by microbes to fulfill the intensive industrial sector demand. Industrial to agricultural wastewater has a higher content of carbohydrate residues (Pandit et al., 2021; O'Flaherty et al., 2010). In a study, alpha-amylase is used for anaerobic digested sludge treatment (Higuchi et al., 2005). The amylase can be utilized in endogenous amylase treatment, which increases the COD by 78.2% (Yu et al., 2013).

1.8 LIPASE

Lipases with high molecular weight are found in sludge and are an important component to be removed. There are many important industries where the lipid has been found to be extremely high as those in oil industries and restaurants. Lipase has been shown to be of importance in the hydrolysis of lipid in activated sludge (Parmar et al., 2001). Studies have shown many bacterial and fungal isolates having lipase secreting activity, which are used in wastewater sludge treatment (De Felice et al., 1997; Palma et al., 2000; Leal et al., 2006). Microbial enzyme utilization seems a better option for the reduction of sludge in wastewater treatment.

1.9 LYSOZYME

Industrial wastewater from food, textile, pulp, restaurant municipal agriculture, etc., is loaded with organic matter, which gives resources for exponential microbial growth. An abundance of microbial cells has been found loaded in the sludge. Lysozyme dissolves the polysaccharide layer of the bacterial cell wall by the lysis of β-1,4 glycosidic bond and also degrades the organic molecules (Yasunori, 1994). In a comparative study of removal of volatile solid, the lysozyme-producing inoculated bacteria reached up to 62% and uninoculated only 9.8% after a five-day shake cultivation (Ogawa, 2003). Another study showed that lysozyme pre-treatment in combination with thermal increases COD up to 95% (Jafari et al., 2020).

1.10 TECHNICAL PROGRESSION AND MECHANISM

The enzymes follow an established pathway before being used in wastewater treatment techniques. The development of commercial enzymes usually follows

- The screening of new or improved enzymes,
- Selection of microorganisms (strains),
- Quality and quantity improvement,
- Fermentation to increase enzyme concentration,
- Large-scale purification, and
- Utilization (Patel et al., 2017; Deckers et al., 2020).

The produced enzymes are not preferably used for treatment due to their reduced reusability, conformational changes (harsh pH and temperature), and effluent stream. So other techniques developed to overcome this issue are immobilization, direct culture utilization, and nanotechnological application.

1.11 IMMOBILIZATION

The process of immobilization is a process to provide phasic support (matrix) to the substrate and its products. The matrix can increase the reusability, stability, strength, enzyme efficiency, and nonspecific adsorption. Mainly inert polymers and inorganic materials are used based on their availability and affordability (Zdarta et al., 2018).

There are different techniques used for the immobilization of enzymes like adsorption, covalent binding, affinity immobilization, entrapment, and fabrication support. For adsorption immobilization, scientists use coconut fibers for enzyme laccase, which increases reduction and oxidation reaction (Ghosh and Ghosh, 2019). The porous biochar/chitosan used in covalent coupling with cellulose has high thermal stability and increases the half-life and can be used in wastewater treatment (Mo et al., 2020).

The affinity matrices use silica beads with agarose-linked concanavalin-A, which provide high enzyme concentration, efficiency, and stability (Sirisha et al., 2016). Entrapment provides increased stability and fewer enzyme leakage, which has been implemented for entrapment of lipases in k-carrageenan, alginate can be used in tannery process, olive oil production, and wastewater treatment (Sharmeen et al., 2019). In the fabrication, the crosslinked alginate with divalent ions and glutaraldehyde improves the activity, reusability, and stability of the enzyme (Mohammadi et al., 2019).

1.12 NANOTECHNOLOGICAL APPROACH

Nanotechnology deals with the study of the object's properties (material size, stability, efficiency) of size ranging from nanometers and their application. Nanoparticles, due to their small size, have unique chemical, physical, biological, and electrical properties (Nasir et al., 2015). With the growing application and demands of nanotechnologies, there are different approaches and methodologies used to synthesize the nanoparticles (Tomar et al., 2014). The nanoscale enzyme immobilization can be utilized in various domestic and commercial uses, such as drug delivery (Rabiei et al., 2021; Delcassian and Patel, 2020), biosensors (Gupta et al., 2020; Juska and Pemble, 2020), lactose hydrolysis (Selvarajan et al., 2019), and as antimicrobial agents (Sahani and Sharma, 2021). Recently in research, it has been found that the zerovalent iron immobilized chitosan (nZVI-chitosan) has a significant role in the removal of bisphenol A (BPA) for a solution with an efficiency of 97% and from medical wastewater with an efficiency of 93.8% (Dehghani et al., 2020).

1.13 CONCLUSION

The world is facing an increasing demand for portable water and the treatment of wastewater is one of the suitable alternatives. Many processes are being used for the treatment of water including physical, chemical, and biological. Preferably the biological method is more cost-effective and eco-friendlier. Research and development has shown enzymatic treatment of wastewater is a reliable technique due to its sustainability, better implementation, and cost-effectiveness. These methods can be applied to several industries, including that of solid wastewater treatment, food and beverages, restaurants, textiles, sugar industries, etc.

REFERENCES

Agarwal, S., Gupta, K.K., Chaturvedi, V.K., Kushwaha, A., Chaurasia, P.K. and Singh, M.P., 2018. The potential application of peroxidase enzyme for the treatment of industry wastes. In Shashi Lata Bharati and Pankaj Kumar Chaurasia (eds) *Research Advancements in Pharmaceutical, Nutritional, and Industrial Enzymology* (pp. 278–293). IGI Global, Hershey.

Aung, Y., Chen, X., Wang, X. et al., 2013. Identification and characterization of a cold-active phthalate esters hydrolase by screening a metagenomic library derived from biofilms of a wastewater treatment plant. *PLoS One*, *8*(10), Article ID e75977.

Ball, P., 2017. Water is an active matrix of life for cell and molecular biology. *Proceedings of the National Academy of Sciences*, *114*(51), pp. 13327–13335.

Chatterjee, S., Dutta, S. and Basu, S., 2013. Removal of cadmium(II) from simulated solution using immobilized papain: experiment, modeling and optimization by response surface methodology, In *Recycling and Reuse of Materials and Their Products* (Vol. 3), Y. Grohens, K. K. Sadasivuni, A. Boudenne (Eds.), Apple Academic Press and CRC Press, Taylor & Francis Group, Toronto, Waretown, NJ.

Chatterjee, S., Dutta, S., Mukherjee, M., Ray, P. and Basu, S., 2014. Studies on removal of lead(II) by Alginate Immobilized Bromelain (AIB). *Desalination and Water Treatment*, *56*, pp. 409–424.

Choudhury, P. and Bhunia, B., 2015. Industrial application of lipase: a review. *BioPharm International*, *1*(2), pp. 41–47.

Dalal, S. and Gupta, M.N., 2007. Treatment of phenolic wastewater by horseradish peroxidase immobilized by bioaffinity layering. *Chemosphere*, *67*(4), pp. 741–747.

Deckers, M., Deforce, D., Fraiture, M.A. and Roosens, N.H., 2020. Genetically modified micro-organisms for industrial food enzyme production: an overview. *Foods*, *9*(3), p. 326.

Dehghani, M.H., Karri, R.R., Alimohammadi, M., Nazmara, S., Zarei, A. and Saeedi, Z., 2020. Insights into endocrine-disrupting Bisphenol-A adsorption from pharmaceutical effluent by chitosan immobilized nanoscale zero-valent iron nanoparticles. *Journal of Molecular Liquids*, *311*, p. 113317.

Delcassian, D. and Patel, A.K., 2020. Nanotechnology and drug delivery. In Sylvain Ladame and Jason Chang (eds) *Bioengineering Innovative Solutions for Cancer* (pp. 197–219). Academic Press, Cambridge, MA,.

De Felice, B., Pontecorvo, G. and Carfagna, M., 1997. Degradation of wastewaters from olive oil mills by Yarrowia lipolytica ATCC 20255 and Pseudomonas putida. *Acta Biotechnologica*, *17*, pp. 231–239.

Dixit, M., Gupta, G.K., Usmani, Z., Sharma, M. and Shukla, P., 2021. Enhanced bioremediation of pulp effluents through improved enzymatic treatment strategies: a greener approach. *Renewable and Sustainable Energy Reviews*, *152*, p. 111664.

Gautam, S.P., Bundela, P.S., et al., 2012. Diversity of cellulolytic microbes and the biodegradation of municipal solid waste by a potential strain. *International Journal of Microbiology*, *2012*, p. 325907.

Ghosh, P. and Ghosh, U., 2019. Immobilization of purified fungal laccase on cost effective green coconut fiber and study of its physical and kinetic characteristics in both free and immobilized form. *Current Biotechnology*, *8*(1), pp. 3–14.

Gowthaman, M., Krishna, C. and Moo-Young, M., 2001. Fungal solid state fermentation – an overview. *Applied Mycology and Biotechnology*, *1*, 305–352.

Gupta, R., Sagar, P., Priyadarshi, N., Kaul, S., Sandhir, R., Rishi, V. and Singhal, N.K., 2020. Nanotechnology-based approaches for the detection of SARS-CoV-2. *Frontiers in Nanotechnology*, *2*, p. 6.

Higuchi, Y., Ohashi, A., Imachi, H. and Harada, H., 2005. Hydrolytic activity of alpha-amylase in anaerobic digested sludge. *Water Science and Technology*, *52*(1–2), pp. 259–266.

Jafari, S., Salehiziri, M., Foroozesh, E., Bardi, M.J. and Rad, H.A., 2020. An evaluation of lysozyme enzyme and thermal pretreatments on dairy sludge digestion and gas production. *Water Science and Technology*, *81*(5), pp. 1052–1062.

Juska, V.B. and Pemble, M.E., 2020. A critical review of electrochemical glucose sensing: evolution of biosensor platforms based on advanced nanosystems. *Sensors*, *20*(21), p. 6013.

Kaila, S., Sejpal, P. and Mona, M.B., 2018. Application of various enzymes in waste treatment. *International Journal of Scientific Research in Chemistry (IJSRCH)*, *3*(1), pp. 60–64.

Kumar, D., Bhardwaj, R., Jassal, S., Goyal, T., Khullar, A. and Gupta, N., 2021. Application of enzymes for an eco-friendly approach to textile processing. *Environmental Science and Pollution Research*, *30*, pp. 1–11.

Kumar, P., Barrett, D.M., Delwiche, M.J. and Stroeve, P., 2009. Methods for pretreatment of lignocellulosic biomass for efficient hydrolysis and biofuel production. *Industrial & Engineering Chemistry Research*, *48*, pp. 3713–3729.

La Rotta, C.E., D'elia, E. and Bon, E.P., 2007. Chloroperoxidase mediated oxidation of chlorinated phenols using electrogenerated hydrogen peroxide. *Electronic Journal of Biotechnology*, *10*(1), pp. 24–36.

Leal Marcia, C.M.R., Freire Denise, M.G., Cammarota Magali, C. and Sant Anna, G.L., 2006. Effect of enzymatic hydrolysis on anaerobic treatment of dairy wastewater. *Process Biochemistry*, *41*, pp. 1173–1178.

Littlechild, J. A., 2015. Archaeal enzymes and applications in industrial biocatalysts. *Biotechnological Uses of Archaeal Proteins*, 147671, pp. 1–10.

Liu, X. and Kokare, C., 2017. Microbial enzymes of use in industry. In Goutam Brahmachari, Arnold L Demain, Jose L Adrio (eds) *Biotechnology of Microbial Enzymes* (pp. 267–298). Academic Press, Cambridge, MA.

Luque-Almagro, V.M., Cabello, P., Sáez, L.P., Olaya-Abril, A., Moreno-Vivián, C. and Roldán, M.D., 2018. Exploring anaerobic environments for cyanide and cyano-derivatives microbial degradation. *Applied Microbiology and Biotechnology*, *102*(3), pp. 1067–1074.

Macaskie, L.E., Bonthrone, K.M. and Rouch, D.A., 1994. Phosphatase-mediated heavy metal accumulation by a Citrobacter sp. and related enterobacteria. *FEMS Microbiology Letters*, *121*(2), pp. 141–146.

Mahdi, B.A., Bhattacharya, A. and Gupta, A., 2012. Enhanced lipase production from Aeromonas sp. S1 using Sal deoiled seed cake as novel natural substrate for potential application in dairy wastewater treatment. *Journal of Chemical Technology & Biotechnology*, *87*(3), 418–426.

Mahendran, R., Sabna, B.S., Thandeeswaran, M., Vijayasarathy, M., Angayarkanni, J. and Muthusamy, G., 2020. Microbial (enzymatic) degradation of cyanide to produce pterins as cofactors. *Current Microbiology*, *77*(4), pp. 578–587.

Meena, V.D., Dotaniya, M.L., Saha, J.K. and Patra, A.K., 2015. Antibiotics and antibiotic resistant bacteria in wastewater: impact on environment, soil microbial activity and human health. *African Journal of Microbiology Research*, *9*(14), pp. 965–978.

Metcalf, L. and Eddy, H.P., 2003. *Wastewater Engineering Treatment and Reuse*, 4th ed., McGraw Hill, NY.

Mo, H., Qiu, J., Yang, C., Zang, L., Sakai, E. and Chen, J., 2020. Porous biochar/chitosan composites for high performance cellulase immobilization by glutaraldehyde. *Enzyme and Microbial Technology*, *138*, p. 109561.

Mohammadi, M., Heshmati, M.K., Sarabandi, K., Fathi, M., Lim, L.T. and Hamishehkar, H., 2019. Activated alginate-montmorillonite beads as an efficient carrier for pectinase immobilization. *International Journal of Biological Macromolecules*, *137*, pp. 253–260.

Nasir, A., Kausar, A. and Younus, A., 2015. A review on preparation, properties and applications of polymeric nanoparticle-based materials. *Polymer-Plastics Technology and Engineering*, *54*(4), pp. 325–341.

Nicell, J.A., Al-Kassim, L., Bewtra, J.K. and Taylor, K.E., 1993. Wastewater treatment by enzyme catalysed polymerization and precipitation. *Biodeterioration Abstracts* *7*(1), pp. 1–8.

O'Flaherty, V., Collins, G. and Mahony, T., 2010. Anaerobic digestion of agricultural residues. *Environmental Microbiology*, *11*, pp. 259–279.

Osuoha, J.O., Abbey, B.W., Egwim, E.C. and Nwaichi, E.O., 2019. Production and characterization of tyrosinase enzyme for enhanced treatment of organic pollutants in petroleum refinery effluent. In *SPE Nigeria Annual International Conference and Exhibition*. OnePetro, Richardson, TX.

Palma, M.B., Pinto, A.L., Gombert, A.K., Seitz, K.H., Kivatinitz, S.C., et al., 2000. Lipase production by *Penicillium restrictum* using solid waste of industrial babassu oil production as substrate. *Applied Biochemistry and Biotechnology*, *84*, pp. 1137–1145.

Pandit, S., Savla, N., Sonawane, J.M., Sani, A.M.D., Gupta, P.K., Mathuriya, A.S., Rai, A.K., Jadhav, D.A., Jung, S.P. and Prasad, R., 2021. Agricultural waste and wastewater as feedstock for bioelectricity generation using microbial fuel cells: recent advances. *Fermentation*, *7*(3), p. 169.

Parmar, N., Singh, A. and Ward, O.P., 2001. Enzyme treatment to reduce solids and improve settling of sewage sludge. *Journal of Industrial Microbiology and Biotechnology*, *26*, pp. 383–386.

Patel, A.K., Singhania, R.R. and Pandey, A., 2017. Production, purification, and application of microbial enzymes. In Goutam Brahmachari, Arnold L Demain, Jose L Adrio (eds) *Biotechnology of Microbial Enzymes* (pp. 13–41). Academic Press, Cambridge, MA.

Patel, H., Ray, S., Patel, A., Patel, K. and Trivedi, U., 2020. Enhanced lipase production from organic solvent tolerant Pseudomonas aeruginosa UKHL1 and its application in oily waste-water treatment. *Biocatalysis and Agricultural Biotechnology*, *28*, p. 101731.

Pei, H.Y., Hu, W.R. and Liu, Q.H., 2010. Effect of protease and cellulase on the characteristic of activated sludge. *Journal of Hazardous Materials*, *178*, pp. 397–403.

Qadir, M., Bahri, A., Sato, T. and Al-Karadsheh, E., 2010. Wastewater production, treatment, and irrigation in Middle East and North Africa. *Irrigation and Drainage*, *24*(*1-2*), pp. 37–51.

Rabiei, M., Kashanian, S., Samavati, S.S., Derakhshankhah, H., Jamasb, S. and McInnes, S.J., 2021. Nanotechnology application in drug delivery to osteoarthritis (OA), rheumatoid arthritis (RA), and osteoporosis (OSP). *Journal of Drug Delivery Science and Technology*, *61*, p. 102011.

Roson, R. and Sartori, M., 2015. A decomposition and comparison analysis of international water footprint time series. *Sustainability*, *7*(5), pp. 5304–5320.

Runajak, R., Chuetor, S., Rodiahwati, W., Sriariyanun, M., Tantayotai, P. and Phornphisutthimas, S., 2020. Analysis of microbial consortia with high cellulolytic activities for Cassava pulp degradation. In E3S Web of Conferences (Vol. 141, p. 03005). EDP Sciences, France.

Sahani, S. and Sharma, Y.C., 2021. Advancements in applications of nanotechnology in global food industry. *Food Chemistry*, *342*, p. 128318.

Saini, R., Rani, V., Sharma, S. and Verma, M.L., 2021. Screening of microbial enzymes and their potential applications in the bioremediation process. In Pankaj Kumar Arora (ed.) *Microbial Products for Health, Environment and Agriculture* (pp. 359–378). Springer, Singapore.

Sarac, N. and Ugur A., 2016. A green alternative for oily wastewater treatment: lipase from Acinetobacter haemolyticus NS02-30. *Desalination and Water Treatment*, *57*(42), pp. 19750–19759.

Selvarajan, E., Nivetha, A., Subathra Devi, C. and Mohanasrinivasan, V., 2019. Nanoimmobilization of β-galactosidase for lactose-free product development. In K M Gothandam, Shivendu Ranjan, Nandita Dasgupta, Eric Lichtfouse (eds) *Nanoscience and Biotechnology for Environmental Applications* (pp. 199–223). Springer, Cham.

Sharmeen, S., Rahman, M.S., Islam, M.M., Islam, M.S., Shahruzzaman, M., Mallik, A.K., Haque, P. and Rahman, M.M., 2019. Application of polysaccharides in enzyme immobilization. In Sabyasachi Maiti and Sougata Jana (eds) *Functional Polysaccharides for Biomedical Applications* (pp. 357–395). Woodhead Publishing, Sawston.

Singh, R.S., Singh, T. and Pandey, A., 2019. Microbial enzymes – an overview. *Advances in Enzyme Technology*, pp. 1–40.

Sirisha, V.L., Jain, A. and Jain, A., 2016. Enzyme immobilization: an overview on methods, support material, and applications of immobilized enzymes. *Advances in Food and Nutrition Research*, *79*, pp. 179–211.

Song, H., Zhou, L., Zhang, L., Gao, B., Wei, D., Shen, Y., Wang, R., Madzak, C. and Jiang, Z., 2011. Construction of a whole-cell catalyst displaying a fungal lipase for effective treatment of oily wastewaters, *Journal of Molecular Catalysis B: Enzymatic*, *71*(3–4), pp. 166–170.

Sonkar, M., Kumar, V., Kumar, P., Shah, M.P., Majumdar, C.B., Biswas, J.K., Dutt, D. and Mishra, P.K., 2021. Bioaugmentation with existing potent microorganisms to accelerate the treatment efficacy of paper industry wastewater pollutants. *Journal of Environmental Chemical Engineering*, *9*(5), p. 105913.

Sun, Y, and Cheng, J., 2002. Hydrolysis of lignocellulosic materials for ethanol production: a review. *Bioresource Technology*, *83*(1), pp. 1–11.

Sutherland, T., Russell, R. and Selleck, M., 2002. Using enzymes to clean up pesticide residues. *Pesticide Outlook*, *13*(4) pp. 149–151.

Tao, Z., Dong, B., Teng, Z. and Zhao, Y., 2020. The classification of enzymes by deep learning. *IEEE Access*, *8*, pp. 89802–89811.

Tomar, R.S., Chauhan, P.S. and Shrivastava, V., 2014. A critical review on nanoparticle synthesis: physicochemical v/s biological approach. *World Journal of Pharmaceutical Research*, *4*(1), pp. 595–620.

Tsuji, M., Yokota, Y., Shimohara, K., Kudoh, S. and Hoshino, T., 2013. An application of wastewater treatment in a cold environment and stable lipase production of Antarctic basidiomycetous yeast Mrakia blollopis. *PLoS One*, *8*(3), p. e59376.

Unuofin, J.O., Okoh, A.I. and Nwodo, U.U., 2019. Aptitude of oxidative enzymes for treatment of wastewater pollutants: a laccase perspective. *Molecules*, *24*(11), p. 2064.

Viswanath, B., Rajesh, B., Janardhan, A., Kumar, A.P. and Narasimha, G., 2014. Fungal laccases and their applications in bioremediation. *Enzyme Research*, *2014*, p. 163242.

Wei, Y. and Liu, J., 2006. *Sludge Reduction with a Novel Combined Worm-Reactor. Aquatic Oligochaete Biology IX* (pp. 213–222). Springer, New York.

Westgate, P.J. and Park, C., 2010. Evaluation of proteins and organic nitrogen in wastewater treatment effluents. *Environmental Science & Technology*, *44*(14), pp. 5352–5357.

Wood, T.M. and Bhat, K.M., 1988. Methods for measuring cellulase activities. *Methods Enzymol*, *160*, pp. 87–112.

Wu, H.S. and Tsai, M.J., 2004. Kinetics of tributyrin hydrolysis by lipase. *Enzyme and Microbial Technology*, *35* (6–7), pp. 488–493.

Yu, B., Cheng, H., Zhuang, W., Zhu, C., Wu, J., Niu, H., Liu, D., Chen, Y. and Ying, H., 2019. Stability and repeatability improvement of horseradish peroxidase by immobilization on amino-functionalized bacterial cellulose. *Process Biochemistry*, *79*, pp. 40–48.

Yu, S., Zhang, G., Li, J., Zhao, Z. and Kang, X., 2013. Effect of endogenous hydrolytic enzymes pretreatment on the anaerobic digestion of sludge. *Bioresource Technology*, *146*, pp. 758–761.

Zarei, Z., Karami, E. and Keshavarz, M., 2020. Co-production of knowledge and adaptation to water scarcity in developing countries. *Journal of Environmental Management*, *262*, p. 110283.

Zdarta, J., Meyer, A.S., Jesionowski, T. and Pinelo, M., 2018. A general overview of support materials for enzyme immobilization: characteristics, properties, practical utility. *Catalysts*, *8*(2), p. 92.

Zhao, S., Xu, W., Zhang, W., Wu, H., Guang, C. and Mu, W., 2021. In-depth biochemical identification of a novel methyl parathion hydrolase from Azohydromonas australica and its high effectiveness in the degradation of various organophosphorus pesticides. *Bioresource Technology*, *323*, p. 124641.

2 Microbial Technologies for the Removal of Pharmaceuticals and Persistent Organic Pollutants from Wastewater

Rameshwari A. Banjara, Ashish Kumar, Roman Kumar Aneshwari, and Nagendra Kumar Chandrawanshi

2.1 INTRODUCTION

The number of industries has increased exponentially as a result of population growth. Rapid resource consumption, economic development, urbanization, and industrialization have caused tremendous environmental, ecosystem, and human health issues in recent years (Bakirtas and Akpolat, 2018). As a result, severe environmental issues and the demand for water are severe. Given the current state of water resources, it is essential to employ novel ideas to improve water cycle management in the public and private sectors. In addition, to recognize the entire worth of water, novel sustainable practices in the water cycle must be implemented. With all that in mind, wastewater recovery can be viewed as a precious resource that can be achieved through advanced emerging technologies . The pharmaceutical sector is the source of the most abrasive wastewater (Gadipelly et al., 2014; Yakubu, 2017). Pharmaceutical wastewater is known to include highly active chemicals, which can be a significant source of concern due to their potential adverse effects on human health and the environment (Virkutyte et al., 2010). Pharmaceutical pollutants have been suggested as potential endocrine disruptors, mimicking growth hormones even when consumed at nanogram quantities. Pharmaceuticals and their metabolites enter aquatic habitats via land-applied sewage sludge effluents from treatment plants and surface water runoff from the industry. Carbamazepine (CBZ) and sulfamethoxazole (SMX), which are resistant substances whose usage has increased dramatically in recent decades, are two of the most often discovered pharmaceutically active compounds in wastewater (García-Espinoza and Nacheva, 2019). It's worth noting that the presence of drugs in

DOI: 10.1201/9781003517238-2

the environment has become widespread, posing a life-threatening hazard to humans and ecosystems worldwide. The concentrations of pharmaceuticals and their residues in the environment vary and are influenced by various factors, including industrial waste, human consumption patterns, wastewater treatment process, population increase, etc. (Mulbry et al., 2008).

In addition persistent organic pollutants (POPs), such as organochlorine insecticides, polychlorinated dibenzodioxins (PCDDs), polychlorinated biphenyls (PCBs), polybrominated diphenyl ethers (PBDEs), and polychlorinated dibenzofurans (PCDFs), are mostly anthropogenic. They have been widely employed in various products (Mato et al., 2001). POPs are harmful organic substances that take a long time to break down in the environment under natural circumstances and can accumulate in living organisms and ecosystems. These POPs have been a hazard to ecosystems because they can enter the environment through agricultural, municipal, and industrial activities (Pariatamby and Kee, 2016). Due to their various environmental impacts, the widespread production and use of hazardous and poisonous POPs have sparked severe concerns (Ashraf, 2017).

The removal of pharmaceutical pollutants and POPs has become a subject of urgency for the water sector and other regulatory authorities due to widespread concern about their toxicity at low doses. In recent years, significant efforts have been made to reduce the concentration of such contaminants in real wastewaters, in response to the current low removal rates achieved in conventional biological treatments for industrial effluents, such as standard anaerobic digestion (Garcia-Gomez et al., 2016; Chtourou et al., 2018). This chapter provides a detailed overview and reports on the current state of biodegradation/bioremediation research to remove pharmaceutical waste and POPs, such as pesticides, polycyclic aromatic hydrocarbons (PAHs), PCBs, and PPCPs, from wastewater.

2.2 PHARMACEUTICAL RESIDUES IN WASTEWATER

Pharmaceuticals are often used in human and veterinary medicine to prevent and treat diseases. Their endurance and negative consequence to the aquatic ecosystem make these biologically active compounds classified as emerging contaminants. These refractory emergent contaminants (RECs) (antibiotics, analgesics, anti-inflammatories, and anti-epileptics) are largely endocrine disruptive substances that enter the aquatic environment in small amounts regularly. They affect the water quality and negatively impact the ecosystem and human health even in low quantities. Antibiotics are the most common and lasting medicinal compounds in the aquatic environment. In hormones used for therapeutic purposes, several studies have established the occurrence of many estrogens in both the influent and effluent municipal wastewater treatment plants, with concentrations ranging from 5 to 188 ng/L and 0.3 to 12.6 ng/L, respectively. Due to the persistence in the aquatic (ground and surface water) environment, drugs related to the 140 analgesic classes, such as naproxen, acetaminophen, ibuprofen, diclofenac, and meprobamate, were classified as serious environmental contaminants (Radjenovic et al., 2009).

Pharmaceutical compounds (PhCs) have accumulated in water and wastewater treatment plants, making the treated effluents non-reusable because the sludge produced is hazardous and highly contaminated with PhCs. Contaminants found in treated effluents can't be eliminated using traditional treatment procedures; thus, they are frequently dumped into the environment without being treated (Edwards et al., 2018; Sophia and Lima, 2018). The global consumption of pharmaceuticals is estimated to be hundreds of thousands of tons (t). For example, millions of anti-inflammatory non-steroidal medications such as paracetamol, aspirin, diclofenac, and ibuprofen are being manufactured. Mirzaei et al (2018) has reported that the abundance of PhCs in aquatic ecosystems varies from country to country and may be influenced by prescription methods. In many countries, the following trace pharmaceuticals were identified in the concentration range of 0.38–3.59 g/L: CBZ, atenolol, propranolol, metronidazole, ranitidine, and paracetamol. However, such concentrations may damage aquatic ecosystems. The presence of various pharmaceutically active chemicals (PACs) in hospital sewage wastewater, wastewater, drinking water, and treated water was investigated by Ayman and Işık (2015). The pesticides atrazine and dimethoate were the most often discovered PACs in sediments, whereas CBZ and metamizole metabolites were regularly detected in water samples. It has been investigated that malaria treatment medicine and paracetamol were found in irrigation water from the garbage. Antibiotics, estrogens, and lipid-lowering residues were found in surface water sewage from a wastewater treatment plant in Lagos, Nigeria (Olarinmoye et al., 2016).

Antibiotics like fluoroquinolones and sulfonamides, for example, may be toxic to aquatic organisms, and their phytotoxic behavior may be harmful to the environment. In drinking water, medications such as acetaminophen, aspirin, ibuprofen, diclofenac, ketoprofen, and naproxen have been detected in tens of milligrams per liter, and this active metabolite has been found in wastewater treatment plants, tap water, and groundwater (Khetan and Collins, 2007). This is most likely due to its long-lasting feature and great mobility in the aquatic environment (Khetan and Collins, 2007). Beta-blockers, such as atenolol, diuretic furosemide, metoprolol, and propranolol, as well as angiotensin-converting enzyme (ACE) inhibitors, angiotensin II receptor antagonists, and calcium channel blockers, have all been found in tens of milligrams per liter in ground, surface, and drinking water (Postigo and Barcelo, 2015; Hapeshi et al., 2015). As a result of the increasing accumulation of persistent organic pharmaceuticals, their removal from various water sources using advanced microbial techniques has become a critical component of water and wastewater treatment.

2.3 POPS IN WASTEWATER

Most POPs are anthropogenic toxic organic compounds; they have been used in a variety of products and enter into the environment through municipal, industrial, and agricultural activities (Mato et al., 2001). It takes a long time to degrade in the environment and can accumulate in living beings and ecosystems (Pariatamby and Kee, 2016). The extensive manufacturing and usage of dangerous and deadly POPs have prompted serious concerns due to their diverse environmental impact (; Ashraf, 2017). Some POPs such as pesticides have been detected in water and soil (for example, drinking water, surface water, and groundwater), harming the environment (Pariatamby and Kee, 2016). Because of their longevity in the ecosystem, bioaccumulation and biomagnification in ecosystems, and considerable adverse effects on the human body, POPs have sparked widespread concern worldwide (Al-Mulali et al., 2015; Bakirtas and Akpolat, 2018). Further, POPs can be classified into three groups.

2.3.1 Industrial Chemicals

This group of compounds consists of hexabromobiphenyl, hexabromocyclododecane, PCBs, decabromodiphenyl ether, polychlorinated naphthalenes, hexachlorobutadiene, pentachlorobenzene, hexachlorobenzene (HCB) and related compounds, short-chain chlorinated paraffins, pentabromodiphenyl ether, tetrabromodiphenyl ether, perfluorooctane sulfonic acid, its salts, and perfluorooctane sulfonyl fluoride. POPs in this category are widely employed in industrial operations of production and manufacture. For example, PCBs were used as industrial lubricants and coolants in the manufacture of transformers, capacitors, and many other electrical devices. Various consumer goods resistant to oil, grease, stains, water, and heat include perfluorooctanoic acid. As observed in many prior studies, the emission of these POPs resulted in environmental contamination in soil, drinking water, and air (Bakirtas and Akpolat, 2018).

2.3.2 Pesticides

The majority of the compounds in this category are organochlorine pesticides, which have been identified for their harmful effects on the human body and their persistence in nature. These chemicals are dicofol, aldrin, chlordane, dieldrin, chlordecone, endosulfan, DDT and its related isomers, heptachlor, HCB, α-hexachlorocyclohexane, β-hexachlorocyclohexane, endrin, lindane, mirex, pentachlorobenzene, pentachlorophenol and its esters and salts, perfluorooctane sulfonic acid, perfluorooctane sulfonyl fluoride and its salts, and toxaphene.

2.3.3 Unintentional Production

These substances are undesirable by-products of the chemical or combustion processes when chlorine compounds are present. PCBs, PCDFs, and PCDDs are some of the most well-known substances in this category. HCB, polychlorinated naphthalene, pentachlorobenzene, and hexachlorobutadiene are some other compounds in this category. To promote a safe and sustainable environment for the community and ecology, these chemicals should be measured to avoid unintended discharges.

For the breakdown and mineralization of pharmaceutical substances and POPs, various physical, chemical, biological, and combination approaches have been proposed. Traditional pharmaceutical substances and POP treatments, such as coagulation-flocculation and adsorption with sedimentation as a post-treatment, however, merely fixate the hazardous substances rather than altogether remove them (Padmanabhan et al., 2006); chemical oxidation has several drawbacks, including exorbitant costs due to significant chemical consumption, incomplete destruction, and a long overall treatment time (Dong et al., 2015). Preferred practical approaches for removing POPs include those that are highly efficient, technologically reliable, environmentally friendly, and relatively of low cost. New, exceptionally cost-effective pharmaceutical substances and POP remediation solutions are urgently needed to handle such compounds' safe disposal and cleanup. In this chapter, microbials including algal, enzymological, and advanced biotechnological techniques are given special attention because of its low cost, effectiveness in pharmaceutical substances, POP degradation, technological feasibility, and cost-efficiency. The biodegradation process is also aided by advances in microbial fuel cells (MFCs), nanomaterials, biofilms, and engineered wetlands. Gene duplication, mutational drift, recombination, and hypermutation are used by various bacteria that carry several plasmids and catabolic genes to adapt to these unfavorable environmental conditions. Some significant pharmaceutical substances and POP catabolic genes, such as biphenyl dioxygenase (bph), hydroxylation, phosphotriesterases, and oxygenase, contribute to the breakdown of organic pollutants through genetic factors. However, with technological advancements in genetic engineering, the role of genetically modified organisms is being considered, as well as metabolomics and metagenomics, to develop effective, low-cost, and reliable methods for detection, determination, and removal of ultra-trace concentrations of pharmaceutical waste and POPs.

2.4 EMERGING MICROBIAL TECHNOLOGIES FOR REMOVAL OF PHARMACEUTICALS AND POPS

Various water treatment procedures, such as adsorption, filtration, precipitation, coagulation, and flocculation, are examples of physical, chemical, and biological wastewater treatment methods (Maletz et al., 2013; Mouele et al., 2015). Comprehensive remediation strategies use MFCs or microbial electrolysis cells (MECs) that focus on the same chemical principles as coagulation, sedimentation, Fenton oxidation, ozonation, chemical oxidation, and adsorption (Brown et al., 2014; Heidrich et al., 2014). These systems produce hydroxyl radicals and other co-species such as sulfates and hydrogen carbonates, which help remove persistent organic contaminants (Martin et al., 2018). Microbes help in electricity generation from organic ingredients in MFCs and MECs, making microbial electrochemical technologies (METs) energy-independent and promising in the future (Christin et al., 2017). The effects of pharmaceuticals on humans and aquatic species have been studied using coagulation/flocculation, biological treatments, ion exchange, reverse osmosis, and adsorption (activated carbon). However, these procedures may necessitate a lot of wet chemistry, and the necessary equipment to scale them up could be costly, limiting their use in the remediation of pharmaceuticals. As a result, more extended, efficient treatment techniques are necessary for wastewater treatment. The overuse of various active pharmaceutical ingredients has resulted in higher direct or indirect pollution of water sources, and the pharmaceutical remains in water sources pose potential risks to genome modification of living organisms and potentially long impacts on human health (Sires and Brillas, 2012).

2.4.1 MFC and METs

Consumption of energy is always a key concern in biological wastewater treatment that harms the environment. In this regard, a MFC is offered as a promising alternative that converts organic waste in wastewater directly into electricity (Liu et al., 2005). To generate electricity, electrons generated by bacteria from various substrates are transmitted to the anode and flow to the cathode. The H^+ ions pass through the semi-permeable barrier to the cathode, where they interact with dissolved oxygen in oxygen reduction reactions (ORRs) induced by biocatalysts to form water. However, most studies found that poor extracellular electron transfer (EET) efficiency is a significant challenge that must be addressed before MFCs can recover enough energy and reduce the environmental consequences of wastewater treatment (Hsu et al., 2018). Various designs of MFC and their operating principles, electrodes, and bacteria species have been studied widely in recent decades to alleviate this limitation. To achieve maximum bacterial colonization, the energy density in MFC is traditionally increased by increasing the surface area of the anode (Figure 2.1). Because of their high specific surface area, carbon-based materials have been extensively investigated to serve as high-performance MFC anodes in this technique.

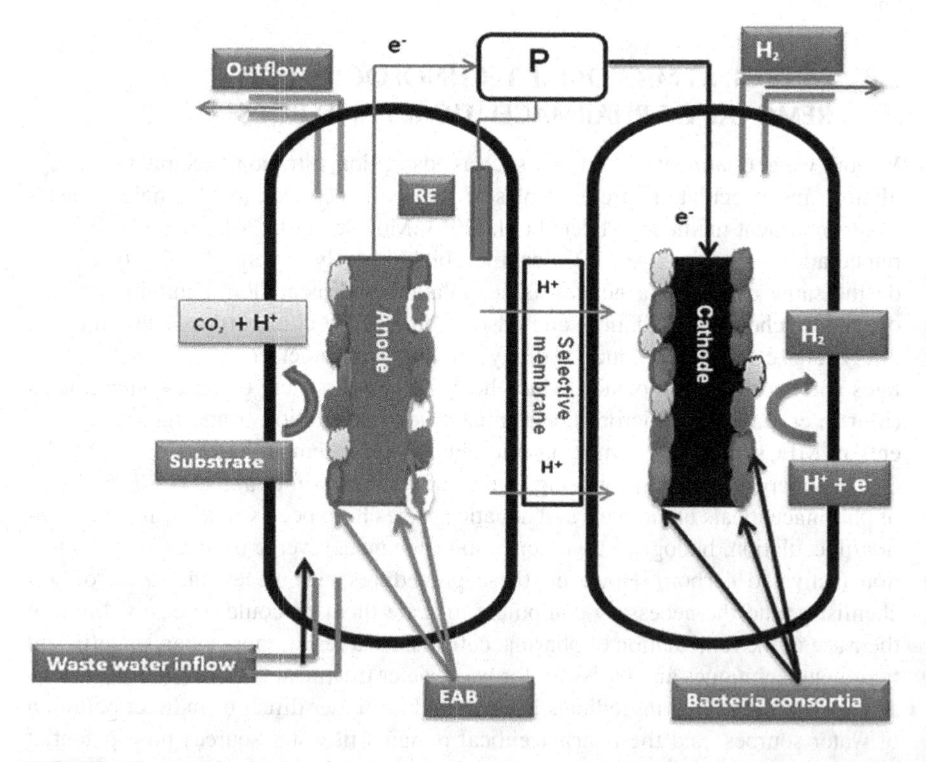

FIGURE 2.1 Schematic diagram of a two-chamber microbial fuel cell reactor for removal of pharmaceuticals from wastewater.

Furthermore, their biocompatibility and conductivity can aid interactions between bacteria and the electrode. Various synthetic ORR catalysts can be used on the cathode of MFCs to improve energy conversion efficiency. Platinum (Pt) is still the most effective ORR catalyst, and it's commonly employed in MFCs for high-energy output (Yuan et al., 2016). The complex compositions in wastewater, on the other hand, frequently pose a threat to the long-term stability of Pt catalysts. Many non-Pt catalysts, such as gold (Kargi and Eker, 2007), palladium (Hosseini and Ahadzadeh, 2012), and porphyrin-related compounds (Birry et al., 2011), have recently been employed in the development of high-performance, long-lasting MFCs. However, nanomaterials with excellent electrical characteristics and tunability can improve EET at both bio-bio and bio-electrode interfaces, hence increasing electricity production in MFCs.

2.4.2 Microbial Electrochemical Fluidized Bed Reactor

The ability of live microbes to relate their metabolism to electrodes has sparked interest in METs in recent years. The liberated electrons from the oxidation of organic matter are transported to electroconductive electrodes that operate as a terminal electron acceptor (TEA), alleviating the need for aeration in traditional aerobic reactors where oxygen is the primary TEA (Pant et al., 2012). These microorganisms, known as active electro bacteria (Tejedor-Sanz et al., 2018), have been extensively explored for various applications, with MET emerging as a promising new technology. Some of these technologies include wastewater treatment and recalcitrant pollutant removal (Pun et al., 2019). METs' wastewater treatment capacity has previously been assessed using three different configurations: MFCs, MECs, and microbial electrochemical snorkel (MES). Typically, MFCs are constructed based on their ability to generate electrical power during wastewater treatment (Gao et al., 2020; Lin et al., 2020; Zhang et al., 2020). MFCs have also been used in desalination to convert the generated power (Ramirez-Moreno et al., 2021).

MECs, on the other hand, have been evaluated in general for developing novel and innovative remediation procedures, such as dechlorination (Ter Heijne et al., 2019). Despite this, MEC's strong capacity to improve wastewater treatment capacity by applying a very tiny potential drop between the anode and a reference electrode has not been thoroughly evaluated on a broad scale (Yao et al., 2019). Optimization of anodic stimulation to accelerate organic matter oxidation could be a first step toward the industrialization of bioelectrochemical technology. Only hybrid designs incorporating MET into built wetlands (CW) have been thoroughly researched and developed up to full scale (about 100 m^3/day). Wetland systems, for example, are equipped with fixed-bed biofilters built of electroconductive coke or charcoal (Prado et al., 2019) instead of the electronic systems used in traditional METs.

More recently, METland has been created to direct electron flow to particular bed sections or even to work in the presence of oxygen to boost nitrification processes (Prado et al., 2020). The high capacity of electroactive bacteria to transport electrons to the electroconductive material improves the organic matter consumption included in real wastewaters, resulting in a sustainable solution. Electroactive bacteria's ability to exchange electrons with electroconductive materials has been studied in a new subject called electro microbiology during the past two decades. This type of microbial metabolism has been shown to improve wastewater pollutant bioremediation. After confirming the pharmaceutical harmful by-product's low adsorption ability on the fluid-bed electrode (7.92 0.05% CBZ and 9.42 0.09% SMX), the microbial electrochemical fluidized bed reactor (ME-FBR) showed a significant capacity to perform better classical alternatives for removal of pollutants from the pharmaceutical companies (more than 80%).

Furthermore, the ME-FBR demonstrated the importance of anode potential selection by successfully eliminating both pollutants at +200 mV. Because of the high total organic carbon (TOC) removal efficiency, electro-stimulation of electroactive bacteria in ME-FBR overcame the expected microbial inhibition caused by CBZ and SMX. Finally, after treating wastewater with a fluid-like anode (+400 mV), *Vibrio fischeri*-based ecotoxicity was reduced by 70%, demonstrating the potential performance of this bioelectrochemical technique. Membrane bioreactors (MBR) have been the most extensively studied technique for removing CBZ and SMX from synthetic hospital wastewaters, reaching more excellent detoxification rates of roughly 37% for CBZ and SMX. Anaerobic MBR (AnMBR) has also been tested, with results showing that it can remove up to 80% of CBZ and SMX (Cheng et al., 2018). Furthermore, operational issues such as membrane biofouling, high membrane prices, and high energy demands for both systems have hampered the scale-up of both techniques for treating pharmaceutically contaminated wastewaters (Garca-Gómez et al., 2016). METs have lately been identified as one of the most promising approaches for achieving long-term bioremediation in a new environmental niche (Wang et al., 2020). In addition, MET has been evaluated in the past for wastewater treatment capabilities using three main approaches: MFCs (Cecconet et al., 2018; Gao et al., 2020), MECs (Bajracharya et al., 2016; Leon-Fernandez et al., 2019), and MES (Viggi et al., 2017). Due to its outstanding performance at lab scale, dual-chamber MEC has been the most examined technology for eliminating pharmaceutical pollutants like CBZ and SMX (Tahir et al., 2019; Rodrigo et al., 2019).

Alternative METs, in this context, have tremendous potential but are underdeveloped, necessitating extensive study to scale up the technologies. This is the situation with ME-FBR. This device uses fluid-like electrodes rather than traditional materials like graphite sheets or carbon fibers to interact with bacteria. The use of electroconductive bed materials as working electrodes is increasingly gaining popularity due to the large mass transfer area, which stimulates both anodic oxidative processes and cathodic reductions such as denitrification (Tejedor-Sanz et al., 2017). Furthermore, such fluid-like materials have a larger active surface area, lowering the technology's operating expenses even more (Xie et al., 2015).

2.4.3 Membrane Bioreactor

MBRs are one among the innovative wastewater treatment technologies. MBRs combine biological processes and membrane filtration. The decomposition of biomass takes place inside the bioreactor tank, while the filtration of treated wastewater from microorganisms is completed in a membrane module in this scenario. MBRs have gotten a lot of attention in the last two decades because of their ability to produce high-quality effluent and are now regarded as an established wastewater treatment technology (Marrot et al., 2004). In general, bioreactor process parameters have a significant impact on microorganism properties such as size, growth rate, filamentous microbe concentration, and so on.

On the other hand, microorganism activities can influence the MBR's efficiency in two ways: the effluent quality and how well the MBR can treat the wastewater pollutant, and the fouling qualities of the membranes. To obtain the optimum operating parameters of the bioreactor and characteristics of MBR plants, a thorough understanding of the principles of biological wastewater treatment, such as microorganism metabolism, microbial stoichiometry, and kinetics in bioreactors, is required (Wolff et al., 2018). The structure and composition of a bioreactor's microbial population differ from one MBR plant to the next and over time for a given MBR unit. The essential qualities of microorganisms in environmental engineering systems, especially MBR plants, are the fundamental reason for this difference. Diverse microorganisms are formed into a range of communities due to the influence of wastewater from the atmosphere, which is supplied into the bioreactor.

2.4.3.1 Types of MBR and Its Configurations

- **Aerobic membrane bioreactor and AnMBR**

  For almost a century, aerobic treatment technology has treated industrial wastewater and effluent. However, the aeration process consumes more energy, produces a lot of sludge, emits greenhouse gases like nitrous oxide (N_2O), has a large footprint, and requires a lot of maintenance. The reduction of necessary energy is essential for the broader implementation of aerobic MBR, and aeration control mechanisms in aeration basins play a vital role in reducing the power usage of the operation. Recent research in full-scale MBRs found that an ammonia-*N*-based aeration management technique reduced aeration and energy usage by 20% and 4%, respectively (Sun et al., 2016). By combining anaerobic digestion treatment with membrane filtration to treat wastewater and address some of the challenges of MBR procedures, an alternative MBR design has been proposed. The energy required for wastewater treatment is less in this situation because organic debris decomposes into methane-rich biogas. Additionally, nutrient recovery is possible after precipitation due to the conversion of nutrients into chemically accessible forms. Membrane stability, fouling, dilution of resources, and salinity buildup are some of the significant challenges to its development (Maaz et al., 2019).

- **Anaerobic fluidized membrane bioreactor (AFMBR)**

  Membrane fouling of AnMBRs, as noted in the previous section, prevents their broad use, despite their promising outcomes in eliminating pharmaceuticals and other benefits associated with them. AFMBR has recently been established to lower the rate of cake layer formation while also increasing the removal of pharmaceuticals from wastewater. It's a hybrid of membrane technology, liquid circulation, and particle sprinkling. Its key advantages over standard AnMBR include increased growth rate of bacteria that can break down pharmaceuticals (i.e., increased antibiotic elimination), lower extracellular polymeric substance (EPS) and soluble microbial product (SMP) concentrations, and more consistent sludge with the larger size. However, evaluating the effectiveness of such a unique model will require further research.

- **Membrane photobioreactor**

  Membrane photobioreactors (MPBRs) are made up of immersed micro- or ultra-filtration membranes such as hollow fibers or flat layers combined with PBRs. Most of the other benefits connected with MPBR include effective microalgae separation, system stability, and improved effluent quality. However, a key impediment to the development of this design is its inefficiency in treating primary raw domestic wastewater with high organic matter levels. This is due to the fact that it contains a low concentration of organic compounds and could be used as a source of food for microalgae in wastewater treatment (Ashadullah et al., 2021).

2.5 MICROALGAE-BASED WASTEWATER TREATMENT TECHNOLOGY

Some researchers have reported that bioaccumulation, biodegradation, and bioadsorption are the key mechanisms by which pharmaceutical pollutants and POPs are removed directly by microalgae (Leng et al., 2020; Hena et al., 2021). Additionally, with the presence of microalgae or the microalgal treatment system itself, certain pharmaceutical pollutants and POPs can be eliminated indirectly by photodegradation and volatilization (Sutherland and Ralph, 2019). Furthermore, this procedure can be broken down into three steps: 1) fast passive adsorption via physicochemical interactions between the cell surface and contaminants, followed by 2) a relatively sluggish transfer of molecules across the cell membrane, and 3) bioaccumulation, biodegradation, or both in the cell (Yu et al., 2017). Following is a more extensive overview of various microalgae-based pharmaceutical removal mechanisms.

2.5.1 Bioadsorption Using Microalgae

Pharmaceuticals such as antibiotics are absorbed into the microalgae's cell walls or onto organic molecules discharged by microalgae into their immediate surroundings, such as EPS (Sutherland and Ralph, 2019). EPS is a mixture of microorganism-derived high-molecular-weight polymers that include proteins, polysaccharides, lipids, nucleic acids, and human compounds that protect cells from the hostile environment (Sheng et al., 2010). As an adaptive mechanism responding to antibiotic toxicity, microorganisms tend to excrete more EPS (Wang et al., 2020). EPSs contribute to distinct functional groups such as amine, carboxyl, hydrophobic, and hydroxyl, due to their complex and diverse composition, consequently offering suitable binding sites for the adsorption of various organic and inorganic substances (More et al., 2014; Hansda et al., 2016). Pharmaceutical pollutants and POPs interact with microalgal cell walls or excretions (both referred to as cell surfaces) in a non-metabolic, passive manner. Adsorption reactions, surface complexation reactions, ion-exchange reactions, micro-precipitation, and chelation are all used to achieve bioadsorption (Tan et al., 2015).

Bioadsorption is one of the key processes for removing certain antibiotics and organic pollutants. The reported bioadsorption capacities of microalgae vary, owing to an increased awareness of antibiotic contamination. The adsorption of 7-amino-cephalosporanic acid by *Chlorella* sp., *Chlamydomonas* sp., and *Mychonastes* sp. was found to be pretty quick at first, taking only 10 minutes, with adsorption capacities of 4.74, 3.09, and 2.95 mg/g, respectively (Guo et al., 2016). Furthermore, the maximal tetracycline adsorption capabilities of *Scenedesmus quadricauda* and *Tetraselmis suecica* were reported to be 295.34 and 56.25 mg/g, respectively (Daneshvar et al., 2018). These findings suggested that microalgae's bioadsorption capacities were significantly dependent on physical and chemical parameters such as surface chemistry and surface area (Norvill et al., 2016). Due to the presence of prominent functional groups such as hydroxyl, carboxyl, and phosphoryl, the cell walls of microalgae and EPS primarily carry negative charges. As a result of electrostatic interactions, antibiotics having a positive charge can be successfully absorbed (Xiong et al., 2018). For example, *Chlorella saccharophila* effectively removed azithromycin (almost 100%), but trimethoprim bioadsorption rate was less than 30% (Gojkovic et al., 2019). Previous research has revealed that the adsorption performance of various antibiotics varies depending on their hydrophilicity, structure, and functional groups (Hena et al., 2021). Due to electrostatic interactions, lipophilic compounds have high bioadsorption affinity values for microalgae, whereas hydrophilic compounds have low bioadsorption affinity values and are more durable in growth medium (Xiong et al., 2018; Sutherland and Ralph, 2019). Antibiotics attach to the microalgal surface on both living and non-living microalgal cell surfaces since bioadsorption is a non-metabolic process. The biomass of non-living microalgae is a tremendous potential biosorbent for removing antibiotics. Cefalexin (at 50 mg/L) was successfully removed from wastewater by non-living *Chlorella* sp. and lipid-extracted *Chlorella* sp., respectively, with adsorption capacities of 129 and 63 mg/g (Angulo et al., 2018).

2.5.2 Bioaccumulation Using Microalgae

Bioaccumulation is an active metabolic process in which antibiotics bind to intracellular proteins or other chemicals in living microalgal cells (Xiong et al., 2018). Although bioadsorption is the first stage in bioaccumulation, not all contaminants that have been adsorbed onto the surface of microalgae may enter the cell and bioaccumulate (Wu et al., 2012). Pharmaceutical waste like antibiotics and organic pollutants can enter the cell through three basic pathways: passive diffusion, passive-facilitated diffusion, and energy-dependent/active absorption (Sutherland and Ralph, 2019). Because some antibiotics can diffuse through the cell membrane from high (external) to low (internal) concentration, passive diffusion of antibiotics requires no energy. Because of its hydrophobicity, antibiotics with low molecular weight, particularly non-polar and lipid-soluble antibiotics, may pass through the cell membrane via passive diffusion. Antibiotics infiltrated microalgal cells via passive diffusion after eliminating SMX, trimethoprim, florfenicol, and CBZ, causing bioaccumulation (Bai and Acharya, 2016; Song et al., 2019). Passive diffusion can also be caused by changes in the permeability of the cell membrane due to antibiotic treatment or a stressful environment, a phenomenon credited to membrane depolarization or hyperpolarization (Sutherland and Ralph, 2019). Passive diffusion of antibiotics may be aided by interfering with the integrity of the cell membrane. For example, adding 1% (w/v) sodium chloride to *Chlorella vulgaris* increased levofloxacin bioaccumulation to 101 g/g, compared to 34 g/g without sodium chloride (Xiong et al., 2017a). Antibiotics pass through the cell membrane with the help of transporter proteins, responsible for facilitating the inflow of polar compounds into the cell. The third pathway is an energy-driven active transport process in which antibiotics move up and down a concentration gradient (Sutherland and Ralph, 2019). CBZ bioaccumulation in *Chlamydomonas mexicana* and *Scenedesmus obliquus* increased when CBZ concentration and cultivation duration increased (Xiong et al., 2016). Optimization of physicochemical parameters such as pH, temperature, contact time, and antibiotic concentrations may influence the rate and amount of antibiotics acquired by microalgae and protect the cells from any toxicity, hence increasing antibiotic clearance. On the other hand, microalgae may drain accumulated antibiotics in cells through metabolism, implying that buildup is a prerequisite for biodegradation. Earlier research has revealed that the simultaneous accumulation and biodegradation of antibiotics in microalgae cells play a prominent role in antibiotic metabolism (Sun et al., 2016; Xiong et al., 2017a; Song et al., 2019). Levofloxacin and sulfamethazine, for example, were reported to be successfully removed by microalgae through accumulation and subsequent intracellular biodegradation (Xiong et al., 2017a).

2.5.3 Biodegradation Using Microalgae

As previously stated, bioadsorption and bioaccumulation by microalgae play a vital role in removing antibiotics from the aqueous phase. Bioadsorption, on the other hand, is a phase transfer phenomenon that functions as a biological filter for concentrating antibiotics of interest (Sutherland and Ralph, 2019). Bioabsorbed antibiotics are likely to be released into the aqueous phase, reversing this process (Oberoi et al., 2019). Biodegradation is the most effective method for removing antibiotics from microalgae (Xiong et al., 2018). Biodegradation is a complex process that involves breaking down organic molecules, either by biotransformation or by complete mineralization to CO_2 and H_2O by pure or mixed microbial cultures (Xiong et al., 2019, 2020). The underlying mechanisms of biodegradation can be divided into two categories:

- metabolic degradation, in which the antibiotic acts as the sole carbon source or electron donor/acceptor for microalgae; and
- co-metabolism, in which additional organic substrates both sustain biomass production and act as an electron donor for the non-growth substrate (Tiwari et al., 2017; Leng et al., 2020).

Extracellular and intracellular biodegradation of antibiotics by microalgae is common or a combination of both, where initial degradation occurs extracellularly and the breakdown products are then metabolized intracellularly (Xiong et al., 2018). Intracellular degradation is based on antibiotics accumulating in microalgal cells, whereas extracellular degradation is based on EPS excreted by microalgae. Kiki et al. (2020) investigated the ability of four microalgal strains, *Haematococcus pluvialis*, *Selenastrum capricornutum*, *S. quadricauda*, and *C. vulgaris*, to remove ten antibiotics and discovered that biodegradation contributed between 23% and 99% of the removal efficiency.

Furthermore, Xie et al. (2020) showed that *Chlamydomonas* sp. Tai-03 had a 100% removal efficiency of ciprofloxacin, with biodegradation accounting for 65.05% of the clearance. Biodegradation is dependent on the microalgae's cellular metabolism, which comprises several complex enzyme activities. Antibiotic biodegradation mediated by microalgae is a two-phase enzymatic catalysis process based on enzyme functions (Leng et al., 2020; Wang et al., 2022). The first phase (phase I) process involves the addition of reactive functional groups by oxidation, reduction, or hydrolysis reactions. Microsomal monooxygenase enzymes or mixed-function oxidases, such as cytochrome b5, cytochrome P450, and nicotinamide adenine dinucleotide phosphate (NADPH) cytochrome P450 reductase, catalyze these processes (Torres et al., 2008). As a result, numerous enzymatic reactions have been reported during phase I activities, including decarboxylation, carboxylation, hydroxylation, oxidation, dehydroxylation, methylation, demethylation, hydrogenation, and ring cracking (Xiong et al., 2019, 2020; Xie et al., 2020). According to Xiong et al. (2020), SMX biodegradation by *Chlorella pyrenoidosa* may involve phase I processes such as amine group oxidation and hydroxylation, followed by phase II reactions such as formulation and pterin-related conjugation. Furthermore, the ability of microalgae to detoxify xenobiotics is similar to that of mammalian livers, implying that microalgae can be used as "green livers" for environmental pollutant detoxification (Torres et al., 2008). Antibiotic biodegradation by microalgae involves enzymes from phases I and phase II enzyme groups. However, the major enzymes that regulate this process and the clarification of their roles have yet to be extensively studied (Xiong et al., 2018).

2.5.4 Microalgae-Mediated Acclimation

Improved microalgae response is critical for wastewater treatment, which can be accomplished by acclimating microalgae to harsh or restricting conditions, such as nutrient shortage and toxicant exposure. Previous research suggests that acclimated microalgae have higher tolerance and biodegradation ability than wild-type microalgae, implying that pollutants are removed more efficiently (Liao et al., 2016; Xiong et al., 2017a). Microalgae acclimation has been widely employed in wastewater treatment to remove nutrients and micro-pollutants. Chen et al. (2015) found that acclimated *C. pyrenoidosa* with 60 mg/L cefradine had a higher removal efficiency. Xiong et al. (2017a) observed that *C. vulgaris* pre-exposed to 200 mg/L of levofloxacin for 11 days had a higher levofloxacin elimination capacity (28%) than wild species (16%). Furthermore, Liao et al. (2016) found that microbial communities incubated at a higher temperature (45°C) for 28 days had increased sulfanilamide biodegradation. Overall, these investigations show that acclimation can help improve microalgae tolerance to contaminants during bioremediation and biomass production. Genetic alterations generated by spontaneous mutation or physiological adaptation are credited with microalgae's ability to adapt to harsh circumstances (Osundeko et al., 2014). Improved capability of acclimated microalgae, on the other hand, could be attributed to improved photosynthesis, carotenoid biosynthesis, antioxidant system activities, or metabolism processes (Cho et al., 2016; Xiong et al., 2018).

2.5.5 Microalgae-Mediated Co-metabolism

The transformation of a non-growth substrate in the presence of a growth substrate or another biodegradable chemical is known as co-metabolism (Tran et al., 2013). Additional organic substrates help maintain biomass production and act as electron donors for non-growth substrate co-metabolism (Xiong et al., 2018). Adding organic substrates or other nutritional items to build a co-metabolic system has been demonstrated to be an effective technique for boosting microalgae-mediated biodegradation of POPs in previous research (Xiong et al., 2020; Vo et al., 2020a). Following the addition of sodium acetate, the elimination efficiency of ciprofloxacin by *C. mexicana* increased from 13% to 56% (Xiong et al., 2017b). Sodium acetate significantly improved the removal efficiency of SMX by *C. pyrenoidosa* (99.3%) compared to the control group (6.05%) in our earlier work (Xiong et al., 2020).

Furthermore, Liang et al. (2019) discovered that acetate's effect on the degradation of 26 pharmaceuticals was dosage and chemical dependent. Acetate concentration was linked to both improved parent chemical removal efficiency and the regulation of transformation products and pathways. Co-metabolism can be influenced by various parameters, including response time, target chemical concentration, microalgal species, and nutritional material type and concentration. The type and concentration of the nutritional substance is an essential aspect among these (Tran et al., 2013). The effects of various carbon sources on the co-metabolism of micro-pollutants by microalgae have been thoroughly investigated. Their findings revealed that sugar-based carbon sources had the highest concentrations of EPS and enzymes and higher removal efficiency of pharmaceuticals and POPs (Vo et al., 2020a, 2020b). However, the presence of some organic substrates can reduce the pollutants' removal effectiveness, presumably due to catabolite repression (Xiong et al., 2017b, 2020). As a result, it is vital to assess the impacts of various carbon sources, select the most appropriate ones, and ensure effective application in wastewater bioremediation. Meanwhile, this procedure will aid biomass recovery and increase revenue for multiple sectors.

2.6 POTENTIAL OF GENETIC ENGINEERING IN REMOVAL OF PHARMACEUTICALS AND POPS

Microorganisms, in general, have species-specific preferences and tolerances for their surroundings, as well as specialized biodegradation performance for various pharmaceuticals and POPs (Sutherland and Ralph, 2019; Leng et al., 2020; Nguyen et al., 2020a,b). Prospecting for good microbial strains is a critical first step in building a stable microorganism-based system for continuous and effective pollution removal from wastewater, as well as a crucial component in the long-term generation of biofuels (Perera et al., 2019; Zhang et al., 2020). The high-throughput sequencing technology allows researchers to examine the structure, functions, and interactions of the microbial population in a microalgae-based system, as well as isolated microbes capable of effective micropollutant biodegradation (Nguyen et al., 2020a,b). Zhang et al. (2020b) studied the biodegradation potential of chloramphenicol by an enhanced bacterial consortium, and the results revealed that the main bacterial taxa *Burkholderia*, *Chryseobacterium*, *Cupriavidus*, *Sphingomonas*, and *Pigmentiphaga* were involved. Furthermore, chloramphenicol was eliminated by the isolated *Sphingomonas* sp. CL5.1 in 48 hours, with a 50.4% mineralization rate (Zhang et al., 2020b).

Furthermore, modern genetic engineering technology has been used to engineer microbial strains or consortia to improve microbe metabolism or enrich bacteria with specific characteristics (Rios Miguel et al., 2020; Khatiwada et al., 2020). Compared to wild bacteria, thermo-tolerant facultative anaerobes altered by genetic engineering displayed much better and efficient degrading capacity on numerous nitroalkanes (Zhang et al., 2021a). Microalgae clones containing functional enzyme genes, such as laccase, have been shown to improve oxidoreductase stability, ensuring efficient bioremediation of pollutants (Subashchandrabose et al., 2013). Laccases are a type of extracellular oxidoreductase found in both bacteria and plants. Because of their poor substrate specificity, they can catalyze a wide range of reactions, including one-electron oxidation of monophenols, diphenols, and polyphenols, as well as aromatic amines and diamines (Bilal et al., 2019). Targeted genome editing, in addition to introducing functional genes into microalgal genomes, is becoming increasingly common in the alteration of microalgal strains (Khatiwada et al., 2020). In *Euglena gracilis*, the CRISPR-based gene targeting technique successfully created a glucan synthase-like 2 gene responsible for paramylon production (Nomura et al., 2019). Furthermore, increased transfer of changed genes may enhance the mobilization of antibiotic resistance genes (ARGs). To widen their application in wastewater bioremediation, more attention should be paid to applying genetic engineering technologies in on-site reactors at wastewater treatment plants and assessing the risk of gene alteration, such as ARG transfer.

2.6.1 Metabolic Engineering for Enhanced Biotransformation

Microorganisms have significant metabolic pathways to use numerous harmful substances as a source of energy for development and growth through respiration, fermentation, and co-metabolism, during the physical and chemical adsorption of wastes employing various microorganisms or derived non-living biocomponents. Biotransformation, in contrast to biosorption, which is based on pollutant concentration and kinetic balance of microbial binding sites, is metabolism-driven and so maybe advantageous when great targeting sensitivity is required at low pollutant concentrations. Aerobic bacteria could convert organic pollutants into non-toxic equivalents, mainly carbon dioxide, in the presence of oxygen. Some contaminants can operate as electron acceptors in the reduction process and be electron donors to be removed. For example, certain anaerobic microbes can digest chlorine found in pollutants by using reductive dehalogenation metabolism (Nelson et al., 2011; Zanaroli et al., 2015). The metabolic activities within the cells highlighted above do not encompass all possibilities. EET pathway has been established by certain electrochemically active bacteria (EAB), such as *Shewanella oneidensis*, *Geobacter sulfurreducens*, and *Pseudomonas aeruginosa*, to "dump" metabolically produced electrons outside the cellular membranes, which exterior electron acceptors or electrode materials could then acquire (Malvankar et al., 2011; Hsu et al., 2018).

When hazardous metal impurities are present in the growing biological environment, soluble pollutants can be converted to non-soluble forms by functioning as electron acceptors, accomplishing the purifying goal. Although EAB utilizes simple substrates such as acetate, lactate, or glucose in most scenarios, the study has shown that this is possible to use more complex substrates such as commercial or domestic water, which would further expand the possibilities of EAB-based wastewater treatment (Velvizhi et al., 2015). Electricity can be generated by using solid electrodes as electron acceptors and recovering chemically stored energy in wastewater when bacteria grow on the electrode used in MFCs (He et al., 2017). Physiology manipulation can be used to stimulate functions of microorganisms involved in contaminant degradation by optimizing conditions of the environment such as pH and temperature and adding necessary nutrients for their metabolic reaction to maximize pollutant treatment efficiency. By regulating the development and metabolism of microorganisms, controlling pH and other environmental parameters provides an easy and efficient technique to improve biotreatment ability.

Engineering metabolic pathways of microorganisms offer an alternative route to tackle the above constraints by designing and establishing new catabolic pathways with enhanced pollution treatment abilities for some fundamentally slow and ineffective metabolic activities. In genetic engineering, the development of whole-genome sequencing and high-throughput screening aids the global picture of gene expression, enzymes, and metabolic pathways in microbes under stress produced by contaminants (Dai et al., 2018). Metabolic engineering can then be used to change existing metabolic pathways or incorporate new catabolic pathways onto specific host cells, resulting in improved biotreatment capabilities by increasing enzymatic activity, widening the spectrum of specified pollutants, or improving biofilm formation (Dangi et al., 2019). In microbial metabolic processes, enzymes, the structural components for powering the biotransformation process, play a significant role. However, in realistic situations, the expression levels of an enzyme's native host may be low, compromising its stability and effectiveness in adverse environmental conditions. By removing and introducing the coding genes into some other expression host, genetic engineering, on the other hand, can improve the output of these enzymes.

Furthermore, desirable enzyme features such as substrate consumption, stress tolerance (temperature, pH, solvents), and the reaction pathway can be modified when paired with genetic manipulation or logical design technology. Coconi-Linares et al. (2015) achieved heterologous co-expression of recombinant enzyme peroxidases and laccases in *Phanerochaete chrysosporium* strains, resulting in a wide range of phenolic/non-phenolic biotransformation and a high percentage of synthetic dye decolorization when compared to the wild strain. Harford-Cross et al. (2000) produced a mutant cytochrome P450 in *Pseudomonas putida*. The two targeted mutations in the enzyme active site increase treatment effectiveness against benzo[a]pyrene, fluoranthene, phenanthrene, and pyrene, among other PAHs. For example, the TCA cycle (also known as the citric acid cycle) is a crucial component of metabolism that generates ATP and NAD(P)H to promote downstream quinone reduction. It is thus possible to boost the EET rate by altering TCA cycle activity. Izallalen et al. (2008) have created an ATP drain in *G. sulfurreducens* to lower the cell's ATP level and contribute to the increased respiration rates of engineered cells.

2.6.2 Surface Binding Engineering

Microorganisms have high non-specific pollutant adsorption abilities due to their high surface area per unit weight and the prevalence of electronegative functional groups on cell surfaces such as hydroxyl groups, sulfhydryl groups, carboxyl in anionic groups, phosphate groups, and nitrogen-containing groups such as amino groups (Ayangbenro and Babalola, 2017). Gram-positive bacteria (*Bacillus, Corynebacterium, Streptomyces, Staphylococcus* sp., etc.), gram-negative bacteria (*Pseudomonas, Enterobacter, Aeromonas* sp., etc.), and cyanobacterium have all shown good adsorption capacity (*Anabaena* sp., etc.). Several studies have detailed recent accomplishments in treating dirty water utilizing natural strains as sorbents that can be accessed elsewhere (Gupta and Diwan, 2016; Nzila et al., 2016).

However, the application of employing microbial cells as biosorbents is limited due to the limited expression in specific strains, restricted binding sites on these produced proteins/peptides, and low selectivity of particular ions in the presence of other compounds (Li and Tao, 2015). As a result, there is a pressing need to create various pollutant-binding proteins/peptides that can be shown on a variety of cells and have increased affinity and specificity for pharmaceutical wastes and POPs. *Escherichia coli, P. putida*, and the yeast *Saccharomyces cerevisiae* stand out among commonly employed host cells for their well-studied genetic engineering paradigms. These microbes have been genetically modified to produce MTs or PCs that bind pharmaceutical wastes and POPs (Shahpiri and Mohammadzadeh, 2018; Li et al., 2015). Metalloregulatory proteins can be expressed when the specific performance for a target pharmaceutical waste and POP ion is needed due to their "pollutant sensing" capabilities, which translates binding events into conformational changes (Reyes-Caballero et al., 2011). Protein engineering that permits novel activities and improved performances by random design or guided evolution could also enrich the pollutant-binding capacities of known proteins/peptides and exhibit existing proteins/peptides. Zhou et al. (2014) have produced a designed uranyl-binding protein that is thermally stable and has excellent uranyl sensitivity and selectivity. Incorporating numerous binding sites inside a single protein/peptide can improve the binding affinity and target multiple pollutant ions simultaneously. Mauro et al., for example, built multidomain polypeptides and expressed them in *E. coli*. Non-living biomass can also be employed as sorbents for pollutant removal because the biosorption process is metabolism-independent due to charged functional groups on the EPS matrix, such as carboxylic and phosphoryl groups. EPS released by microbial cells throughout biofilm development has received a lot of attention (Sheng et al., 2010; Dash and Das, 2016). Rather than using the entire EPS matrix, which includes polysaccharides, proteins, lipids, and DNA, adsorption only requires single components to be extracted and immobilized onto a supporting material to act as a "living filter," as opposed to the traditional non-living filtration technologies like reverse osmosis and nanofiltration. Another key EPS component often employed is *E. coli* amyloid protein nanofibers, which may be efficiently genetically modified to provide a range of functionalities, including particular pollutant ion binding (Tay et al., 2017).

2.6.3 Microbial and Engineered Microbial Consortium

Recent research has revealed the potential of microalgal consortia in various applications, including biomass generation and biotechnologically valuable metabolites and cost-effective and environmentally friendly wastewater treatment (Perera et al., 2019; Nguyen et al., 2020a). A mixed culture of two or more microalgal species, or an interaction of microalgae with other microbes such as bacteria or fungi, is referred to as a consortium or co-cultivation culture (Liu et al., 2017). Microalgae-microalgae consortia and microalgae-bacteria consortia are the most widely employed for wastewater treatment among the various consortia that can be developed. Photosynthetic microorganisms (eukaryotic and prokaryotic) make up the microalgae-microalgae consortium, whereas photosynthetic microorganisms and heterotrophic bacteria make up the microalgae-bacteria consortium (Gonçalves et al., 2017). The use of consortia in wastewater bioremediation has two advantages over using individual microorganisms.

To begin with, cooperative interactions between co-cultivated microorganisms improve total nutrient and micropollutant absorption, and these systems are more robust to environmental oscillations (Gonçalves et al., 2017). A microalgae-microalgae consortium made up of *Chlorella* sp. and *Scenedesmus* sp. increased the biodegradation removal effectiveness of ibuprofen by 40% and cut the lag phase of caffeine by three days (Matamoros et al., 2016). Although the reason for the microalgae co-cultures' higher pollutant removal efficiency is still unknown, researchers have linked it to competitive/inhibitive and cooperative interactions between the different microalgal strains (Hena et al., 2021). Mendes and Vermelho (2013) proposed that the exchange of metabolites between microorganisms during cooperative interactions improves biomass accumulation and, as a result, nutrient and micropollutant removal efficiency. Xiong et al. (2017c) investigated the ecotoxicity and removal of enrofloxacin by five microalgae species: *C. vulgaris*, *Micractinium reisseri*, *S. obliquus*, *Ourococcus multisporus*, and *C. mexicana*, as well as their consortia.

Microalgae-bacteria consortium has been widely used for wastewater treatment than the microalgae-microalgae consortium because of its less energy demand, cost-effectiveness, and possible resource recovery (Perera et al., 2019; Zhang et al., 2020a). In the microalgae-bacteria consortium, interactions between microalgae and bacteria encompass a broad spectrum of connections, from collaboration to rivalry (Figure 2.2).

In general, heterotrophic bacteria can use the oxygen created by microalgae for oxidation reactions of organic carbons. The CO_2 emitted during bacterial mineralization can be supplied back to microalgae for photosynthesis (Quijano et al., 2017). Microalgae offer dissolved organic carbon to bacteria, whereas bacteria provide fixed nitrogen, vitamin B12, and siderophores for microalgal development (Tong et al., 2023). The direct or indirect cooperative interactions between microalgae and bacteria enable the microalgae-bacteria consortium to remove pollutants more effectively and efficiently than the microalgae-microalgae consortium (Gonçalves et al., 2017; Zhang et al., 2020a). Although various interactions between microalgae and bacteria have been discovered, little is known about their inner molecular mechanisms. The majority of prior research has characterized microalgae-bacteria interactions from

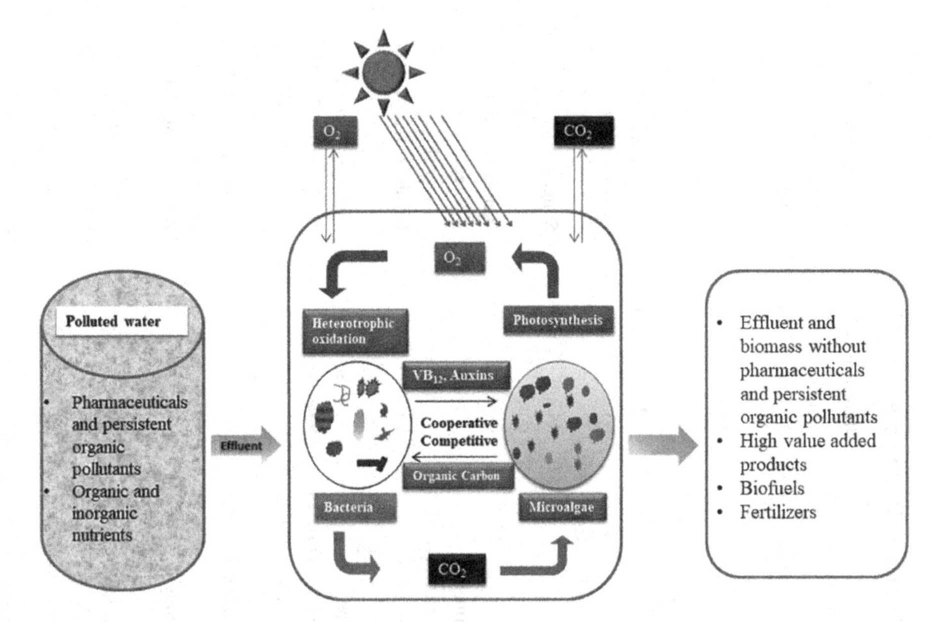

FIGURE 2.2 Interaction between bacteria and microalgae for removal of pharmaceuticals and POPs from wastewater.

biological and physical viewpoints, with the only limited molecular and biochemical explanation of these interactions. A thorough understanding of the mechanisms underlying the interactions between microalgae and bacteria is required to inspire large-scale uses of the microalgae-bacteria consortia. Detail about emerging microbial technologies for removal of pharmaceuticals and POPs from wastewater is given in Table 2.1.

2.6.4 Engineering Consortia

Unlike previous discussions, which have focused on monocultures, microbial consortia, which consist of multiple microorganisms working together, have distinct benefits, like the ability to perform complicated tasks that individual populations would otherwise be incapable of, as well as a higher tolerance to changing environments (Brenner et al., 2008; Bernstein and Carlson, 2012). Wastewater comprises a wide range of contaminants and certain pollutants with complex chemical structures. As a result, using a single strain to eliminate all wastes simultaneously is difficult. Furthermore, more efficient and productive treatment can be expected when co-metabolic processes inside microbial communities complement each other. Researchers have effectively reduced certain wastes, such as phenol, by constructing particular consortia (Azaizeh et al., 2015), organic acid (Ogbonna et al., 2000), nitrates and phosphate (Dahiya and Venkkatamohan, 2016), and cellulose (Henske et al., 2018). Consortia of cyanobacteria/microalgae and bacteria are one example (Gonçalves et al., 2017). Photosynthesizing microorganisms, in particular, supply oxygen, which

TABLE 2.1
Different Microbial Technologies for Removal of Pharmaceuticals and POPs from Wastewater

S. No.	Pharmaceuticals/ POPs	Methods/Reactors	Removal Efficiency	Reference
1.	Ibuprofen and diclofenac	Batch experiments (microcosms)	Complete removal	Langenhoff et al. (2013)
2.	Antibiotics carbamazepine and sulfamethoxazole	Microbial electrochemical fluidized bed reactor (ME-FBR)	More than 80%	Yeray et al. (2021)
3.	Antibiotic compounds; sulfonamides, beta-lactam, and chloramphenicol	*MFC bacterial oxidation reactor*	More than 75%	Zhang et al. (2020),
4.	Chloramphenicol (CAP)	Two-chambered MFC reactor	84%	Zhang et al. (2017)
5.	Cefazolin sodium	Two-chambered MFC; METs	1.2–6.8 mg/L/h	Wen et al. (2011)
6.	Bisphenol A and ibuprofen	Upflow Microbial fuel cell-coupled constructed wetland	63.2%–78.7%	Li et al. (2019)
7.	Antibiotic metronidazole	MFC reactor	85.4%	Song et al. (2013)
8.	Antibiotic oxytetracycline	Two-chambered MFC reactor	99%	Yan et al. (2018)
9.	Ceftriaxone sodium	Single-chamber MFC	98%	Wen et al. (2011)
10.	Azithromycin	Activated sludge followed by chlorination	47.90	Blair et al. (2015)
11.	Ranitidine	Activated sludge followed by biologic filtration	More than 66%	Gros et al. (2010)
12.	Mixture of analgesics (ketoprofen, paracetamol, and aspirin)	Artificial microalgal-bacterial consortium	95%	Ismail et al. (2017)
13.	Levofloxacin	Microalgae *Chlorella vulgari* consortia	70%–80%	Ndlela et al., (2023)
14.	PoP; lindane (γ-hexachlorocyclohexane)	Microbial metabolization by *Kocuria* sp. DAB-1Y, *Staphylococcus* sp. DAB-1W, and *Sphingobium japonicum*	94%–98%	Nagata et al., (2019), Kumar et al. (2016)
15.	PoP; endosulfan	Microbial metabolization by *Bacillus subtilis*	94%	Kumar et al. (2014), Ahmad et al. (2020)
16.	PoP; pentachlorophenol	Microbial metabolization by *Pseudomonas fluorescens*	99%	Ammeri et al. (2017)
17.	PoP; 4-chlorophenol	Microbial metabolization by *Bacillus subtilis* MF447840.1	100%	Sandhibigraha et al. (2019)
18.	PoP; 1,1,1-trichloro-2,2-bis(4-chlorophenyl) ethane (DDT)	Microbial metabolization by *Pseudoxanthomonas* sp.	95%	Wang et al. (2010)

is required by pollutant-degrading heterotrophic bacteria. Bacterial mineralization produces carbon dioxide in exchange, which completes the photosynthetic cycle. This symbiotic connection can be considered an ideal self-sustaining system that is superior to traditional engineering methods for oxygen addition. When biomass degrader-anaerobic fungi are co-cultured with methanogenic archaea, the effectiveness of waste breakdown is improved by strengthening synergistic interactions between them, according to the researchers (Swift et al., 2019). Co-cultivation of *Neocallimastix* strain N1 with *Methanobacterium formicicum* strains, for example, boosted cellulose digesting rate while simultaneously shifting reaction products from less valuable molecules like lactate and ethanol to more valuable fuel energy—methane (Teunissen et al., 1992). Following the lead of natural symbiotic strains, rationally selecting and controlling desired cell-cell communications among engineering microorganisms to form a coordinated cellular network could provide a way to reduce metabolic burden in microbial cells or degrade complex organic compounds that are difficult to construct a whole degradation pathway in one strain.

Even though genetically designed biological machinery has exhibited promising therapeutic outcomes due to its remarkable adsorption and catalytic capabilities, its use is still restricted to appropriate laboratory circumstances (Ezezika and Singer, 2010). Before modified microorganisms may work in a real-world environment, there are primarily two hurdles that must be overcome. The first difficulty is the durability of designed cells in polluted environments with considerable variations from laboratory values in pH, salinity, temperature, dissolved oxygen, redox potential, radioactivity, and overall cleanliness. Upregulation of genes in the plasmids of modified bacteria could also place excessive pressure on the cells, inhibiting their growth and reproduction (Rosano and Ceccarelli, 2014).

2.7 NOVEL STRATEGIES FOR ENHANCED REMOVAL OF PHARMACEUTICALS AND POPS

Over the last few decades, omics technologies combined with biochemical and microbial analysis have proven to be promising molecular tools, providing detailed information on gene, transcriptome, metabolic, and protein expression changes, as well as the regulation of these changes in the biodegradation of micro-pollutants (Mishra et al., 2019). Even though omics technologies have emerged as potential molecular tools for studying micropollutant biodegradation mechanisms, their application in antibiotic biodegradation is still in its infancy, and more research is needed. Genomics, proteomics, and metagenomics methods can be used to uncover the structure and function of genes, proteins, enzymes, and metabolic pathways involved in the degradation process and screen and experimentally validate gene functions (Ufarte et al., 2015; Jhariya and Pal, 2022). Transcriptomics, proteomics, and metabolomics have also been used to decipher microalgae's complex molecular and cellular responses to various pollutants (Nagarajan et al. 2022; Wang et al., 2017). Wang et al. (2017) used transcriptome profiling to uncover the molecular response of *Desmodesmus* sp. to bisphenol A (BPA) exposure. Their findings revealed that genes encoding oxidoreductases and glycosyltransferases were significantly upregulated in microalgae, with oxidoreductases identified to increase BPA oxidative biodegradation and glycosyltransferases linked with detoxification (Wang et al., 2017). Patel et al. (2015) used label-free shotgun proteome analysis to look at *Chlamydomonas reinhardtii*'s nitrogen absorption and metabolism in response to wastewater processing. Their findings demonstrated increased enzymes and proteins involved in nitrogen metabolism and assimilation, carbohydrate synthesis and utilization, amino acid recycling, oxidative stress, and lipid biosynthesis (Patel et al., 2015). In combination with data analytics, multi-omics has aided in gaining a better understanding of the inner molecular mechanisms of micropollutant biodegradation (Mishra et al., 2019). Zhang et al. (2021b) integrated transcriptome, metabolomic, and free radical investigations to inquire into the functional groups and detailed molecular mechanisms of sulfonamide biodegradation. In the biodegradation of sulfonamides, carbohydrate-active enzymes, oxidoreductases, and OH played critical roles (Zhang et al., 2021b).

2.8 FUTURE CHALLENGES IN REMOVAL OF PHARMACEUTICALS AND POPS

The action of enzymes secreted by microorganisms breaks down complex, dangerous chemical compounds into more specific, less harmful molecules, thus microorganisms play a crucial role in eliminating pharmaceuticals and POPs from wastewater. The solubility of pharmaceutical contaminants in wastewater determines their biodegradation effectiveness. If pollutant solubility is low especially in case of hydrophobic compound it may be retained in sewage sludge. The retention of these molecules in sludge allows for additional time for microbial breakdown, in which the micro-pollutants are either destroyed by catabolic microbial enzymes or used as a carbon source by microorganisms. Hydrophilic micro-pollutants, on the other hand, may escape without biodegradation and permeate, evading the biodegradation process. In both aerobic and anaerobic digestion, the speed and efficiency of degradation differ from compound to compound, depending on the structure and functional group of the molecules. Sludge retention time (SRT) is another element that has a substantial impact on biodegradation rates. The rate of biodegradation is determined by operating conditions (pH, temperature, and retention time), complexity, microbial population, and bioavailability of micro-pollutants. The hydrodynamic condition of the membrane module/cassette/tank arrangements should be ideal to ensure the best fouling management.

2.9 CONCLUSION

Microorganisms have showed great promise in recovering contaminated wastewater in a cost-effective and long-term manner. The fact that several microorganisms, particularly microalgae, play an important role in the removal of pharmaceuticals and POPs promises a bright future for microorganism-based wastewater treatment. Recent advances in synthetic biology have offered viable platforms for specifically engineering their structure and function, improving adsorption and catalytic capabilities, and increasing biotreatment efficiency and specificity (EPS). Furthermore, there is a lot of evidence suggesting an extracellular polymeric material is important for antibiotic bioadsorption and extracellular biodegradation. Previous research has demonstrated that co-metabolism is a successful technique for boosting antibiotic biodegradation by microalgae, with the type and concentration of carbon sources playing the most important role in antibiotic co-metabolism. The role of EPS in antibiotic elimination and the underlying processes that drive the complex interactions between EPS and antibiotics are unknown. Advanced genetic engineering approaches and omics technologies, such as genomics, proteomics, metabolomics, and transcriptomics, are shedding light on functional genes and metabolic pathway, identifying potential proteins that mediate electron transport, and enzymes that catalyze pharmaceutical and POP metabolic reactions.

REFERENCES

Ahmad, K. S. 2020. Remedial potential of bacterial and fungal strains (*Bacillus subtilis, Aspergillus niger, Aspergillus flavus* and *Penicillium chrysogenum*) against organochlorine insecticide Endosulfan. *Folia Microbiol.* 65, 801–810.

Al-Mulali, U., Ozturk, I., Lean, H. H. 2015. The influence of economic growth, urbanization, trade openness, financial development, and renewable energy on pollution in Europe. Nat. *Hazards.* 79(1), 621–644.

Ammeri, R. W., Mehri, I., Badi, S., Hassen, W., Hassen, A. 2017. Pentachlorophenol degradation by *Pseudomonas fluorescens. Water Qual. Res. J.* 52, 99–108.

Angulo, E., Bula, L., Mercado, I., Montano, A., Cubillan, N. 2018. Bioremediation of cephalexin with non-living *Chlorella* sp., biomass after lipid extraction. *Bioresour. Technol.* 257, 17–22.

Ashadullah, A. K. M., Shafiquzzaman, M., Haider, H., Alresheedi, M., Azam, M. S., Ghumman, A. R. 2021. Wastewater treatment by microalgal membrane bioreactor: evaluating the effect of organic loading rate and hydraulic residence time. *J. Environ. Manag.* 278, 111548.

Ashraf, M. A. 2017. Persistent organic pollutants (POPs): a global issue, a global challenge. *Environ. Sci. Pollut. Res.* 24(5), 4223–4227.

Ayangbenro, S. A., Babalola, O. O. 2017. A new strategy for heavy metal polluted environments: a review of microbial biosorbents. *Int. J Env. Res. and Pub. Health.* 14(1), 94.

Ayman, Z., Işık, M. 2015. Pharmaceutically active compounds in water, Aksaray, Turkey. *Clean Soil Air Water.* 43, 1381–1388.

Azaizeh, H., Kurzbaum, E., Said, O., Jaradat, H., Menashe, O. 2015. The potential of autochthonous microbial culture encapsulation in a confined environment for phenol biodegradation. *Env. Sci. Poll. Res.* 22(19), 15179–15187.

Bai, X. L., Acharya, K., 2016. Removal of trimethoprim, sulfamethoxazole, and triclosan by the green alga *Nannochloris* sp. *J. Hazard. Mater.* 315, 70–75.

Bajracharya, S., Sharma, M., Mohanakrishna, G., Dominguez Benneton, X., Strik, D. P., Sarma, P. M., Pant, D. 2016. An overview on emerging bioelectrochemical systems (BESs): technology for sustainable electricity, waste remediation, resource recovery, chemical production and beyond. *Renew. Energy.* 98, 153–170.

Bakirtas, T., Akpolat, A. G. 2018. The relationship between energy consumption, urbanization, and economic growth in new emerging-market countries *Energy.* 147, 110–121.

Bernstein, H. C., Carlson, R. P. 2012. Microbial consortia engineering for cellular factories: in vitro to *in silico* systems. *Comp. Struc. Biotech. J.*, 3(4), e201210017.

Bilal, M., Adeel, M., Rasheed, T., Zhao, Y., Iqbal, H. M. N. 2019. Emerging contaminants of high concern and their enzyme-assisted biodegradation – a review. *Environ. Int.* 124, 336–353.

Birry, L., Mehta, P., Jaouen, F., Dodelet, J. P., Guiot, S. R., Tartakovsky, B. 2011. Application of iron-based cathode catalysts in a microbial fuel cell. *Electrochimica Acta*. 56(3), 1505–1511.

Blair, B., Nikolaus, A., Hedman, C., Klaper, R., Grundl, T. (2015). Evaluating the degradation, sorption, and negative mass balances of pharmaceuticals and personal care products during wastewater treatment. *Chemosphere*. 134, 395–401.

Brenner, K., You, L., Arnold, F. H. 2008. Engineering microbial consortia: a new frontier in synthetic biology. *Trends Biotechnol*. 26(9), 483–489.

Brown, R. K., Harnisch, F., Wirth, S., Wahlandt, H., Dockhorn, T., Dichtl, N., Schroder, U. 2014. Evaluating the effects of scaling up on the performance of bioelectrochemical systems using a technical scale microbial electrolysis cell. *Bioresour. Technol*. 163, 206–213.

Cecconet, D., Molognoni, D., Callegari, A., Capodaglio, A. G. 2018. Agro-food industry wastewater treatment with microbial fuel cells: energetic recovery issues. *Int. J. Hydrogen Energy*. 43, 500–511.

Chen, J. Q., Zheng, F. Z., Guo, R. X., 2015. Algal feedback and removal efficiency in a sequencing batch reactor algae process (SBAR) to treat the antibiotic cefradine. *PLoS One*. 10, e0133273.

Cheng, D., Ngo, H. H., Guo, W., Liu, Y., Chang, S. W., Nguyen, D. D. 2018. Anaerobic membrane bioreactors for antibiotic wastewater treatment: performance and membrane fouling issues. *Bioresour. Technol*. 267, 714–724.

Cho, K., Lee, C. H., Ko, K., Lee, Y. J., Kim, K. N., Kim, M. K., Chung, Y. H., Kim, D., Yeo, I. K., Oda, T. 2016. Use of phenol-induced oxidative stress acclimation to stimulate cell growth and biodiesel production by the oceanic microalga *Dunaliella salina*. *Algal Res*. 17, 61–66.

Christin, K., Benjamin, K., Falk, H. 2017. Microbial ecology-based engineering of microbial electrochemical technologies. *Microb. Biotechnol*. 11, 22–38.

Chtourou, M., Mallek, M., Dalmau, M., Mamo, J., Santos-Clotas, E., Salah, A. 2018. Triclosan, carbamazepine and caffeine removal by activated sludge system focusing on membrane bioreactor. *Process. Saf. Environ. Prot*. 118, 1–9.

Coconi-Linares, N., Ortiz-Vázquez, E., Fernandez, F., Loske, A. M., Gomez-Lim, M. A., 2015. Recombinant expression of four oxidoreductases in *Phanerochaete chrysosporium* improves degradation of phenolic and non-phenolic substrates. *J. Biotechnol*. 209, 76–84.

Dahiya, S., Venkata Mohan, S. 2016. Strategic design of synthetic consortium with embedded wastewater treatment potential: deciphering the competence of isolates from diverse Microbiome. *Front. Environ. Sci*. 4,1 1–15.

Dai, Z., Zhang, S., Yang, Q., Zhang, W., Qian, X., Dong, W., Jiang, M., Xin, F. 2018. Genetic tool development and systemic regulation in biosynthetic technology. *Biotechnol Biofuels*. 11, 152–152.

Daneshvar, E., Zarrinmehr, M. J., Hashtjin, A. M., Farhadian, O., Bhatnagar, A. 2018. Versatile applications of freshwater and marine water microalgae in dairy wastewater treatment, lipid extraction and tetracycline biosorption. *Bioresour. Technol*. 268, 523–530.

Dangi, A. K., Sharma, B., Hill, R. T., Shukla, P. 2019. Bioremediation through microbes: systems biology and metabolic engineering approach. *Crit. Rev. Biotech*. 39(1), 79–98.

Dash, H. R., Das, S. 2016. Interaction between mercuric chloride and extracellular polymers of biofilm-forming mercury resistant marine bacterium *Bacillus thuringiensis* PW-05. *RSC Adv*. 6(111), 109793–109802.

Dong, H., Zeng, G., Tang, L., Fan, C., Zhang, C., He, X., He, Y. 2015. An overview on limitations of TiO2-based particles for photocatalytic degradation of organic pollutants and the corresponding countermeasures. *Water Res*. 79, 128–146.

Edwards, Q. A., Sultana, T., Kulikov, S. M., Garner-O'Neale, L. D., Yargeau, V., Metcalfe, C. D. 2018. Contaminants of emerging concern in wastewaters in Barbados, West Indies. *Bull. Environ. Contam. Toxicol.* 101, 1–6.

Ezezika, O. C., Singer, P. A. 2010. Genetically engineered oil-eating microbes for bioremediation: prospects and regulatory challenges. *Technol. Soc.* 32(4), 331–335.

Gadipelly, C., Perez-Gonzalez, A., Yadav, G., Ortiz, I., Ibanez, R., Rathod, V. 2014. Pharmaceutical industry wastewater: review of the technologies for water treatment and reuse. *Ind. Eng. Chem. Res.* 53, 11571–11592.

Gao, N., Fan, Y., Long, F., Qiu, Y., Geier, W., Liu, H. 2020. Novel trickling microbial fuel cells for electricity generation from wastewater. *Chemosphere*. 248, 126058.

García-Espinoza, J. D., Nacheva, P. M. 2019. Degradation of pharmaceutical compounds in water by oxygenated electrochemical oxidation: parametric optimization, kinetic studies and toxicity assessment. *Sci. Total Environ.* 691, 417–429.

Garcia-Gomez, C., Drogui, P., Seyhi, B., Gortares-Moroyoqui, P., Buelna, G., Estrada-Alvgarado, M., Alvarez, L. H. 2016. Combined membrane bioreactor and electrochemical oxidation using Ti/PbO2 anode for the removal of carbamazepine. *J. Taiwan Inst. Chem. Eng.* 64, 211–219.

Gojkovic, Z., Lindberg, R. H., Tysklind, M., Funk, C. 2019. Northern green algae have the capacity to remove active pharmaceutical ingredients. *Ecotoxicol. Environ. Saf.* 170, 644–656.

Gonçalves, A. L., Pires, J. C. M., Simoes, M. 2017. A review on the use of microalgal consortia for wastewater treatment. *Algal Res.* 24, 403–415.

Gros, M., Petrovic, M., Ginebreda, A., Barcelo, D. 2010. Removal of pharmaceuticals during wastewater treatment and environmental risk assessment using hazard indexes. *Env. Inter.* 36, 15–26.

Guo, Wan-Qian, He-Shan Zheng, Shuo Li, Juan-Shan Du, Xiao-Chi Feng, Ren-Li Yin, Qing-Lian Wu, Nan-Qi Ren, and Jo-Shu Chang. 2016. Removal of cephalosporin antibiotics 7-ACA from wastewater during the cultivation of lipid-accumulating microalgae. Bioresource Technology 221: 284–290.

Gupta, P., Diwan, B. 2016. Bacterial exopolysaccharide mediated heavy metal removal: a Review on biosynthesis, mechanism and remediation strategies. *Biotechnol. Rep. (Amst).* 13, 58–71.

Gupta, V. K., Nayak, A., Agarwal, S. 2015. Bioadsorbents for remediation of heavy metals: current status and their future prospects. *Env. Eng. Res.* 20(1), 1–18.

Hansda, A., Kumar, V., Anshumali. 2016. A comparative review towards potential of microbial cells for heavy metal removal with emphasis on biosorption and bioaccumulation. *World J. Microb. Biot.* 32, 170.

Hapeshi, E., Gros, M., Lopez-Serna, R., Boleda, M. R., Ventura, F., Petrovic, M., Barcelo, D., Fatta-Kassinos, D. 2015. Licit and illicit drugs in urban wastewater in Cyprus. *Clean Soil Air Water.* 43, 1272–1278.

Harford-Cross, C. F., Carmichael, A. B., Allan, F. K., England, P. A., Rouch, D. A., Wong, L. L. 2000. Protein engineering of cytochrome P450cam (CYP101) for the oxidation of polycyclic aromatic hydrocarbons. *Protein Eng. Des. Sel.* 13(2), 121–128.

He, L., Du, P., Chen, Y., Lu, H., Cheng, X., Chang, B., Wang, Z. 2017. Advances in microbial fuel cells for wastewater treatment. *Renew. Sustain. Energy Rev.* 71, 388–403.

Heidrich, E. S., Edwards, S. R., Dolfing, J., Cotterill, S. E., Curtis, T. P. 2014. Performance of a pilot scale microbial electrolysis cell fed on domestic wastewater at ambient temperatures for a 12-month period. *Bioresour. Technol.* 173, 87–95.

Hena, S., Gutierrez, L., Croue, J. P. 2021. Removal of pharmaceutical and personal care products (PPCPs) from wastewater using microalgae: a review. *J. Hazard. Mater.* 403, 124041.

Henske, J. K., Wilken, S. E., Solomon, K. V., Smallwood, C. R., Shutthanandan, V., Evans, J. E., Theodorou, M. K., O'Malley, M. A. 2018. Metabolic characterization of anaerobic fungi provides a path forward for bioprocessing of crude lignocellulose. *Biotech. Bioeng.* 115(4), 874–884.

Hosseini, M. G., Ahadzadeh, I. 2012 A dual-chambered microbial fuel cell with Ti/nano-TiO_2/Pd nano-structure cathode. *J Power Sources,* 220, 292–297.

Hsu, L., Deng, P., Zhang, Y., Nguyen, H. N., Jiang, X. 2018. Nanostructured interfaces for probing and facilitating extracellular electron transfer. *J Mate. Chem.* B 6(44), 7144–7158.

Ismail, M. M., Essam, T. M., Ragab, Y. M., El-Sayed, A. E., Khair, B., Mourad, F. E. 2017. Remediation of a mixture of analgesics in a stirred-tank photobioreactor using microalgal-bacterial consortium coupled with attempt to valorise the harvested biomass. *Biores. Tech.* 232, 364–371.

Izallalen, M., Mahadevan, R., Burgard, A., Postier, B., Didonato, R., Sun, J., Schilling, C. H., Lovley, D. R. 2008. *Geobacter sulfurreducens* strain engineered for increased rates of respiration. *Metabolic Eng.* 10(5), 267–275.

Jhariya, U., & Pal, S. (2022). Proteomic, Genomic, and Metabolomic Understanding and Designing for Bioremediation of Environmental Contaminants. In *Omics Insights in Environmental Bioremediation* (pp. 415–435). Singapore: Springer Nature Singapore.

Kargi, F., Eker, S. 2007. Electricity generation with simultaneous wastewater treatment by a microbial fuel cell (MFC) with Cu and Cu-Au electrodes. *J. Chem. Tech. Biotech.* 82(7), 658–662.

Khatiwada, B., Sunna, A., Nevalainen, H. 2020. Molecular tools and applications of *Euglena gracilis*: from biorefineries to bioremediation. *Biotechnol. Bioeng.* 117, 3952–3967.

Khetan, S. K., Collins, T. J. 2007. Human pharmaceuticals in the aquatic environment: a challenge to green chemistry. *Chem. Rev.* 107, 2319–2364.

Kiki, C., Rashid, A., Wang, Y.W., Li, Y., Zeng, Q. T., Yu, C. P., Sun, Q. 2020. Dissipation of antibiotics by microalgae: kinetics, identification of transformation products and pathways. *J. Hazard. Mater.* 387, 121985.

Kumar, A., Bhoot, N., Soni, I., John, P. J. 2014. Isolation and characterization of a *Bacillus subtilis* strain that degrades endosulfan and endosulfan sulfate. *3 Biotech*, 4, 467–475.

Kumar, D., Kumar, A., Sharma, J. 2016. Degradation study of lindane by novel strains *Kocuria* sp. DAB-1Y and *Staphylococcus* sp. DAB-1W. *Bioresour. Bioprocess.* 3, 53.

Mirzaei, R., Yunesian, M., Nasseri, S., Gholami, M., Jalilzadeh, E., Shoeibi, S., & Mesdaghinia, A. 2018. Occurrence and fate of most prescribed antibiotics in different water environments of Tehran, Iran. *Science of the total environment*, 619, 446–459.

Langenhoff, A., Nadia, I., Teun, V., Gosse, S., Marco, B., Katarzyna, K. R., Huub, R. 2013. Microbial removal of the pharmaceutical compounds Ibuprofen and Diclofenac from Wastewater, *BioMed. Res. Int.* 9, 325806.

Leng, L. J., Wei, L., Xiong, Q., Xu, S. Y., Li, W. T., Lv, S., Lu, Q., Wan, L.P., Wen, Z. Y., Zhou, W. G. 2020. Use of microalgae based technology for the removal of antibiotics from wastewater: a review. *Chemosphere,* 238, 124680.

Leon-Fernandez, L. F., Villaseñor, J., Rodriguez, L., Cañizares, P., Rodrigo, M. A., Fernández-Morales, F. J. 2019. Dehalogenation of 2,4-dichlorophenoxyacetic acid by means of bioelectrochemical systems. *J. Electroanal. Chem.* 854:113564.

Li, H., Cong, Y., Lin, J., Chang, Y. 2015. Enhanced tolerance and accumulation of heavy metal ions by engineered *Escherichia coli* expressing *Pyrus calleryana* phytochelatin synthase. *J. Basic Microbiol.* 55(3), 398–405.

Li, H., Zhang, S., Yang, X. L., Yang, Y. L., Xu, H., Li, X. N., Song, H. L. 2019. Enhanced degradation of bisphenol A and ibuprofen by an up-flow microbial fuel cell-coupled constructed wetland and analysis of bacterial community structure. *Chemosphere.* 217, 599–608.

Li, P. S., Tao, H. C. 2015. Cell surface engineering of microorganisms towards adsorption of heavy metals. *Crit. Rev. Microbiol.* 41(2), 140–149.

Liao, X. B., Li, B. X., Zou, R. S., Xie, S. G., Yuan, B. L., 2016. Antibiotic sulfanilamide biodegradation by acclimated microbial populations. *Appl. Microbiol. Biotechnol.* 100, 2439–2447.

Liang, C., Zhang, L., Nord, N. B., Carvalho, P. N., & Bester, K. 2019. Dose-dependent effects of acetate on the biodegradation of pharmaceuticals in moving bed biofilm reactors. *Water Research*, 159, 302–312.

Lin, C. W., Chen, J., Zhao, J., Liu, S. H., Lin, L. C. 2020. Enhancement of power generation with concomitant removal of toluene from artificial groundwater using a mini microbial fuel cell with a packed-composite anode. *J. Hazard. Mater.* 387, 121717.

Liu, H., Grot, S., Logan B. E. 2005. Electrochemically assisted microbial production of hydrogen from acetate. *Environ. Sci. Technol.* 39, 4317–4320.

Liu, H., Ramnarayanan, R., Logan, B. E. 2004. Production of electricity during wastewater treatment using a single chamber microbial fuel cell. *Env. Sci. Tech.* 38(7), 2281–2285.

Liu, S. Y., Charles, W., Ho, G., Cord-Ruwisch, R., Cheng, K. Y. 2017. Bioelectrochemical enhancement of anaerobic digestion: comparing single- and two-chamber reactor configurations at thermophilic conditions. *Bioresour. Technol.* 245, 1168–1175.

Maletz, S., Floehr, T., Beier, S., Klumper, C., Brouwer, A., Behnisch, P., Higley, E., Giesy, J. P., Hecker, M., Gebhardt, W. 2013. In vitro characterization of the effectiveness of enhanced sewage treatment processes to eliminate endocrine activity of hospital effluents. *Water Res.* 47, 1545–1557.

Malvankar, N. S., Vargas, M., Nevin, K. P., Franks, A. E., Leang, C., Kim, B. C., Inoue, K., Mester, T., Covalla, S. F., Johnson, J. P., Rotello, V. M., Tuominen, M. T., Lovley, D. R. 2011. Tunable metallic-like conductivity in microbial nanowire networks. *Nat. Nanotechnol.* 6(9), 573–579.

Marrot, B., Barrios-Martinez, A., Moulin, P., Roche, N. 2004. Industrial wastewater treatment in a membrane bioreactor: a review. *Environ. Prog.* 23(1) 59–68.

Martin, A. L., Satjaritanun, P., Shimpalee, S., Devivo, A.B., Weidner, J., Greenway, S., Henson, M.J., Turick, E. C. 2018. In-situ electrochemical analysis of microbial activity. *AMB Express.* 8, 162.

Matamoros, V., Uggetti, E., Garcia, J., Bayona, J. M. 2016. Assessment of the mechanisms involved in the removal of emerging contaminants by microalgae from wastewater: a laboratory scale study. *J. Hazard. Mater.* 301, 197–205.

Mato, Y., Isobe, T., Takada, H., Kanehiro, Ohtake, H. C., Kaminuma T. 2001. Plastic resin pellets as a transport medium for toxic chemicals in the marine environment *Environ. Sci. Technol.* 35(2), 318–324.

Maaz, M., Yasin, M., Aslam, M., Kumar, G., Atabani, A. E., Idrees, M., ... & Kim, J. 2019. Anaerobic membrane bioreactors for wastewater treatment: Novel configurations, fouling control and energy considerations. *Bioresource Technology*, 283, 358–372.

Mendes, L. B. B., Vermelho, A. B. 2013. Allelopathy as a potential strategy to improve microalgae cultivation. *Biotechnol. Biofuels.* 6, 152–165.

Mishra, A., Medhi, K., Malaviya, P., Thakur, I. S. 2019. Omics approaches for microalgal applications: prospects and challenges. *Bioresour. Technol.* 291, 121890.

More, T. T., Yadav, J. S., Yan, S., Tyagi, R. D., Surampalli, R. Y. 2014. Extracellular polymeric substances of bacteria and their potential environmental applications. *J. Environ. Manage.* 144, 1–25.

Mouele, E. S. M., Tijani, O. J., Fatoba, O. O., Petrik, L. F. 2015. Degradation of organic pollutants and microorganisms from wastewater using different dielectric barrier discharge configurations-a critical review. *Environ. Sci. Pollut. Res.* 22, 18345–18362.

Mulbry, W., Kondrad, S., Pizarro, C., Kebede-Westhead, E. 2008. Treatment of dairy manure effluent using freshwater algae: algal productivity and recovery of manure nutrients using pilot-scale algal turf scrubbers. *Bioresour. Technol.* 99, 8137–8142.

Nagata, Y., Kato, H., Ohtsubo, Y., Tsuda, M. 2019. Lessons from the genomes of lindane-degrading sphingomonads. *Environ. Microbiol. Rep.* 11, 630–644.

Nagarajan, D., Lee, D. J., Varjani, S., Lam, S. S., Allakhverdiev, S. I., & Chang, J. S. (2022). Microalgae-based wastewater treatment–microalgae-bacteria consortia, multi-omics approaches and algal stress response. *Science of The Total Environment*, *845*, 157110.

Nelson, J. L., Fung, J. M., Cadillo-Quiroz, H., Cheng, X., Zinder, S. H. 2011. A role for *Dehalobacter* spp. in the reductive dehalogenation of dichlorobenzenes and monochlorobenzene. *Env. Sci. Tech.* 45(16), 6806–6813.

Ndlela, L. L., Schroeder, P., Genthe, B., & Cruzeiro, C. (2023). Removal of antibiotics using an Algae-Algae consortium (Chlorella protothecoides and Chlorella vulgaris). *Toxics*, *11*(7), 588.

Nguyen, L. N., Commault, A. S., Tim, K., Peter, J., Galil, R. 2020a. Genome sequencing as a new window into the microbial community of membrane bioreactors – a critical review. *Sci. Total Environ.* 704, 135279.

Nguyen, V. H., Smith, S. M., Wantala, K., Kajitvichyanukul, P. 2020b. Photocatalytic remediation of persistent organic pollutants (POPs): a review, *Arab. J. Chem.* 13(11), 8309–8337.

Nomura, T., Inoue, K., Uehara-Yamaguchi, Y., Yamada, K., Iwata, O., Suzuki, K., Mochida, K. 2019. Highly efficient transgene-free targeted mutagenesis and single-stranded oligodeoxynucleotidemediated precise knock-in in the industrial microalga *Euglena gracilis* using Cas9 ribonucleoproteins. *Plant Biotechnol. J.* 17, 2032–2034.

Norvill, Z. N., Shilton, A., Guieysse, B. 2016. Emerging contaminant degradation and removal in algal wastewater treatment ponds: identifying the research gaps. *J. Hazard. Mater.* 313, 291–309.

Nzila, A., Razzak, S. A., Zhu, J. 2016. Bioaugmentation: an emerging strategy of industrial wastewater treatment for reuse and discharge. *Int. J. Env. Res. Public Health.* 13(9), 846.

Oberoi, A. S., Jia, Y., Zhang, H. Q., Khanal, S. K., Lu, H. 2019. Insights into the fate and removal of antibiotics in engineered biological treatment systems: a critical review. *Environ. Sci. Technol.* 53, 7234–7264.

Ogbonna, J. C., Yoshizawa, H., Tanaka, H. 2000. Treatment of high strength organic wastewater by a mixed culture of photosynthetic microorganisms. *J. Appl. Phycol.* 12(3), 277–284.

Olarinmoye, O., Bakare, A., Ugwumba, O., Hein, A. 2016. Quantification of pharmaceutical residues in wastewater impacted surface waters and sewage sludge from Lagos, Nigeria. *J. Environ. Chem. Ecotoxicol.* 8, 14–24.

Osundeko, O., Dean, A. P., Davies, H., Pittman, J. K. 2014. Acclimation of microalgae to wastewater environments involves increased oxidative stress tolerance activity. *Plant Cell Physiol.* 55, 1848–1857.

Padmanabhan, P. V. A., Sreekumar, K. P., Thiyagarajan, T. K., Satpute, R. U., Bhanumurthy, K., Sengupta P., Dey, G. K., Warrier, K. G. K. 2006. Nano-crystalline titanium dioxide formed by reactive plasma synthesis. *Vacuum.* 80(11), 1252–1255.

Pant, D., Singh, A., Van Bogaert, G., Irving Olsen, S., Singh Nigam, P., Diels, L., Vanbroekhoven, K. 2012. Bioelectrochemical systems (BES) for sustainable energy production and product recovery from organic wastes and industrial wastewaters. *RSC Adv.* 2, 1248–1263.

Pariatamby, A., Kee, Y. L. 2016. Persistent organic pollutants management and remediation. *Procedia Environ. Sci.* 31, 842–848.

Patel, A. K., Huang, E. L., Low-Decarie, E., Lefsrud, M. G. 2015. Comparative shotgun proteomic analysis of wastewater-cultured microalgae: nitrogen sensing and carbon fixation for growth and nutrient removal in *Chlamydomonas reinhardtii*. *J. Proteome Res.* 14, 3051–3067.

Perera, I. A., Abinandan, S., Subashchandrabose, S. R., Venkateswarlu, K., Naidu, R., Megharaj, M. 2019. Advances in the technologies for studying consortia of bacteria and cyanobacteria/microalgae in wastewaters. *Crit. Rev. Biotechnol.* 39, 709–731.

Postigo, D., Barcelo, C. 2015. Synthetic organic compounds and their transformation products in groundwater: occurrence, fate and mitigation. *Sci. Total Environ.* 503–504, 32–47.

Prado, A., Berenguer, R., Esteve-Nunez, A. 2019. Electroactive biochar outperforms highly conductive carbon materials for biodegrading pollutants by enhancing microbial extracellular electron transfer. *Carbon.* 146, 597–609.

Prado, A., Ramirez-Vargas, C. A., Arias, C. A., Esteve-Nunez, A. 2020. Novel bioelectrochemical strategies for domesticating the electron flow in constructed wetlands. *Sci. Total Environ.* 735, 139522.

Pun, Á., Boltes, K., Letón, P., & Esteve-Nuñez, A. 2019. Detoxification of wastewater containing pharmaceuticals using horizontal flow bioelectrochemical filter. *Bioresource Technology Reports*, 7, 100296.

Quijano, G., Arcila, J. S., Buitron, G. 2017. Microalgal-bacterial aggregates: applications and perspectives for wastewater treatment. *Biotechnol. Adv.* 35, 772–781.

Radjenovic, J., Petrovic, M., Barcelo, D. 2009. Fate and distribution of pharmaceuticals in wastewater and sewage sludge of the conventional activated sludge (CAS) and advanced membrane bioreactor (MBR) treatment. *Water Res.* 43(3), 831–841.

Ramirez-Moreno, M., Esteve-Núñez, A., Ortiz, J. M. 2021. Desalination of brackish water using a microbial desalination cell: analysis of the electrochemical behavior. *Electrochim. Acta.* 388, 138570.

Reyes-Caballero, H., Campanello, G. C., Giedroc, D. P. 2011. Metalloregulatory proteins: metal selectivity and allosteric switching. *Biophys. Chem.* 156(2), 103–114.

Rios Miguel, A. B., Jetten, M. S. M., Welte, C. U. 2020. The role of mobile genetic elements in organic micropollutant degradation during biological wastewater treatment. *Water Res. X.* 9, 100065.

Rodrigo, Q. J., Tejedor-Sanz, S., Schroll, R., Esteve-Nunez, A. 2019. Electrodes boost microbial metabolism to mineralize antibiotics in manure. *Bioelectrochem.* 128, 283–290.

Rosano, G. L., Ceccarelli, E. A. 2014. Recombinant protein expression in Escherichia coli: advances and challenges. *Front.Microbiol.* 5, 172–172.

Sandhibigraha, S., Chakraborty, S., Bandyopadhyay, T., Bhunia, B. A. 2019. Kinetic study of 4-chlorophenol biodegradation by the novel isolated *Bacillus subtilis* in batch shake flask. *Environ. Eng. Res.* 25, 62–70.

Shahpiri, A., Mohammadzadeh, A. 2018. Mercury removal by engineered *Escherichia coli* cells expressing different rice metallothionein isoforms. *Annal. Microbiol.* 68(3), 145–152.

Sheng, G. P., Yu, H. Q., Li, X. Y. 2010. Extracellular polymeric substances (EPS) of microbial aggregates in biological wastewater treatment systems: a review. *Biotech. Adv.* 28(6), 882–894.

Sires, I., Brillas, E. 2012. Remediation of water pollution caused by pharmaceutical residues based on electrochemical separation and degradation technologies: a review. *Environ. Int.* 40, 212–229.

Song, C. F., Wei, Y. L., Qiu, Y. T., Qi, Y., Li, Y., Kitamura, Y. 2019. Biodegradability and mechanism of florfenicol via *Chlorella* sp. UTEX1602 and L38: experimental study. *Bioresour. Technol.* 272, 529–534.

Song, H., Guo, W., Liu, M., Sun, J. 2013. Performance of microbial fuel cells on removal of metronidazole. *Water Sci. Technol.* 68, 2599–2604.

Sophia, A. C., Lima, E. C. 2018. Removal of emerging contaminants from the environment by adsorption. *Ecotoxicol. Environ. Saf.* 150, 1–17.

Subashchandrabose, S. R., Ramakrishnan, B., Megharaj, M., Venkateswarlu, K., Naidu, R. 2013. Mixotrophic cyanobacteria and microalgae as distinctive biological agents for organic pollutant degradation. *Environ. Int.* 51, 59–72.

Sun, J., Liang, P., Yan, X., Zuo, K., Xiao, K., Xia, J., Qiu, Y., Wu, Q., Wu, S., Huang, X., Qi, M., Wen, X. 2016. Reducing aeration energy consumption in a large-scale membrane bioreactor: process simulation and engineering application. *Water Res.* 93, 205–213.

Sutherland, D. L., Ralph, P. J. 2019. Microalgal bioremediation of emerging contaminants - opportunities and challenges. *Water Res.* 164, 114921.

Swift, C. L., Brown, J. L., Seppala, S., O'Malley, M. A. 2019. Co-cultivation of the anaerobic fungus *Anaeromyces robustus* with *Methanobacterium bryantii* enhances transcription of carbohydrate active enzymes. *J. Ind. Microbiol & Biotech.* 46(9), 1427–1433.

Tahir, K., Miran, W., Nawaz, M., Jang, J., Shahzad, A., Moztahida, M. 2019. Investigating the role of anodic potential in the biodegradation of carbamazepine in bioelectrochemical systems. *Sci. Total Environ.* 688, 56–64.

Tan, X. F., Liu, Y. G., Zeng, G. M., Wang, X., Hu, X. J., Gu, Y. L., Yang, Z. Z. 2015. Application of biochar for the removal of pollutants from aqueous solutions. *Chemosphere*. 125, 70–85.

Tay, P. K. R., Nguyen, P. Q., Joshi, N. S. A. 2017. Synthetic circuit for mercury bioremediation using self-assembling functional amyloids. *ACS Syn. Biol.* 6(10), 1841–1850.

Tejedor-Sanz, S., Ortiz, J. M., Esteve-Nunez, A. 2017. Merging microbial electrochemical systems with electrocoagulation pretreatment for achieving a complete treatment of brewery wastewater. *Chem. Eng. J.* 330, 1068–1074.

Tejedor-Sanz, S., Fernández-Labrador, P., Hart, S., Torres, C. I., Esteve-Nunez, A. 2018. Geobacter dominates the inner layers of a stratified biofilm on a fluidized anode during brewery wastewater treatment. *Front. Microbiol.* 9, 378.

Ter Heijne, A., Geppert, F., Sleutels, T. H., Batlle-Vilanova, P., Liu, D., Puig, S. 2019. Mixed culture biocathodes for production of hydrogen, methane, and carboxylates In F. Harnisch, & D. Holtmann (Eds.), *Bioelectrosynthesis Advances in Biochemical Engineering/Biotechnology, vol 167*. Springer, Cham. (pp. 203–229). https://doi.org/10.1007/10_2017_15

Teunissen, M. J., Kets, E. P. W., Op den Camp, H. J. M., Huis in't Veld, J. H. J., Vogels, G. D. 1992. Effect of coculture of anaerobic fungi isolated from ruminants and non-ruminants with methanogenic bacteria on cellulolytic and xylanolytic enzyme activities. *Arch. Microbiol.* 157(2), 176–182.

Tiwari, B., Sellamuthu, B., Ouarda, Y., Drogui, P., Tyagi, R.D., Buelna, G. 2017. Review on fate and mechanism of removal of pharmaceutical pollutants from wastewater using biological approach. *Bioresour. Technol.* 224, 1–12.

Tong, C. Y., Honda, K., & Derek, C. J. C. 2023. A review on microalgal-bacterial co-culture: The multifaceted role of beneficial bacteria towards enhancement of microalgal metabolite production. *Environmental Research*, 115872.

Torres, M. A., Barros, M. P., Campos, S. C., Pinto, E., Rajamani, S., Sayre, R. T., Colepicolo, P. 2008. Biochemical biomarkers in algae and marine pollution: a review. *Ecotoxicol. Environ. Saf.* 71, 1–15.

Tran, N. H., Urase, T., Ngo, H. H., Hu, J. Y., Ong, S. L. 2013. Insight into metabolic and cometabolic activities of autotrophic and heterotrophic microorganisms in the biodegradation of emerging trace organic contaminants. *Bioresour. Technol.* 146, 721–731.

Ufarte, L., Laville, E., Duquesne, S., Potocki-Veronese, G. 2015. Metagenomics for the discovery of pollutant degrading enzymes. *Biotechnol. Adv.* 33, 1845–1854.

Velvizhi, G., Venkata Mohan, S. 2015. Bioelectrogenic role of anoxic microbial anode in the treatment of chemical wastewater: microbial dynamics with bioelectro-characterization. *Water Res.* 70, 52–63.

Viggi, C. C., Matturro, B., Frascadore, E., Insogna, S., Mezzi, A., Kaciulis, S. 2017. Bridging spatially segregated redox zones with a microbial electrochemical snorkel triggers biogeochemical cycles in oil-contaminated River Tyne (UK) sediments. *Water Res.* 127, 11–21.

Virkutyte, J., Varma, R., Jegatheesan, V. 2010. *Treatment of Micropollutants in Water and Wastewater.* New York: IWA Publishing.

Vo, H. N. P., Ngo, H. H., Guo, W. S., Liu, Y. W., Chang, S. W., Nguyen, D. D., Zhang, X. B., Liang, H., Xue, S. 2020a. Selective carbon sources and salinities enhance enzymes and extracellular polymeric substances extrusion of *Chlorella* sp. for potential cometabolism. *Bioresour. Technol.* 303, 122877.

Vo, H. N. P., Ngo, H. H., Guo, W., Nguyen, K. H., Chang, S. W., Nguyen, D. D., Liu, Y.W., Liu, Y., Ding, A., Bui, X. T. 2020b. Micropollutants cometabolism of microalgae for wastewater remediation: effect of carbon sources to cometabolism and degradation products. *Water Res.* 183, 115974.

Wang, G., Zhang, J., Wang, L., Liang, B., Chen, K., Li, S., Jiang, J. 2010 Cometabolism of DDT by the newly isolated bacterium, *Pseudoxanthomonas* sp. wax. *Braz. J. Microbiol.* 41, 431–438.

Wang, R., Diao, P., Chen, Q., Wu, H., Xu, N., & Duan, S. 2017. Identification of novel pathways for biodegradation of bisphenol A by the green alga Desmodesmus sp. WR1, combined with mechanistic analysis at the transcriptome level. Chemical Engineering Journal, 321, 424–431.Wang, X., Aulenta, F., Puig, S., Esteve-Núñez, A., He, Y., Mu, Y. 2020. Microbial electrochemistry for bioremediation. *Environ. Sci. Ecotechnol.* 1, 100013.

Wang, Yue, Yuanyuan He, Xiaoqiang Li, Dillirani Nagarajan, and Jo-Shu Chang. 2022. Enhanced biodegradation of chlortetracycline via a microalgae-bacteria consortium. *Bioresource technology.* 343: 126149.

Wen, Q., Kong, F., Zheng, H., Yin, J., Cao, D., Ren, Y., Wang, G. 2011. Simultaneous processes of electricity generation and ceftriaxone sodium degradation in an air-cathode single chamber microbial fuel cell. *J. Power Sources.* 196(5), 2567–2572.Wolff, D., Krah, D. D., Ghattas, A. K., Wick, A., Ternes, T. A. 2018. Insights into the variability of microbial community composition and micropollutant degradation in diverse biological wastewater treatment systems. *Water Res.* 143, 313–324.

Wu, Y. H., Li, T. L., Yang, L. Z. 2012. Mechanisms of removing pollutants from aqueous solutions by microorganisms and their aggregates: a review. *Bioresour. Technol.* 107, 10–18.

Xie, X., Criddle, C., Cui, Y. 2015. Design and fabrication of bioelectrodes for microbial bioelectrochemical systems. *Energy Environ. Sci.* 8, 3418–3441.

Xie, Peng, Shih-Hsin Ho, Qing-Yang Xiao, Xi-Jun Xu, Lei Zhao, Xu Zhou, Duu-Jong Lee, Nan-Qi Ren, and Chuan Chen. 2020. Revealing the role of nitrate on sulfide removal coupled with bioenergy production in Chlamydomonas sp. Tai-03: Metabolic pathways and mechanisms. Journal of Hazardous Materials 399: 123115.

Xiong, J. Q., Kim, S. J., Kurade, M. B., Govindwar, S., Abou-Shanab, R. A. I., Kim, J. R., Roh, H. S., Khan, M. A., Jeon, B. H. 2019. Combined effects of sulfamethazine and sulfamethoxazole on a freshwater microalga, *Scenedesmus obliquus*: toxicity, biodegradation, and metabolic fate. *J. Hazard. Mater.* 370, 138–146.

Xiong, J. Q., Kurade, M. B., Abou-Shanab, R. A., Ji, M. K., Choi, J., Kim, J. O., Jeon, B. H., 2016. Biodegradation of carbamazepine using freshwater microalgae *Chlamydomonas mexicana* and *Scenedesmus obliquus* and the determination of its metabolic fate. *Bioresour. Technol.* 205, 183–190.

Xiong, J. Q., Kurade, M. B., Jeon, B. H. 2017a. Biodegradation of levofloxacin by an acclimated freshwater microalga, *Chlorella vulgaris. Chem. Eng.* J. 313, 1251–1257.

Xiong, J. Q., Kurade, M. B., Jeon, B. H. 2018. Can microalgae remove pharmaceutical contaminants from water? *Trends Biotechnol.* 36, 30–44.

Xiong, J. Q., Kurade, M. B., Kim, J. R., Roh, H. S., Jeon, B. H. 2017b. Ciprofloxacin toxicity and its co-metabolic removal by a freshwater microalga *Chlamydomonas mexicana. J. Hazard. Mater.* 323, 212–219.

Xiong, J. Q., Kurade, M. B., Patil, D. V., Jang, M., Paeng, K. J., Jeon, B. H. 2017c. Biodegradation and metabolic fate of levofloxacin via a freshwater green alga, *Scenedesmus obliquus* in synthetic saline wastewater. *Algal Res.* 25, 54–61.

Xiong, J. Q., Liu, Y.S., Hu, L. X., Shi, Z. Q., Cai, W. W., He, L. Y., Ying, G. G. 2020. Cometabolism of sulfamethoxazole by a freshwater microalga *Chlorella pyrenoidosa. Water Res.* 175, 115656.

Yakubu, O. H. 2017. Pharmaceutical wastewater effluent-source of contaminants of emerging concern: phytotoxicity of metronidazole to soybean (*Glycine max*). *Toxics.* 5, 10.

Yan, W., Guo, Y., Xiao, Y., Wang, S., Ding, R., Jiang, J., Gang, H., Wang, H., Yang, J., Zhao, F. 2018. The changes of bacterial communities and antibiotic resistance genes in microbial fuel cells during long-term oxytetracycline processing. *Water Res.* 142, 105–114.

Yao, J., Pan, B., Shen, R., Yuan, T., Wang, J. 2019. Differential control of anode/cathode potentials of paired electrolysis for simultaneous removal of chemical oxygen demand and total nitrogen. *Sci. Total Environ.* 687, 198–205.

Yeray, A., Llorente, M., Sánchez-Gómez, A., Manchon, C., Boltes, K., Esteve-Nunez, A. 2021. Microbial electrochemical fluidized bed reactor: a promising solution for removing pollutants from pharmaceutical industrial wastewater. *Front. Microbiol.* 12, 737112.

Yu, Y., Zhou, Y. Y., Wang, Z.L., Torres, O. L., Guo, R. X., Chen, J. Q. 2017. Investigation of the removal mechanism of antibiotic ceftazidime by green algae and subsequent microbic impact assessment. *Sci. Rep.* 7, 4168.

Yuan, H., Hou, Y., Abu-Reesh, I. M., Chen, J., He, Z. 2016. Oxygen reduction reaction catalysts used in microbial fuel cells for energy efficient wastewater treatment: a review. *Mater. Horiz.* 3(5), 382–401.

Zanaroli, G., Negroni, A., Hoggblom, M. M., Fava, F. 2015. Microbial dehalogenation of organohalides in marine and estuarine environments. *Curr. Opin. Biotechnol.* 33, 287–295.

Zhang, J., Cao, X., Wang, H., Long, X., Li, X. 2020. Simultaneous enhancement of heavy metal removal and electricity generation in soil microbial fuel cell. *Ecotoxicol. Environ. Saf.* 192, 110314.

Zhang, Q., Zhang, Y., Li, D. 2017. Cometabolic degradation of chloramphenicol via a meta-cleavage pathway in a microbial fuel cell and its microbial community. *Bioresour. Technol.* 229, 104–110.

Zhou, L., Bosscher, M., Zhang, C., Ozçubukçu, S., Zhang, L., Zhang, W., Li, C. J., Liu, J., Jensen, M. P., Lai, L., He, C. 2014. A protein engineered to bind uranyl selectively and with femtomolar affinity. *Nat. Chem.* 6(3), 236–241.

3 Nano-carriers for Enzyme/Microbe Immobilization for Wastewater Treatment

Saba Ghattavi and Ahmad Homaei

3.1 INTRODUCTION

Water is a necessary component of life (Wang et al., 2019b); today, the production of vast volumes of wastewater from various manufacturing operations poses a serious threat to most communities and endangers human health and the environment. Contamination of water has become a significant point of concern and a top priority for society and government (Crini, 2005; Crini and Lichtfouse, 2019; Sonune and Ghate, 2004); the removal of these compounds from water and wastewater prior to disposal into the atmosphere is critical due to their challenging biodegradability and possible long-term harmful effects on ecosystem protection and even human health (Ufarté et al., 2015; Zangeneh et al., 2015; Zhou et al., 2020). Table 3.1 shows typical resistant contaminants in wastewater and their categorization (Feng et al., 2021; Varga et al., 2019; Wang and Wang, 2016; Zhang et al., 2020). These chemicals

TABLE 3.1
Typical Resistant Contaminants in Wastewater and their Categorization

Resistant contaminants	Oil, fat, grease	–	–
	Organic micropollutants	Pharmaceuticals	Hormones
			Anti-infectives
			Metabolic agent
		Pesticides	Organophosphorus pesticides
			Carbamate pesticides
		Products for personal care	Sunscreen agents
			Disinfectant
		Chemicals used in industry	Plastics
			Dye
			Additive

DOI: 10.1201/9781003517238-3

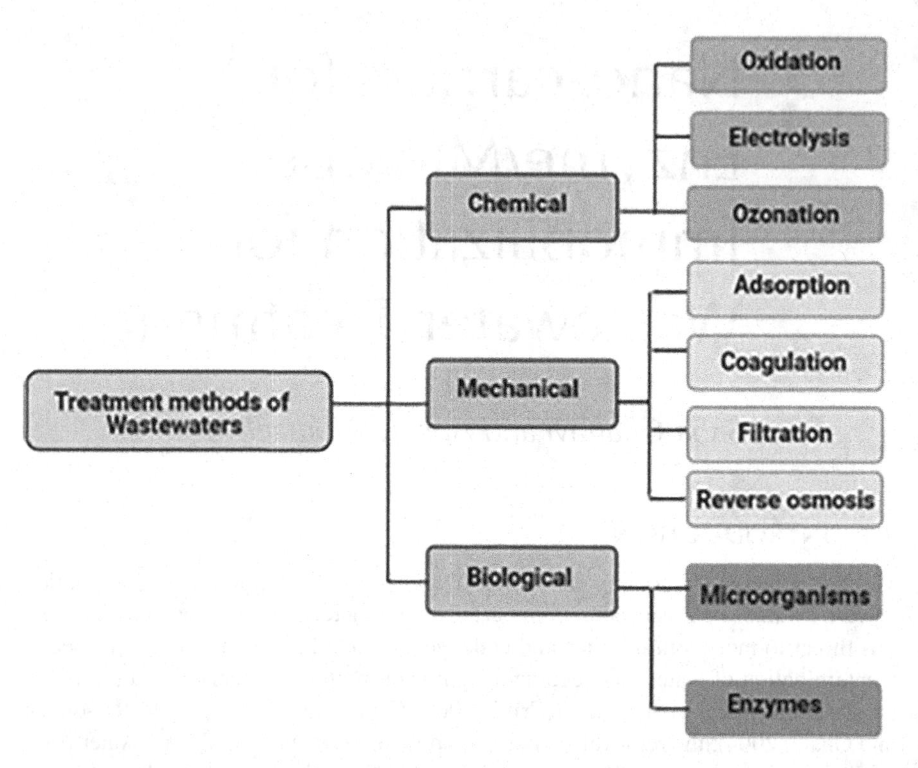

FIGURE 3.1 Summary of different wastewater treatment methods.

are difficult to detect or manage (Mishra et al., 2020) due to small nanograms or micrograms per litre quantities in the environment (Virkutyte and Varma, 2010). Wastewater can be treated in a variety of ways, including biological, mechanical and chemical processes (Alshabib et al., 2019; Zhang et al., 2018), as shown in Figure 3.1 (Saratale et al., 2011). These methods are good for treatment but they are extremely costly and may result in additional pollution and harm (Bhatnagar and Anastopoulos, 2017; Villegas et al., 2016).

Biological techniques are more environmentally friendly and can remove the majority of contaminants from wastewater (Alshabib et al., 2019). Biotechnological methods, which include the use of microorganisms with the goal of removing wastewater pollutants, have also been commonly used for wastewater treatment (Wang et al., 2019b; Zhang et al., 2018). During the process, organisms used in these technologies might break down or absorb pollutants (Ebele et al., 2017; Sharma et al., 2019; Wilkinson et al., 2016). Besides these methods, enzyme is a powerful biocatalyst that can destroy compounds in a controlled environment (Yao et al., 2020). Many enzymes for the elimination of contaminants and pathogens have been discovered in recent decades (Sharma et al., 2018; Stadlmair et al., 2018; Thallinger et al., 2013). Additionally, enzymes may also reduce the time it takes for substrates to enter cells, making these processes more efficient (Mishra et al., 2020). Enzymatic methods have better reaction kinetics and use less water and energy than chemical processes, which means enzymes are not consumed by reactions and can be reused (Feng et al.,

2021; Kalia et al., 2013; Summerscales and Manufacturing, 2021). Recent biotechnological advances have allowed the production of cheaper and more readily available enzymes through better isolation and purification procedure (Durán, 1997) and also many researches have shown that it outperforms contemporary technology (Sharma et al., 2018). Studies have shown that the benefits of enzymatic treatment over traditional treatments include: operation in a broad pH range in high- and low-contaminant concentrations, application in a range of temperatures and salinity, and the simple control procedure (Durán and Esposito, 2000; Karam and Nicell, 1997).

Recent developments in nanotechnology have created highly active and stable enzyme nanomaterial composites that outperform free enzymes in many ways (Cipolatti et al., 2016). Nanomaterials are beneficial for immobilized carriers so they can satisfy their needs (Luo et al., 2005) and also they have special features such as small sizes and physicochemical properties that they can use in various applications (Dasgupta et al., 2017). Nano-carriers help immobilized enzymes maintain their stability, activity and increase the scope of their application. Nano-carrier enzymes have been studied for many fields such as manufacturing of biorenewables in the industrial sector, food industry, medical uses (therapy delivery and disease diagnosis), cosmetics (Cipolatti et al., 2016; Sharma et al., 2018; Stadlmair et al., 2018) and wastewater treatment (Bouabidi et al., 2019; Wang et al., 2019b). The carrier or support material chosen in an immobilization process has a significant impact on the viability of immobilized microorganisms and, as a result, on the productivity of the wastewater treatment process in which the immobilized system will be used (Bouabidi et al., 2019). In this chapter, we provide an overview of the strengths of nano-carrier enzyme immobilization expressed in wastewater treatment and review nano-carriers that are used for enzyme immobilization and also we discuss the application of immobilized enzymes for use in the treatment of water.

3.2 WASTEWATER TREATMENT ENZYMES

At the present time, the removal of contaminants from wastewater is inadequate and results in continuous discharge into the aquatic environment. Bioremediation procedures have risen in prominence in recent years as a potential solution to this problem, as they may have a lower environmental impact than chemical or physical treatment techniques. Enzyme-based technologies appear to be a promising option because they can target individual molecules (Stadlmair et al., 2018). Biomolecules obtained from natural materials play a significant role in the production of everyday items. Enzymes are exceptionally effective at catalysing certain processes that may be described with the induced fit model or using the lock and key model (Gholami-Borujeni et al., 2011; Torres et al., 2017).

Enzymes are one of the compounds that are well known around the world for their many industrial applications (Jayasekara and Ratnayake, 2019). Many enzymes have been tested to remove refractory contaminants in wastewater, and many of them have proven effective (Feng et al., 2021). The efficiency of enzymes in wastewater treatment is highly dependent on enzyme types, as well as their sources (Cummins et al., 2007). Examples of some enzymes that are involved in wastewater treatment are presented in Figure 3.2.

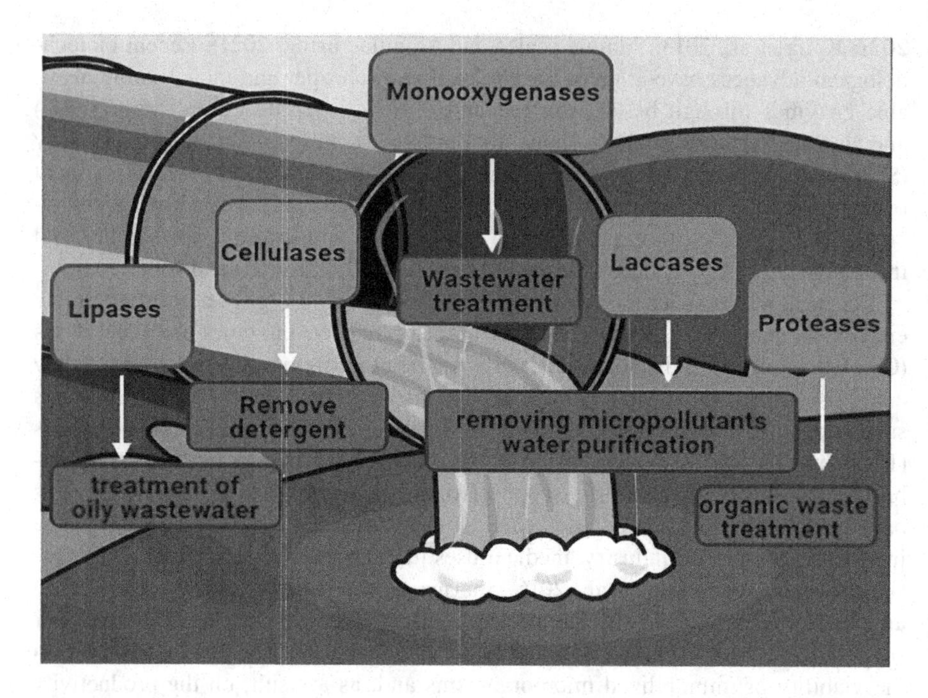

FIGURE 3.2 Some wastewater treatment enzymes and their application in wastewater treatment.

3.2.1 Lipases

Lipases are ubiquitous enzymes that belong to the triacylglycerol acylhydrolase (class with Enzyme Commission number 3.1.1.3) (Arpigny and Jaeger, 1999; Chioke et al., 2018) and have attracted the attention of many industries (Chandra et al., 2020), particularly microbial lipases (Bharathi et al., 2019). Microbial lipases are unquestionably important biocatalysts due to their wide range of catalytic processes in both aqueous and non-aqueous conditions. Due to their great specificity and stability, lipases are favoured over chemical catalysts (Nimkande and Bafana, 2022). Lipases have been isolated from a wide variety of organisms, including yeast, bacteria, fungi, plants and animals. Microorganism-derived enzymes are used in a number of sectors and various industries (Jaeger and Eggert, 2002; Saxena, 1999) such as cosmetic, food, chemical, manufacturing of detergents (Hasan et al., 2010) and water reconditioning (Novak et al., 1990). Aside from industrial uses, microbial lipases have the potential to provide significant benefits over chemical and mechanical wastewater treatment approaches (Nimkande and Bafana, 2022). Lipases are often used in the treatment of oily wastewater because they hydrolyse oil and fat into long chain fatty acids and glycerol (Jamie et al., 2016).

3.2.2 Cellulases

Cellulases are the most often used industrial enzymes, and they have been commercially accessible for more than 30 years (Kuhad et al., 2011). Despite the fact that cellulases have been used in the industry sector for more than three decades, this enzyme remains a subject of interest for both academic research and industrial applications (Chang et al., 2016; Ejaz et al., 2020). Cellulases are a diverse group of enzymes generated by a variety of microorganisms such as bacteria, fungi and actinomycetes (Ejaz et al., 2021; Jayasekara and Ratnayake, 2019) and also it is the most abundant polysaccharide that exists on the earth (O'sullivan, 1997). Cellulases have been shown to be more heat resistant than other enzymes that degrade plant cell walls (Chang et al., 2016); as a result, they are a superior option for industrial applications (Ejaz et al., 2020).

Cellulases have shown that they can be used in a wide range of industries such as food, agriculture, detergent, textile (Jayasekara and Ratnayake, 2019). Cellulases are enzymes that break down cellulose microfibrils that form when cotton-based cloths are washed and it can remove ink from recycled paper (Jayasekara and Ratnayake, 2019; Sun and Cheng, 2002).

3.2.3 Monooxygenases

Enzymes that catalyse the incorporation of a single oxygen atom from O_2 into an organic substrate are known as monooxygenases (Pazmino et al., 2010). In most situations, monooxygenases need binding cofactors such as pterin, haem, metal ions and flavin to aid in the activation of molecular oxygen (Pazmino et al., 2010). The monooxygenase enzymes catalyse oxidative reactions with different substrates of alkanes to complex molecules. These enzymes only need molecular oxygen to work and use the substrate as a reducing agent (Arora et al., 2009). As a result, they are appealing biocatalysts (Pazmino et al., 2010). The monooxygenase enzyme is used in several industries, one of its important applications in wastewater treatment (Todorova et al., 2019).

3.2.4 Laccases

Laccases are blue multicopper oxidase enzymes that catalyse the oxidation of a variety of substrates, including phenols and aromatic or aliphatic amines, converting molecular oxygen to water (Riva, 2006). Laccases are often glycosylated in eukaryotes, and the carbohydrate component appears to promote laccase conformational stability and to protect enzymes against radical and proteolytic inactivation (Arregui et al., 2019). These enzymes come in a variety of forms. In reality, while the majority are monomeric, some have been shown to be multimeric, homodimeric and heterodimeric with molecular masses ranging from 50 to 140kDa (with their sugar component) (Arregui et al., 2019; Janusz et al., 2020). Laccases are present in a wide variety of species, including bacteria, fungi, insects and higher plants (Gianfreda et al., 1999). Laccase is a type of oxidoreductase enzyme that contains multicopper atoms which in the presence of molecular oxygen can catalyse the oxidation of substrates by a single electron (Ramírez-Montoya et al., 2015; Su et al., 2018). Laccase is a promising potential water purification alternative due to its excellent catalytic properties and wide substrate range (Strong and Claus, 2011). The commercial involvement in laccases is also clearly demonstrated by the high number of patents issued in recent times (Zerva et al., 2019). Laccases' ability to be used in synthetic or degradative processes, as well as their broad substrate specificity, makes them ideal "green tools" for a wide variety of applications in industries (Bassanini et al.,

2021; Couto and Herrera, 2006; Pezzella et al., 2015). Their principal technological uses are in the textile business, for example, for delignification of woody fibres in the paper production, enhancing textile properties and procedures related to dyeing and fibre bleaching, as well as in the food sector for food enhancement (Bassanini et al., 2021; Kunamneni et al., 2008; Pezzella et al., 2015; Riva, 2006). Laccase is an interesting biocatalyst for removing micropollutants, purifying water (Zhou et al., 2020) and wastewater treatment (Ba et al., 2013); it can only perform its important catalytic function after being effectively immobilized (Zhou et al., 2020). However, laccases are limited to users in the industries because of their low degree of stability, high price, short service life and difficult reusability (Mokhtar et al., 2019; Zdarta et al., 2019).

3.2.5 Proteases

Proteases are a global entity found in all plants, animals and microorganisms (Gurumallesh et al., 2019; Razzaq et al., 2019) but the majority of protease enzymes are derived through the bacterial fermentation process (Gurumallesh et al., 2019). Proteases are categorized into six types: threonine, glutamate, cysteine, metallo, aspartate and serine (López-Otín and Bond, 2008). Peptide bonds are found in polypeptide chains of amino acids that have been degraded by proteases (Barrett and McDonald, 1986; Razzaq et al., 2019). Protease is a degradative enzyme that modifies proteins with specificity and selectivity (Rao et al., 1998). Microorganism-derived proteases are one of the three most important types of industrial enzymes, responsible for 60% enzyme sales worldwide (Kasana, 2010). Proteases are degradative enzymes that can modify proteins with a high degree of specificity and selectivity (Rao et al., 1998) and also they have proven to be a successful solution to chemicals and an environmentally sustainable symbol for nature or the environment (Razzaq et al., 2019). Proteases play an important role in biotechnology, and they are a useful method for removing proteins (Rao et al., 1998). For industrial usage, proteases must frequently sustain high activity under non-physiological circumstances such as high temperatures and pH, intense calcium chelating agents and detergents (Bryan and Enzymology, 2000; Li et al., 2013). Proteases have been used in the food and detergent industry for a long time; however, their use in the leather industry and for the disposal of organic waste treatment is a recent biotechnological invention (Dambmann and Aunstrup, 1981).

3.3 IMMOBILIZATION OF ENZYMES

In 1916, the method of immobilization of enzymes was invented (Nelson and Griffin, 1916). An enzyme that has been immobilized is one that has been attached to an organic, inorganic or inert material like silica, calcium or alginate. In addition, attaching an enzyme to a solid support can improve its resistance to environmental changes such as pH and temperature (Homaei et al., 2013a). Four methods commonly used for enzyme immobilization including cross-linking, adsorption, covalent bonding and entrapment (Figure 3.3) (Brena et al., 2013; Wong et al., 2019). Each of these strategies performs differently. Adsorption and entrapment are two ways of physical

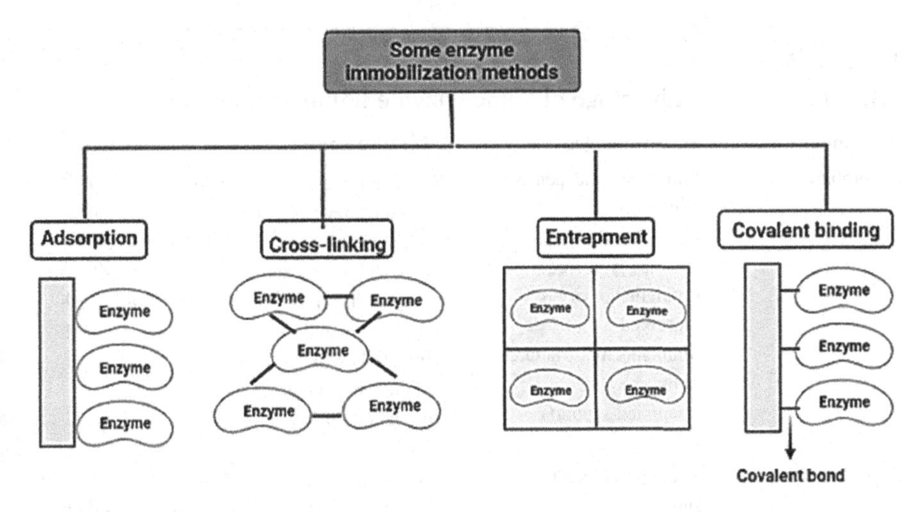

FIGURE 3.3 Schematic of some enzyme immobilization methods.

immobilization that rely on weak interactions that include van der Waals forces, ion binding and physical restriction of the enzyme inside the support (Bilal et al., 2016; Gupta and Mattiasson, 1992; Quilles Junior et al., 2016). The physical immobilization approach has an advantage of preserving the enzyme's catalytic activity since the native structure of the enzyme is not changed (Torres et al., 2017). Adsorption is the most scalable of all the established strategies utilized to carry out enzyme immobilization in the contemporary industry because of its high efficiency, simplified mechanism and affordability (Homaei et al., 2013b; Jesionowski et al., 2014). Enzyme immobilization has some benefits, including improved stability, the ability to use the enzyme in solutions where the enzyme is insoluble and ease of separation (Cao, 2005). Advantages and disadvantages of some enzyme immobilization methods are shown in Table 3.2.

Immobilized enzymes are commonly used in the medicinal, food and research industries (Homaei, 2015; Homaei et al., 2013a) as well as for organic synthesis on a laboratory scale, medical and analytical applications (DiCosimo et al., 2013), and for wastewater treatment (Bouabidi et al., 2019; Chen et al., 2007; El-Naas et al., 2016; Gupta and Balomajumder, 2015). To avoid washout of enzymes in a continuous wastewater treatment procedure, enzymes must be immobilized on a suitably insoluble substrate. Immobilization techniques are commonly classified into three types that include cross-linking, binding to a carrier and entrapment. In recent years, several alternative immobilization strategies have been considered, and different carrier substances have been effectively used (Stadlmair et al., 2018). For the immobilization of enzymes, porous and non-porous compounds have been utilized (Sheldon and van Pelt, 2013a; Stadlmair et al., 2018). The use of immobilization of enzyme technology in the field of wastewater treatment has many advantages over biological decomposition using free cells. The enzyme immobilization technology can provide high biomass, high mechanical sturdiness, high resistance to chemicals and enhancing genetic consistency (Kadimpati et al., 2013).

TABLE 3.2
Advantage and Disadvantage of Some Enzyme Immobilization Methods

Technique	Advantages	Disadvantages	References
Adsorption	– Simple and inexpensive to carry out – Has not been chemically altered – Catalytic activity is preserved – Maintains the structure of the enzyme – Reapplied supports	– Binding forces between enzymes and their substrates are weak and reversible – Changes in pH, temperature and ionic strength have a negative impact on the stability of the solution	Kim et al. (2015); Liu et al. (2018); Torres et al. (2018); Wong et al. (2019); Zucca and Sanjust (2014)
Cross-linking	– There is no need for support – Simple and dependable – Leakage of catalyst is maintained to a minimum – Maintains the catalytic activity of the enzyme at its original state	– Experimentation processes may be difficult and complicated – Problems in mass transfer After immobilization, catalytic activity may be lost	Jiang et al. (2014); Sun et al. (2017); Zdarta et al. (2018); Zucca and Sanjust (2014)
Entrapment	– Inexpensive to carry out – Has not been chemically altered – The enzyme preserves catalytic activity – Conformation is maintained – Prevents the aggregation of enzymes – Enzyme leaching is reduced	– Leakage of enzymes – Problems in mass transfer – Macromolecular substrates are not allowed to pass through	Bilal et al. (2017); Jamal et al. (2013); Liu et al. (2018); Shaheen et al. (2017); Wong et al. (2019)
Covalent binding	– The binding's strength – Enzyme leaching is reduced – Improves the stability of enzymes – Stable covalent bonds are formed	– Enzyme sterical alterations are a possibility – Enzymatic activity may be reduced – Chemical changes to the support are required – In most cases, the connection is permanent – Prohibiting the support from being reused	Ai et al. (2014); Bilal et al. (2018); Liu et al. (2018); Vineh et al. (2018)

3.3.1 Nano-carriers for Immobilization of Enzymes

Aside from the immobilization method, the immobilization carrier is also an important consideration for enzyme immobilization (Sirisha et al., 2016). Enzymes that have been immobilized on nano-carriers are more stable and robust, they may be recycled and renewed (Liu and Dong, 2020), they cost less, they may be plentiful, and they are suitable for the environment (eco-friendly) (Liu and Dong, 2020; Sirisha et al., 2016). As nanotechnology progresses, nanomaterials have been widely used in industry and medicine. Because nanomaterials may satisfy the requirements of immobilization carriers, they are predicted to be desirable carriers (Liu and Dong, 2020). Some nanomaterials in the enzymes of immobilization and their applications are summarized in the following text.

3.3.1.1 Metallic Nano-carriers

A group of nanomaterials containing a metal element is known as metal nanomaterials (Liu and Dong, 2020). Metal-organic frameworks (MOFs) are porous, ordered metallic nanomaterials made by cross-linking metal ions and organic ligands. Physical adsorption or covalent binding among enzyme molecules and built-in functional groups on MOFs can be used to immobilize enzymes on MOFs (Wu et al., 2015).

Protein-inorganic hybrid nanoflowers are a different type of metallic nano support that has been published. The first was created by coprecipitating copper salts and proteins in a single step (Ge et al., 2012), and it was later expanded to include other metal ions like iron (Ocsoy et al., 2015). Hybrid nanoflowers and MOFs will create a metal-ion-rich microenvironment as one of their most important benefits (Ge et al., 2012; Lyu et al., 2014; Ocsoy et al., 2015). Gold nanoparticles are the most widely utilized as metal nanomaterials for enzyme immobilization due to their enormous surface area, strong thermal and mechanical stability, simplicity of functionalization and favourable biocompatibility (Liu and Dong, 2020). Chemical (Luo et al., 2005) or physical adsorption (Venditti et al., 2015), cross-linking (Xu et al., 2010) and covalent attachment (Zhang et al., 2005) have been used to immobilize enzymes on bare or functionalized gold nanoparticles. Furthermore, for enzyme immobilization, gold nanoparticles are always attached on other nano-carriers (Jeong et al., 2006).

Magnetic nanoparticles have been used for enzyme immobilization, making it easier to separate and recover enzymes from reaction media (Cipolatti et al., 2016). Immobilization of enzymes on noble metal nanoparticles has also been attempted. Despite the fact that noble metals are less widely used in water treatment due to their high cost, their accumulation of optical properties combined with enzyme activities can be used to fabricate sensors for the identification of trace pollutants (Rome and Kickhoefer, 2013).

3.3.1.2 Silica Nano-carriers

Silica (SiO_2) is a solid nano-carrier used for the immobilization of enzymes that are abundant, chemically and thermally inert material. Surfactant templated polymerization processes are used to make mesoporous silica materials from silicon alkoxide precursors (Beck et al., 1992; Magner, 2013; Zhou and Hartmann, 2013). The enzymes may be covalently bound to the bare mesoporous silica or chemically grafted with functional groups like silanols, thiols and amines. Enzymes can be entrapped during in situ silica synthesis in the presence of enzymes. Surfactant is often carried out under hard conditions such as high temperature, organic solvents and severe pH that can reduce the activity of enzymes. Biosilicification, or the use of biomolecules to produce silica deposition, is an alternative enzyme method in water solutions with a neutral pH (Betancor and Luckarift, 2008). In fact, silica is an excellent option for removing pollution and water disinfection due to easy synthesis (Betancor and Luckarift, 2008; Gasser et al., 2014b; Luckarift et al., 2006).

3.3.1.3 Carbon Nano-carriers

Carbon nanotubes (CNTs) are one-dimensional hollow carbon atom tubes that contain single-wall nanotubes and multi-walled nanotubes (Liu and Dong, 2020; Wang et al., 2003). In general, carbonaceous nanomaterials for enzyme immobilization include reduced graphene oxide, graphene and CNTs (Beg et al., 2011; Liu and Dong, 2020). Physical adsorption and surface grafting, in which enzymes are covalently attached to carbon nanomaterials, immobilize enzymes (Cipolatti et al., 2016). Enzyme immobilization is also facilitated by mesoporous carbon materials (Zhou and Hartmann, 2013). They are made by polymerizing and carbonizing carbon precursors on templates including mesoporous silica and block copolymers, then removing the template. Carbon materials have exceptional electrical conductivity as compared to silica carriers, which can facilitate electron transfer reactions on electrodes (Cipolatti et al., 2016; Pan et al., 2015; Wang et al., 2003; Zhou and Hartmann, 2013) and also CNTs have good thermal, mechanical and biocompatibility (Wang et al., 2003).

3.3.1.4 Biomaterials as Nano-carriers

Biomaterials are less toxic than synthetic nano-carriers and they are biocompatible and environmentally friendly (Azuma et al., 2018; Wang et al., 2015; Wang et al., 2019a; Wang et al., 2019b).

Protein cages, a type of biomaterial, are crucial in immobilizing enzymes due to their ability to maintain a large nuclear volume and protect and stabilize the immobilized enzymes (Pieters et al., 2016; Wang et al., 2019b). DNA Origami is another biomaterial nano-carrier that the enzymes on them are immobilized and support the supplementary motifs on DNA. Its advantages are biocompatibility and increase in the overall activity of enzyme couples that are used in removing pollution (Fu et al., 2012; Saccà and Niemeyer, 2011). Some nano-carriers for immobilization of enzymes and their application in wastewater treatment are shown in Figure 3.4.

3.4 NANO-CARRIER-IMMOBILIZED ENZYMES FOR USE IN THE TREATMENT OF WASTEWATER

There are different methods and techniques to remove contaminants; nanotechnology is one of the important techniques used in wastewater treatment. Nanomaterials are ideal for wastewater treatment and water purification applications due to their wide area of a particular surface, size-dependent properties, high level of reactivity, etc (Bora and Dutta, 2014), and also the enzymatic approach is a more efficient and versatile method, needs fewer chemicals and resources, and produces less toxic waste products (Sharma et al., 2018; Stadlmair et al., 2018; Wang et al., 2019a).

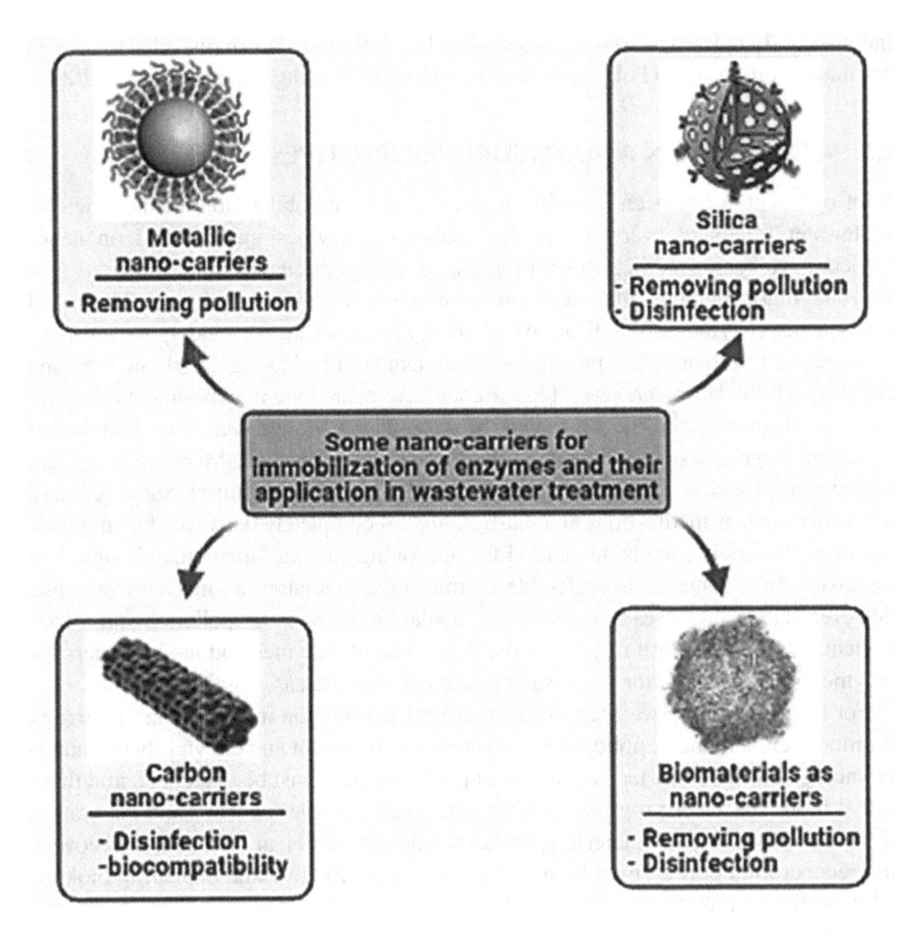

FIGURE 3.4 Some nano-carriers for immobilization of enzymes and their application in wastewater treatment.

Many enzymes have been recognized and prepared to control various contaminants, including lipases, cellulases, monooxygenases, laccases, proteases, etc (Bilal et al., 2019; Gasser et al., 2014a; Sharma et al., 2018). The advantage of using enzymatic methods is that free enzymes can be easily inactivated because they are sensitive to ligands and metal ions in natural waters, as well as products from their own activities. Nano-carrier-immobilized enzymes increase their overall stability and expand their usable life with limited activity loss (Wei et al., 2013) and also allow the reuse of enzymes and reduce costs (Gasser et al., 2014b). Enzymatic degradation is a process for treating organic pollutant compounds that are environmentally friendly. For wastewater treatment, enzymes have been immobilized on nanomaterials to increase their stability in harsh environments and make the most of their catalytic activities (Liu and Dong, 2020). Disinfection of water is another advantage of using enzymes where no harmful by-products are produced (Luckarift et al., 2006; Pangule et al., 2010). Antimicrobial enzymes that have been immobilized increase their durability and protect them from becoming digested by bacteria, and also are used for fouling in simulated Point of Use (PoU) processes to control biofouling (Luckarift et al., 2006).

3.5 CONCLUSION AND FUTURE PERSPECTIVES

A lot of research has been done in the area of cell immobilization technologies for wastewater treatment over the last few decades. Enzymes immobilized on nano-carriers have been used in a range of applications. As discussed in the previous section, enzymatic methods offer enormous potential in wastewater treatment. The use of nano-carrier enzyme immobilization technology to eliminate and biodegrade a variety of wastewater pollutants has proven to be efficient (Liu and Dong, 2020; Sheldon and van Pelt, 2013b). However, several researches have been done in experimental/laboratory circumstances, and the wastewater was synthetic. More research on real wastewater and larger-scale experiments is therefore urgently needed. More study on real wastewater, as well as larger-scale studies, are thus critically required. Some resistant pollutants, such as medications and plastics, are not completely destroyed by enzymes. Use of mediators is one viable method for improving enzyme function, although they are toxic. As a result, it is preferable to minimize mediators as much as possible. However, due to increases in the world's population, more water pollution and a consequent decline in natural resources, the future use of enzymes and nano-carriers for enzyme immobilization for wastewater treatment may increase significantly.

For future research, we suggest that there is investigation into more nano-carriers in order to expand their application in wastewater treatment and enzyme-based nanotechnologies which may face a number of problems that must be overcome, and there is still more space for progress. In addition, studies on the recovery and generation of biologically extracted materials, especially heavy metals, are needed, as recovery and regeneration play a key role in providing a cost-effective and appealing biological wastewater treatment process.

REFERENCES

Ai, Q., Yang, D., Li, Y., Shi, J., Wang, X., Jiang, Z.J.B., 2014. Highly efficient covalent immobilization of catalase on titanate nanotubes. *Biochemical Engineering Journal* 83, 8–15.

Alshabib, M., Onaizi, S.A.J.S., Technology, P., 2019. A review on phenolic wastewater remediation using homogeneous and heterogeneous enzymatic processes. *Current Status and Potential Challenges*. 219, 186–207.

Arora, P.K., Kumar, M., Chauhan, A., Raghava, G.P., Jain, R.K., 2009. OxDBase: a database of oxygenases involved in biodegradation. *BMC Research Notes* 2, 1–8.

Arpigny, J.L., Jaeger, K.-E., 1999. Bacterial lipolytic enzymes: classification and properties. *Biochemical Journal* 343, 177–183.

Arregui, L., Ayala, M., Gómez-Gil, X., Gutiérrez-Soto, G., Hernández-Luna, C.E., Herrera de Los Santos, M., Levin, L., Rojo-Domínguez, A., Romero-Martínez, D., Saparrat, M.C., 2019. Laccases: structure, function, and potential application in water bioremediation. *Journal of Molecular Catalysis B: Enzymatic* 18, 1–33.

Azuma, Y., Bader, D.L., Hilvert, D., 2018. Substrate sorting by a supercharged nanoreactor. *Journal of the American Chemical Society* 140, 860–863.

Ba, S., Arsenault, A., Hassani, T., Jones, J.P., Cabana, H., 2013. Laccase immobilization and insolubilization: from fundamentals to applications for the elimination of emerging contaminants in wastewater treatment. *Critical reviews in biotechnology* 33, 404–418.

Barrett, A.J., McDonald, J.K. 1986. Nomenclature: protease, proteinase and peptidase. *Biochemical Journal* 237, 935.

Bassanini, I., Ferrandi, E.E., Riva, S., Monti, D.J.C., 2021. Biocatalysis with laccases: an updated overview. *Catalysts* 11, 26.

Beck, J.S., Vartuli, J., Roth, W.J., Leonowicz, M., Kresge, C., Schmitt, K., Chu, C., Olson, D.H., Sheppard, E., McCullen, S., 1992. A new family of mesoporous molecular sieves prepared with liquid crystal templates. *Journal of the American Chemical Society* 114, 10834–10843.

Beg, S., Rizwan, M., Sheikh, A.M., Hasnain, M.S., Anwer, K., Kohli, K.J., 2011. Advancement in carbon nanotubes: basics, biomedical applications and toxicity. *Journal of Pharmacy and Pharmacology* 63, 141–163.

Betancor, L., Luckarift, H.R., 2008. Bioinspired enzyme encapsulation for biocatalysis. *Trends in Biotechnology* 26, 566–572.

Bharathi, D., Rajalakshmi, G.J.B., Biotechnology, A., 2019. Microbial lipases: an overview of screening, *production and Purification*. 22, 101368.

Bhatnagar, A., Anastopoulos, I.J.C., 2017. Adsorptive removal of bisphenol A (BPA) from aqueous solution: a review. *Chemical Engineering Journal* 168, 885–902.

Bilal, M., Adeel, M., Rasheed, T., Zhao, Y., Iqbal, H.M., 2019. Emerging contaminants of high concern and their enzyme-assisted biodegradation-a review. *Environment International* 124, 336–353.

Bilal, M., Asgher, M., Shahid, M., Bhatti, H.N., 2016. Characteristic features and dye degrading capability of agar agar gel immobilized manganese peroxidase. *International Journal of Biological Macromolecules* 86, 728–740.

Bilal, M., Iqbal, H.M., Hu, H., Wang, W., Zhang, X., 2017. Enhanced bio-catalytic performance and dye degradation potential of chitosan-encapsulated horseradish peroxidase in a packed bed reactor system. *Journal of the Science of Food and Agriculture* 575, 1352–1360.

Bilal, M., Rasheed, T., Zhao, Y., Iqbal, H.M., Cui, J., 2018. "Smart" chemistry and its application in peroxidase immobilization using different support materials. *International Journal of Biological Macromolecules* 119, 278–290.

Bora, T., Dutta, J., 2014. Applications of nanotechnology in wastewater treatment-a review. *Journal of Nanoscience and Nanotechnology* 14, 613–626.

Bouabidi, Z.B., El-Naas, M.H., Zhang, Z., 2019. Immobilization of microbial cells for the biotreatment of wastewater: a review. *Environmental Chemistry Letters* 17, 241–257.

Brena, B., González-Pombo, P., Batista-Viera, F., 2013. Immobilization of enzymes: a literature survey. *International Journal of Molecular Sciences* 15–31.

Bryan, P.N. 2000. Protein engineering of subtilisin. *Journal of Biotechnology and Applied Biochemistry* 1543, 203–222.

Cao, L., 2005. Introduction: immobilized enzymes: past, present and prospects. *Carrier-Bound Immobilized Enzymes: Principles, Application and Design*. Wiley-VCH,
Weinheim, (Vol. 1, pp. 1–52).

Chandra, P., Singh, R., Arora, P.K.J., 2020. Microbial lipases and their industrial applications: a comprehensive review. *Microbial Cell Factories* 19, 1–42.

Chang, C.-J., Lee, C.-C., Chan, Y.-T., Trudeau, D.L., Wu, M.-H., Tsai, C.-H., Yu, S.-M., Ho, T.-H.D., Wang, A.H.-J., Hsiao, C.-D., 2016. Exploring the mechanism responsible for cellulase thermostability by structure-guided recombination. *PLoS One* 11, e0147485.

Chen, Y.-M., Lin, T.-F., Huang, C., Lin, J.-C., Hsieh, F.-M., 2007. Degradation of phenol and TCE using suspended and chitosan-bead immobilized Pseudomonas putida. *Journal of Hazardous Materials* 148, 660–670.

Chioke, O.J., Ogbonna, C.N., Onwusi, C., Ogbonna, J.C., 2018. Lipase in biodiesel production. *African Journal of Biotechnology Research* 12, 73–85.

Cipolatti, E.P., Valerio, A., Henriques, R.O., Moritz, D.E., Ninow, J.L., Freire, D.M., Manoel, E.A., Fernandez-Lafuente, R., de Oliveira, D., 2016. Nanomaterials for biocatalyst immobilization-state of the art and future trends. *RSC Advances* 6, 104675–104692.

Couto, S.R., Herrera, J.L.T, 2006. Industrial and biotechnological applications of laccases: a review. *Biotechnology Advances* 24, 500–513.
Crini, G., 2005. Recent developments in polysaccharide-based materials used as adsorbents in wastewater treatment. *Progress in Polymer Science* 30, 38–70.
Crini, G., Lichtfouse, E., 2019. Advantages and disadvantages of techniques used for wastewater treatment. *Environmental Chemistry Letters* 17, 145–155.
Cummins, I., Landrum, M., Steel, P.G., Edwards, R.J., 2007. Structure activity studies with xenobiotic substrates using carboxylesterases isolated from Arabidopsis thaliana. *Phytochemistry* 68, 811–818.
Dambmann, C., Aunstrup, K., 1981. The variety of serine proteases and their industrial significance, proteinases and their inhibitors. Elsevier, pp. 231–244.
Dasgupta, N., Ranjan, S., Ramalingam, C., 2017. Applications of nanotechnology in agriculture and water quality management. *Environmental Chemistry Letters* 15, 591–605.
DiCosimo, R., McAuliffe, J., Poulose, A.J., Bohlmann, G., 2013. Industrial use of immobilized enzymes. *Chemical Society Reviews* 42, 6437–6474.
Durán, N., 1997. The impact of biotechnology in the pulp and paper industry: state of art. *Progress in Ecology*, 543.
Durán, N., Esposito, E., 2000. Potential applications of oxidative enzymes and phenoloxidase-like compounds in wastewater and soil treatment: a review. *Applied catalysis B: environmental* 28, 83–99.
Ebele, A.J., Abdallah, M.A.-E., Harrad, S., 2017. Pharmaceuticals and personal care products (PPCPs) in the freshwater aquatic environment. *Environmental Chemistry* 3, 1–16.
Ejaz, U., Muhammad, S., Hashmi, I.A., Ali, F.I., Sohail, M., 2020. Utilization of methyltrioctylammonium chloride as new ionic liquid in pretreatment of sugarcane bagasse for production of cellulase by novel thermophilic bacteria. *Journal of Bioscience* 317, 34–38.
Ejaz, U., Sohail, M., Ghanemi, A.J.B., 2021. Cellulases: from bioactivity to a variety of industrial applications. *Biocatalysis* 6, 44.
El-Naas, M.H., Surkatti, R., Al-Zuhair, S., 2016. Petroleum refinery wastewater treatment: a pilot scale study. *Journal of Water Process Engineering* 14, 71–76.
Feng, S., Ngo, H.H., Guo, W., Chang, S.W., Nguyen, D.D., Cheng, D., Varjani, S., Lei, Z., Liu, Y., 2021. Roles and applications of enzymes for resistant pollutants removal in wastewater treatment. *Bioresource Technology* 335, 125278.
Fu, J., Liu, M., Liu, Y., Woodbury, N.W., Yan, H., 2012. Interenzyme substrate diffusion for an enzyme cascade organized on spatially addressable DNA nanostructures. *Journal of the American Chemical Society* 134, 5516–5519.
Gasser, C.A., Ammann, E.M., Shahgaldian, P., Corvini, P.F.-X., 2014a. Laccases to take on the challenge of emerging organic contaminants in wastewater. *Applied microbiology and biotechnology* 98, 9931–9952.
Gasser, C.A., Yu, L., Svojitka, J., Wintgens, T., Ammann, E.M., Shahgaldian, P., Corvini, P.F.-X., Hommes, G., 2014b. Advanced enzymatic elimination of phenolic contaminants in wastewater: a nano approach at field scale. *Applied microbiology and biotechnology* 98, 3305–3316.
Ge, J., Lei, J., Zare, R.N., 2012. Protein-inorganic hybrid nanoflowers. *Nature nanotechnology* 7, 428–432.
Gholami-Borujeni, F., Mahvi, A.H., Naseri, S., Faramarzi, M.A., Nabizadeh, R., Alimohammadi, M.J.R.J.C.E., 2011. Application of immobilized horseradish peroxidase for removal and detoxification of azo dye from aqueous solution. *Journal of Chemical Engineering* 15, 217–222.
Gianfreda, L., Xu, F., Bollag, J.-M., 1999. Laccases: a useful group of oxidoreductive enzymes. *Bioremediation journal* 3, 1–26.

Gupta, A., Balomajumder, C., 2015. Simultaneous removal of Cr (VI) and phenol from binary solution using Bacillus sp. immobilized onto tea waste biomass. *Journal of Water Process Engineering* 6, 1–10.

Gupta, M., Mattiasson, B.., 1992. Unique applications of immobilized proteins in bioanalytical systems. *Methods in Enzymology* 36, 1–34.

Gurumallesh, P., Alagu, K., Ramakrishnan, B., Muthusamy, S., 2019. A systematic reconsideration on proteases. *International Journal of Biological Macromolecules* 128, 254–267.

Hasan, F., Shah, A.A., Javed, S., Hameed, A., 2010. Enzymes used in detergents: lipases. *African journal of biotechnology* 9, 4836–4844.

Homaei, A., 2015. Enzyme immobilization and its application in the food industry. *Advances in Food Biotechnology* 9, 145–164.

Homaei, A.A., Sariri, R., Vianello, F., Stevanato, R., 2013a. Enzyme immobilization: an update. *Journal of chemical biology* 6, 185–205.

Homaei, A.A., Sariri, R., Vianello, F., Stevanato, R., 2013b. Enzyme immobilization: an update. *Journal of Chemical Biology* 6, 185–205.

Jaeger, K.-E., Eggert, T., 2002. Lipases for biotechnology. *Current Opinion in Biotechnology* 13, 390–397.

Jamal, F., Singh, S., Qidwai, T., Singh, D., Pandey, P., Pandey, G., Khan, M.J.B., Biotechnology, A., 2013. Catalytic activity of soluble versus immobilized cauliflower (Brassica oleracea) bud peroxidase-concanavalin a complex and its application in dye color removal. *Journal of Biotechnology and Biomaterials* 2, 311–321.

Jamie, A., Alshami, A.S., Maliabari, Z.O., Ali Ateih, M., Al Hamouz, O.C.S., 2016. Immobilization and enhanced catalytic activity of lipase on modified MWCNT for oily wastewater treatment. *Environmental Progress & Sustainable Energy* 35, 1441–1449.

Janusz, G., Pawlik, A., Świderska-Burek, U., Polak, J., Sulej, J., Jarosz-Wilkołazka, A., Paszczyński, A., 2020. Laccase properties, physiological functions, and evolution. *International Journal of Molecular Sciences* 21, 966.

Jayasekara, S., Ratnayake, R., 2019. Microbial cellulases: an overview and applications. *Cellulose* 22, pp.10-5772.

Jeong, J., Ha, T.H., Chung, B.H., 2006. Enhanced reusability of hexa-arginine-tagged esterase immobilized on gold-coated magnetic nanoparticles. *Analytical and Bioanalytical Chemistry* 569, 203–209.

Jesionowski, T., Zdarta, J., Krajewska, B.J.A., 2014. Enzyme immobilization by adsorption: a review. *Adsorption* 20, 801–821.

Jiang, Y., Cui, C., Zhou, L., He, Y., Gao, J., 2014. Preparation and characterization of porous horseradish peroxidase microspheres for the removal of phenolic compound and dye. *Journal of Industrial and Engineering Chemistry* 53, 7591–7597.

Kadimpati, K.K., Mondithoka, K.P., Bheemaraju, S., Challa, V.R.M., 2013. Entrapment of marine microalga, Isochrysis galbana, for biosorption of Cr (III) from aqueous solution: isotherms and spectroscopic characterization. *Applied Water Science* 3, 85–92.

Kalia, S., Thakur, K., Celli, A., Kiechel, M.A., Schauer, C.L., 2013. Surface modification of plant fibers using environment friendly methods for their application in polymer composites, textile industry and antimicrobial activities: a review. *Journal of Environmental Chemical Engineering* 1, 97–112.

Karam, J., Nicell, J.A., 1997. Potential applications of enzymes in waste treatment. *Journal of Chemical Technology & Biotechnology: International Research in Process, Environmental AND Clean Technology* 69, 141–153.

Kasana, R.C., 2010. Proteases from psychrotrophs: an overview. *Critical reviews in microbiology* 36, 134–145.

Kim, H.J., Park, S., Kim, S.H., Kim, J.H., Yu, H., Kim, H.J., Yang, Y.-H., Kan, E., Kim, Y.H., Lee, S.H., 2015. Biocompatible cellulose nanocrystals as supports to immobilize lipase. *Journal of Molecular Catalysis B: Enzymatic* 122, 170–178.

Kuhad, R.C., Gupta, R., Singh, A, 2011. Microbial cellulases and their industrial applications. *Enzyme Research* 2011.

Kunamneni, A., Plou, F.J., Ballesteros, A., Alcalde, M., 2008. Laccases and their applications: a patent review. *Recent Patents on Biotechnology* 2, 10–24.

Li, Q., Yi, L., Marek, P., Iverson, B.L., 2013. Commercial proteases: present and future. *Journal of Fluorescence* 587, 1155–1163.

Liu, D.-M., Chen, J., Shi, Y.-P., 2018. Advances on methods and easy separated support materials for enzymes immobilization. *Trends in Analytical Chemistry* 102, 332–342.

Liu, D.-M., Dong, C., 2020. Recent advances in nano-carrier immobilized enzymes and their applications. *Process Biochemistry* 92, 464–475.

López-Otín, C., Bond, J.S.., 2008. Proteases: multifunctional enzymes in life and disease. *Journal of Biological Chemistry* 283, 30433–30437.

Luckarift, H.R., Dickerson, M.B., Sandhage, K.H., Spain, J.C., 2006. Rapid, room-temperature synthesis of antibacterial bionanocomposites of lysozyme with amorphous silica or titania. *Small* 2, 640–643.

Luo, X.-L., Xu, J.-J., Zhang, Q., Yang, G.-J., Chen, H.-Y., 2005. Electrochemically deposited chitosan hydrogel for horseradish peroxidase immobilization through gold nanoparticles self-assembly. *Biosensors and Bioelectronics* 21, 190–196.

Lyu, F., Zhang, Y., Zare, R.N., Ge, J., Liu, Z., 2014. One-pot synthesis of protein-embedded metal-organic frameworks with enhanced biological activities. *Nano letters* 14, 5761–5765.

Magner, E., 2013. Immobilisation of enzymes on mesoporous silicate materials. *Chemical Society Reviews* 42, 6213–6222.

Mishra, B., Varjani, S., Agrawal, D.C., Mandal, S.K., Ngo, H.H., Taherzadeh, M.J., Chang, J.-S., You, S., Guo, W.J.E.T., Innovation, 2020. Engineering biocatalytic material for the remediation of pollutants: a comprehensive review. *Environmental Technology & Innovation* 20, 101063.

Mokhtar, A., Nishioka, T., Matsumoto, H., Kitada, S., Ryuno, N., Okobira, T., 2019. Novel biodegradation system for bisphenol A using laccase-immobilized hollow fiber membranes. *International journal of biological macromolecules* 130, 737–744.

Nelson, J., Griffin, E.G., 1916. Adsorption of invertase. *Journal of the American Chemical Society* 38, 1109–1115.

Nimkande, V.D., Bafana, A.J.J.o .W.P.E., 2022. A review on the utility of microbial lipases in wastewater treatment. *Journal of Water Process Engineering* 46, 102591.

Novak, J., Kralova, B., Demnerova, K., Prochazka, K., Vodrazka, Z., Tolman, J., Rysova, D., Smidrkal, J., Lopata, V., 1990. Enzyme agent based on lipases and oxidoreductases for washing, degreasing and water reconditioning. *European Patent* 355, 228.

O'sullivan, A.C., 1997. Cellulose: the structure slowly unravels. *Cellulose* 4, 173–207.

Ocsoy, I., Dogru, E., Usta, S., 2015. A new generation of flowerlike horseradish peroxides as a nanobiocatalyst for superior enzymatic activity. *Enzyme and microbial technology* 75, 25–29.

Pan, D., Gu, Y., Lan, H., Sun, Y., Gao, H., 2015. Functional graphene-gold nano-composite fabricated electrochemical biosensor for direct and rapid detection of bisphenol A. *Analytica chimica acta* 853, 297–302.

Pangule, R.C., Brooks, S.J., Dinu, C.Z., Bale, S.S., Salmon, S.L., Zhu, G., Metzger, D.W., Kane, R.S., Dordick, J.S., 2010. Antistaphylococcal nanocomposite films based on enzyme– nanotube conjugates. *ACS nano* 4, 3993–4000.

Pazmino, D.T., Winkler, M., Glieder, A., Fraaije, M.W., 2010. Monooxygenases as biocatalysts: classification, mechanistic aspects and biotechnological applications. *Journal of biotechnology* 146, 9–24.

Pezzella, C., Guarino, L., Piscitelli, A., 2015. How to enjoy laccases. *Chemical Sciences* 72, 923–940.

Pieters, B.J., Van Eldijk, M.B., Nolte, R.J., Mecinović, J., 2016. Natural supramolecular protein assemblies. *Chemical Society Reviews* 45, 24–39.

Quilles Junior, J.C., Ferrarezi, A.L., Borges, J.P., Brito, R.R., Gomes, E., Da Silva, R., Guisán, J.M., Boscolo, M.J.B., engineering, b., 2016. Hydrophobic adsorption in ionic medium improves the catalytic properties of lipases applied in the triacylglycerol hydrolysis by synergism. *Journal of Biotechnology* 39, 1933–1943.

Ramírez-Montoya, L.A., Hernández-Montoya, V., Montes-Morán, M.A., Cervantes, F.J., 2015. Correlation between mesopore volume of carbon supports and the immobilization of laccase from Trametes versicolor for the decolorization of Acid Orange 7. *Journal of environmental management* 162, 206–214.

Rao, M.B., Tanksale, A.M., Ghatge, M.S., Deshpande, V.V., 1998. Molecular and biotechnological aspects of microbial proteases. *Microbiology and Molecular Biology Reviews* 62, 597–635.

Razzaq, A., Shamsi, S., Ali, A., Ali, Q., Sajjad, M., Malik, A., Ashraf, M., 2019. Microbial proteases applications. *Frontiers in Bioengineering and Biotechnology* 7, 110.

Riva, S., 2006. Laccases: blue enzymes for green chemistry. *Trends in Biotechnology* 24, 219–226.

Rome, L.H., Kickhoefer, V.A., 2013. Development of the vault particle as a platform technology. *Acs Nano* 7, 889–902.

Saccà, B., Niemeyer, C.M., 2011. Functionalization of DNA nanostructures with proteins. *Chemical Society Reviews* 40, 5910–5921.

Saratale, R.G., Saratale, G.D., Chang, J.-S., Govindwar, S., 2011. Bacterial decolorization and degradation of azo dyes: a review. *Journal of the Taiwan Institute of Chemical Engineers* 42, 138–157.

Saxena, R., 1999. Microbial lipases: potential biocatalysts for the future industry. *Current Science*. 77, 101–115.

Shaheen, R., Asgher, M., Hussain, F., Bhatti, H.N., 2017. Immobilized lignin peroxidase from Ganoderma lucidum IBL-05 with improved dye decolorization and cytotoxicity reduction properties. *International Journal of Biological Macromolecules* 103, 57–64.

Sharma, B., Dangi, A.K., Shukla, P., 2018. Contemporary enzyme based technologies for bioremediation: a review. *Journal of Environmental Management* 210, 10–22.

Sharma, R.S., Karmakar, S., Kumar, P., Mishra, V. 2019. Application of filamentous phages in environment: a tectonic shift in the science and practice of ecorestoration. *Evolutionary Ecology* 9, 2263–2304.

Sheldon, R., van Pelt, S., 2013a. Enzyme immobilisation in biocatalysis: why, what and how. *Chemical Society Reviews* 42: 6223–6235.

Sheldon, R., van Pelt, S., 2013b. Evaluation of immobilized enzymes for industrial applications. *Chemical Society Reviews* 15, 6223–6235.

Sirisha, V.L., Jain, A., Jain, A. 2016. Enzyme immobilization: an overview on methods, support material, and applications of immobilized enzymes. *Advances in Food and Nutrition Research* 79, 179–211.

Sonune, A., Ghate, R., 2004. Developments in wastewater treatment methods. *Desalination* 167, 55–63.

Stadlmair, L.F., Letzel, T., Drewes, J.E., Grassmann, J., 2018. Enzymes in removal of pharmaceuticals from wastewater: a critical review of challenges, applications and screening methods for their selection. *Chemosphere* 205, 649–661.

Strong, P., Claus, H., 2011. Laccase: a review of its past and its future in bioremediation. *Critical Reviews in Environmental Science and Technology* 41, 373–434.

Su, J., Fu, J., Wang, Q., Silva, C., Cavaco-Paulo, A., 2018. Laccase: a green catalyst for the biosynthesis of poly-phenols. *Critical reviews in Biotechnology* 38, 294–307.

Summerscales, J. 2021. A review of bast fibres and their composites: Part 4~ organisms and enzyme processes. *Composites Part A: Applied Science and Manufacturing* 140, 106149.
Sun, H., Jin, X., Long, N., Zhang, R., 2017. Improved biodegradation of synthetic azo dye by horseradish peroxidase cross-linked on nano-composite support. *International Journal of Biological Macromolecules* 95, 1049–1055.
Sun, Y., Cheng, J., 2002. Hydrolysis of lignocellulosic materials for ethanol production: a review. *Bioresource Technology* 83, 1–11.
Thallinger, B., Prasetyo, E.N., Nyanhongo, G.S., Guebitz, G.M., 2013. Antimicrobial enzymes: an emerging strategy to fight microbes and microbial biofilms. *Biotechnology Journal* 8, 97–109.
Todorova, Y., Yotinov, I., Topalova, Y., Benova, E., Marinova, P., Tsonev, I., Bogdanov, T., 2019. Evaluation of the effect of cold atmospheric plasma on oxygenases' activities for application in water treatment technologies. *Environmental Technology* 40, 3783–3792.
Torres, J., Nogueira, F., Silva, M., Lopes, J., Tavares, T., Ramalho, T., Corrêa, A, 2017. Novel eco-friendly biocatalyst: soybean peroxidase immobilized onto activated carbon obtained from agricultural waste. *RSC Advances* 7, 16460–16466.
Torres, J., Silva, M., Lopes, J., Nogueira, A., Nogueira, F., Corrêa, A., 2018. Development of a reusable and sustainable biocatalyst by immobilization of soybean peroxidase onto magnetic adsorbent. *Journal of Biotechnology and Microbiology* 114, 1279–1287.
Ufarté, L., Laville, É., Duquesne, S., Potocki-Veronese, G., 2015. Metagenomics for the discovery of pollutant degrading enzymes. *Biotechnology Advances* 33, 1845–1854.
Varga, B., Somogyi, V., Meiczinger, M., Kováts, N., Domokos, E., 2019. Enzymatic treatment and subsequent toxicity of organic micropollutants using oxidoreductases-A review. *Journal of Chemical Technology & Biotechnology* 221, 306–322.
Venditti, I., Palocci, C., Chronopoulou, L., Fratoddi, I., Fontana, L., Diociaiuti, M., Russo, M.V., 2015. Candida rugosa lipase immobilization on hydrophilic charged gold nanoparticles as promising biocatalysts. *Activity and Stability Investigations.* 131, 93–101.
Villegas, L.G.C., Mashhadi, N., Chen, M., Mukherjee, D., Taylor, K.E., Biswas, N., 2016. A short review of techniques for phenol removal from wastewater. *Current Pollution Reports* 2, 157–167.
Vineh, M.B., Saboury, A.A., Poostchi, A.A., Rashidi, A.M., Parivar, K.J., 2018. Stability and activity improvement of horseradish peroxidase by covalent immobilization on functionalized reduced graphene oxide and biodegradation of high phenol concentration. *International Journal of Biological Macromolecules* 106, 1314–1322.
Virkutyte, J., Varma, R., 2010. *Treatment of micropollutants in water and wastewater.* IWA Publishing.
Wang, J., Wang, S., 2016. Removal of pharmaceuticals and personal care products (PPCPs) from wastewater: a review. *Journal of Environmental Management* 182, 620–640.
Wang, M., Abad, D., Kickhoefer, V.A., Rome, L.H., Mahendra, S., 2015. Vault nanoparticles packaged with enzymes as an efficient pollutant biodegradation technology. *ACS Nano* 9, 10931–10940.
Wang, M., Chen, Y., Kickhoefer, V.A., Rome, L.H., Allard, P., Mahendra, S., 2019a. A vault-encapsulated enzyme approach for efficient degradation and detoxification of Bisphenol A and its analogues. *ACS Sustainable Chemistry & Engineering* 7, 5808–5817.
Wang, M., Mohanty, S.K., Mahendra, S., 2019b. Nanomaterial-supported enzymes for water purification and monitoring in point-of-use water supply systems. *Accounts of Chemical Research* 52, 876–885.
Wang, S., Humphreys, E.S., Chung, S.-Y., Delduco, D.F., Lustig, S.R., Wang, H., Parker, K.N., Rizzo, N.W., Subramoney, S., Chiang, Y.-M., 2003. Peptides with selective affinity for carbon nanotubes. *Nature Materials* 2, 196–200.

Wei, W., Du, J., Li, J., Yan, M., Zhu, Q., Jin, X., Zhu, X., Hu, Z., Tang, Y., Lu, Y., 2013. Construction of robust enzyme nanocapsules for effective organophosphate decontamination, detoxification, and protection. *Advanced Materials* 25, 2212–2218.

Wilkinson, J.L., Hooda, P.S., Barker, J., Barton, S., Swinden, J., 2016. Ecotoxic pharmaceuticals, personal care products, and other emerging contaminants: a review of environmental, receptor-mediated, developmental, and epigenetic toxicity with discussion of proposed toxicity to humans. *Critical Reviews in Environmental Science and Technology* 46, 336–381.

Wong, J.K.H., Tan, H.K., Lau, S.Y., Yap, P.-S., Danquah, M.K., 2019. Potential and challenges of enzyme incorporated nanotechnology in dye wastewater treatment: a review. *Journal of Environmental Chemical Engineering* 7, 103261.

Wu, X., Hou, M., Ge, J., 2015. Metal-organic frameworks and inorganic nanoflowers: a type of emerging inorganic crystal nanocarrier for enzyme immobilization. *Catalysis Science & Technology* 5, 5077–5085.

Xu, J., Shang, F., Luong, J.H., Razeeb, K.M., Glennon, J.D. 2010. Direct electrochemistry of horseradish peroxidase immobilized on a monolayer modified nanowire array electrode. *Biosensors and Bioelectronics* 25, 1313–1318.

Yao, Y., Wang, M., Liu, Y., Han, L., Liu, X.J.F.c., 2020. Insights into the improvement of the enzymatic hydrolysis of bovine bone protein using lipase pretreatment. *Food Chemistry* 302, 125199.

Zangeneh, H., Zinatizadeh, A., Habibi, M., Akia, M., Isa, M.H., 2015. Photocatalytic oxidation of organic dyes and pollutants in wastewater using different modified titanium dioxides: a comparative review. *Journal of Industrial and Engineering Chemistry* 26, 1–36.

Zdarta, J., Meyer, A.S., Jesionowski, T., Pinelo, M., 2019. Multi-faceted strategy based on enzyme immobilization with reactant adsorption and membrane technology for biocatalytic removal of pollutants: a critical review. *Biotechnology Advances* 37, 107401.

Zdarta, J., Meyer, A.S., Jesionowski, T., Pinelo, M.J.C., 2018. A general overview of support materials for enzyme immobilization: characteristics, properties, *Practical Utility*. 8, 92.

Zerva, A., Simić, S., Topakas, E., Nikodinovic-Runic, J.., 2019. Applications of microbial laccases: patent review of the past decade (2009-2019). *Catalysts* 9, 1023.

Zhang, S., Wang, N., Niu, Y., Sun, C., 2005. Immobilization of glucose oxidase on gold nanoparticles modified Au electrode for the construction of biosensor. *Sensors and Actuators B: Chemical* 109, 367–374.

Zhang, W., Grimi, N., Jaffrin, M.Y., Ding, L., Tang, B., Zhang, Z., 2018. Optimization of RDM-UF for alfalfa wastewater treatment using RSM. *Environmental Science and Pollution Research* 25, 1439–1447.

Zhang, Y., Xu, Z., Chen, Z., Wang, G.J.C., 2020. Simultaneous degradation of triazophos, methamidophos and carbofuran pesticides in wastewater using an Enterobacter bacterial bioreactor and analysis of toxicity and biosafety. *Journal of Hazardous Materials* 261, 128054.

Zhou, W., Zhang, W., Cai, Y., 2020. Laccase immobilization for water purification: a comprehensive review. *Chemical Engineering Journal*, 126272.

Zhou, Z., Hartmann, M., 2013. Progress in enzyme immobilization in ordered mesoporous materials and related applications. *Chemical Society Reviews* 42, 3894–3912.

Zucca, P., Sanjust, E.J.M., 2014. Inorganic materials as supports for covalent enzyme immobilization: methods and mechanisms. *Journal of Materials Chemistry B* 19, 14139–14194.

4 Accumulation and Detoxification of Aqueous Pollutants by Microbes/Enzymes

Saumik Panja, Rusha Pal, and Abhishek RoyChowdhury

4.1 INTRODUCTION

The declining freshwater resources all over the world have triggered the interest of reusing wastewater and reclaiming physical and chemical resources from the wastewater with the aim of recycling them. The recovery of potable water from industrial and domestic wastewater streams has been the primary focus for various urban sectors worldwide. The key factor to maximize the conversion lies between the freshwater demand and innovative wastewater treatment technology. Currently, another vital objective of the wastewater treatment focuses on reducing environmental impacts due to the presence of emerging contaminants in urban and industrial wastewater. Due to the increasing growth of population along with urban sprawl, the number and variety of wastewater contaminants have changed drastically in the past few decades. Along with the urban and industrial pressure, the wastewater streams also get highly influenced due to the different agricultural practices in the rural environment. While the urban wastewater carries tons of anthropogenic substances and contaminants, the rural sector generates pesticides, nutrient runoff that influences the wastewater streams. The management of wastewater at both sectors faces numerous challenges due to the lack of proper treatment technology and to some extent the lack of proper rules and regulations. Recently, both the academic and the industrial research initiatives have turned their focus on developing efficient treatment techniques and resource recovery from wastewater from different sectors. Considering freshwater as the most valuable resource, the other resources include nutrients, organic matter, metals, etc. When comparing the wastewater treatment technology between developing and developed nations, most of the valuable resources slip through the treatment checkpoints in developing nations due to the lack of efficient treatment technology. With the rise of new technologies over the past few decades, we have also observed an emergence of a new generation of contaminants. Not only is there a lack of proper knowledge about each of these new series of contaminant materials but also we do not have appropriate treatment infrastructure to screen them out of the wastewater

 DOI: 10.1201/9781003517238-4

streams. The emergence of contaminants like microplastics and pharmaceuticals in the natural waterbodies puts the human population at a huge unknown risk. Organic micropollutants (OMPs) are the new generation of wastewater contaminants that are known for their prolonged persistence in the ecosystem. A huge number of contaminants fall into the category of OMPs that primarily includes pharmaceutical compounds (PhCs), industrial chemicals, pesticides and personal care products.[1] Chemotherapeutic agents like antibiotics are frequently detected in the wastewater and these biologically active compounds can easily pass through different levels of wastewater treatment without any significant alteration in their structure. The presence of antibiotics in soil and aquatic ecosystem can trigger antibiotic resistance in microorganisms. Antimicrobial resistance is a serious threat to the global public health system which is significantly reliant on antibiotics. Except triggering resistance in microorganisms, the presence of PhCs like acetaminophen, carbamazepine, gemfibrozil, and venlafaxine has been reported to reduce fecundity in female zebrafish (*Danio rerio*).[1,2] After passing through several checkpoints of conventional wastewater treatment plant (WWTP), these OMPs primarily end up in the aquatic ecosystem and are eventually partitioned in sediments and soil.

Microorganisms, due to their unique adaptive capacity, utilize a huge variety of substances for sustaining their life cycle. While conventional chemical and physical treatments demand complex infrastructure and financial requirement, we can use microbial technology to address the complexity of the remediation process in an easier and sustainable manner. Such an approach, more popularly known as bioremediation, involves the use of microorganisms to degrade the xenobiotic and organic pollutants present in the wastewater. One of the major objectives of running a microbe-oriented remediation operation is to provide a favorable environment for their optimum growth. For centuries, both dairy and brewery industries have developed efficient bioreactor to support the microorganisms to reach their maximum growth potential. When it comes to designing a reactor for wastewater treatment, a couple of considerations need to be taken, such as providing the microorganism of choice the opportunity to gain dominance within the reactor containment as the wastewater contains a preoccupied load of other microorganisms. The choice of microorganisms is also a critical factor as many microorganisms exhibit promising outcome in a controlled environment but they fail to prove their potential in the real-life scenario that comes with numerous variables.

4.1.1 Different Types of Wastewater

Depending on the types of the industrial sectors, the chemical and the physical characteristics of wastewater vary significantly. The wastewater effluent generated from the food and food processing industries contains significant amounts of biological oxygen demand (BOD).[3,4] The elevated level of BOD provides an excellent ground for microorganisms to thrive within it. The wastewater stream from brewery industries contains a number of organic matters such as sugar starch and proteins, a perfect environment for microbial proliferation.[5,6] Recent studies indicate that wastewater generated from brewery industry contains chemical oxygen demand (COD) almost 10 times higher than the domestic wastewater.[7] Many researchers are investigating the feasibility of using this organic matter-rich wastewater effluent to develop microbial

TABLE 4.1
Wastewater Generated from Different Sector and Typical Concentrations of Most Common Constituents, COD, Phosphorus (P), and NH_4-N

Source	COD (mg/L)	P (mg/L)	NH_4-N (mg N/L)
Municipal/domestic	60–111,600	ND	25.7
Fruit processing	2400	ND	ND
Petroleum refinery	1040	ND	ND
Synthetic	620–1563	10	16.8–38.5
Swine	3300	545–563	1000–1300
Winery	6850	0.95	18.3
Combined industrial	376	51	Below Detection Limit
Tannery	1100	ND	431
Rice mill	2775	ND	BDL
Fecal	1500	ND	110
Brewery	510	ND	BDL
Caustic	400	ND	BDL
Paper mill	6300–6500	1.5	38–42
Seafood industry	5600	ND	BDL
Retting	530	7.5	BDL
Yogurt	800–14,000	ND	4–60
Landfill leachate	1938	3.63	2537.6

Source: Munoz-Cupa et al. [4].

fuel cells. In spite of containing a significant load of organic matter, some of the industrial effluents contain other hazardous materials and therefore do not provide an optimal ground for microbial growth and proliferation, for example, wastewater generated from a distillery facility contains a significant amount of salts alongside high organic loads. The presence of nutrients such as nitrate (NO_3^-), phosphate (PO_4^{3-}), and non-oxidized nitrogen significantly affects the removal of COD by microorganisms.[8] On the other hand, the presence of nutrients in the wastewater stream supports the growth of algae and generates a competitive environment for microbial population. Recent investigations showed that in the absence of nutrients, microorganisms can remove up to 94% of COD from dairy wastewater (Table 4.1).

4.2 ENZYME-MEDIATED WASTEWATER REMEDIATION

Conventional WWTPs have a long history of using microorganisms as an integral part of treatment methods commonly known as activated sludge treatment. Sometimes depending on the physical and chemical characteristics of influent wastewater streams, the overall efficiency of the microorganisms gets affected. Microorganisms are highly dependent on their cellular enzymes for metabolizing the organic content as well as detoxifying the toxic substances present in the wastewater matrix. The supercritical role of these enzymes drew the attention of the scientific communities dedicated to developing a cost-effective, sustainable, and green wastewater remediation/reclamation technology. For the past few decades, the researchers have successfully isolated and purified the enzymes that are literally the catalytic component of cellular physiological system. Recently, scientific investigations have reported the efficiency of certain enzymes to degrade and detoxify wastewater contaminants. Although enzymes are extremely efficient in catalyzing biochemical reactions, they are sensitive toward the physical and chemical environment. Some of these critical factors include pH and temperature of the surrounding environment. Sometimes enzymes also need vital cellular components such as cofactors as an essential functional factor. On the contrary, the enzyme-based remediation approach has minimal impact on ecosystems.[9] Several oxidative enzymes have been reported for their potential to remove toxic contaminants from the environment.[9] Dioxygenase and monooxygenase are some of the potential candidates for environmental remediation, but they have limitations for their requirement of cofactor. Alternatively, peroxidases and laccases are the ones that fall in the category of enzymes with environmental implications. Both of these enzymes do not require cofactors to retain their functional ability. Oxidative enzymes like peroxidases have broad spectrum of substrate specificity and possess the transformational ability to detoxify a wide range of environmental contaminants. These enzymes are found in several organisms that are widely distributed in various ecosystems. The application of these enzymes as biocatalysts is found in many industrial sectors such as pulp and paper bleaching. Recently, the application of oxidative enzymes has been reported in wastewater treatment, in situ remediation, and medical diagnostics.

Microorganisms have a diverse pool of enzymes, one of the vital biomolecules that catalyze a number of critical biochemical reactions. While a single microorganism species synthesizes specific categories of enzymes to sustain under a normal growth condition, microbial consortia produce diverse classes of enzymes to support a symbiotic relationship. In a community structure, microbial species also alter the dynamics of the overall growth conditions. Recent scientific investigations focus on exploiting the presence of diverse enzymes in microbial consortia to remove organic micropollutants from wastewater streams.[10] It has been identified that the presence of reactive oxygen species (ROS) in high concentrations within cellular physiological system results in oxidative stress and induces oxidative enzyme gene expression. ROS that are commonly found inside microorganisms include superoxide (O_2^-), hydrogen peroxide (H_2O_2), and hydroxyl radicals ($OH^•$). Induction of oxidative enzyme gene expression triggers the synthesis of antioxidative enzymes such as oxidoreductases (peroxidase and cytochromes) to alleviate the oxidative stress.[11] Manipulation of cellular exposure to dissolved oxygen has been reported to induce the synthesis of antioxidative enzymes that are capable of catalyzing reaction toward the degradation of wastewater contaminants like polycyclic aromatic hydrocarbons and organophosphorus contaminants.[12,13] The intracellular metabolic pathways are disrupted due to the varying concentrations of dissolved oxygen in the surrounding environment that ultimately results in the formation of excess ROS inside the cellular boundary. As a defense mechanism, cells upregulate genetic expression responsible for the synthesis of ROS scavenging enzymes. In a wastewater treatment setting, the manipulation of dissolved oxygen can be easily simulated during the activated sludge secondary treatment phase where oxygen/air is supplied to the reactor chamber from an external source. Under limited oxygen supply, microorganisms belonging to phylum Proteobacteria predominate in the bioreactor. Proteobacteria are commonly found in the wastewater and there are multiple evidence that these microorganisms are capable of transforming emerging contaminants that are commonly detected in the influent stream.

4.3 MICROALGAE-BASED WASTEWATER REMEDIATION

Application of microalgae for wastewater treatment is a relatively new technology and is rapidly gaining popularity in water treatment industries because of its efficiency. The presence of microalgae in the wastewater streams has been widely reported in multiple literatures. The only reason that supports the microalgal population to thrive is the presence of nutrients in the wastewater. The algal population is responsible for triggering eutrophication in the surface water downstream of WWTPs that doesn't have an efficient system to remove nutrients. Eutrophication is defined by massive algal growth in the surface water that eventually creates an anoxic environment. Under controlled environment, the rapid growth potential of microalgae can be utilized for removing nutrients from wastewater matrices. It has been widely reported in many recent studies that microalgae not only remove nutrients from wastewater but also detoxify or metabolize toxic compounds such as heavy metals. Nutrients have always been an integral part of the wastewater regardless of the generating sectors such as urban, agriculture, industries. Microalgae is found in abundance in both marine and freshwater ecosystems. They are unicellular autotrophs and the primary producers in most of the aquatic ecosystem. Recently, microalgae-mediated wastewater treatment received the primary focus not only because of their performance but also for the production of oil in their cells, commonly known as algal biofuel. The focus of the bioenergy production industries is currently shifting toward microorganisms as food crop-based energy production was affected, which addresses the issue of food security for the global population. Under controlled environment, microalgae exhibit rapid growth potential and can be grown and harvested all year around even in suboptimal growth conditions. Runoff generated from agricultural sectors that are largely dependent on traditional inorganic agricultural practices significantly raises the trophic level of surface water by adding massive amounts of nutrients (primarily nitrogen and phosphorus).

Conventional treatment methods for removing nutrients such as activated sludge treatment and physicochemical processes for phosphorus removal are found to be inefficient to meet the stringent regulatory standards in terms of both feasibility and cost. Particularly, these conventional nutrient removal techniques are time consuming, and have high energy and carbon footprint.[14,15] Due to the lack of strict governance and enforcement, the issue of nutrient discharge from various industries in the developing nations results in the massive eutrophication rate in surface water. For example, the average rate of removing ammonium-nitrogen (NH_4-N) in China was less than 40% in 2012.[16] In the past decade, it has been found that approximately 48% of the lakes in North America are suffering from eutrophication.[17] In most cases, WWTPs act as a point source contributor for nutrients to the surface water and provides an optimal environment that promotes massive algal infestation. The application of microalgae for nutrient removal has led toward the establishment of different types of algal reactors and raceway reactors are one of the popular ones for growing algal species (Figure 4.1).

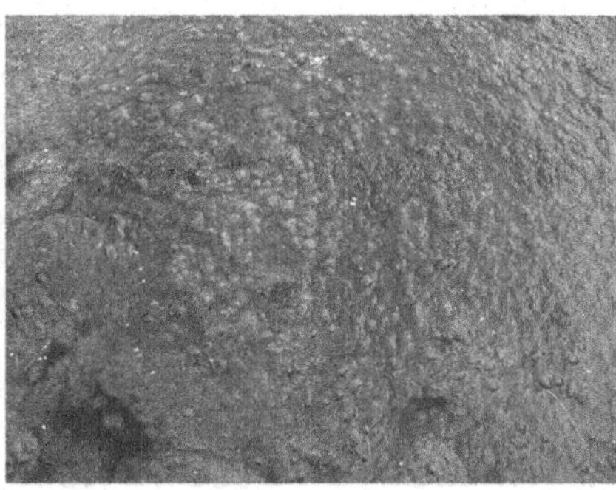

FIGURE 4.1 Algal infestation in Lake Wapalanne, New Jersey, due to nutrient loading from non-point sources. (Adapted from Akerman-Sanchez and Rojas-Jimenez [25].)

Municipal wastewater contains a balanced load of organic and inorganic substances, hence providing an ideal environment for microalgal growth. Along with providing nutritional benefits, municipal wastewater contains much lower concentrations of toxic substances compared to industrial wastewater discharge. The absence of toxic loads eliminates the probabilities of growth inhibition of the microalgal population. WWTPs utilize the microalgae in various stages of the treatment process. Sometimes, the algal cells are introduced at the stage of primary treatment. It has been observed that the algal biomass productivity reached almost 25% during primary treatment compared to the productivity at algal growth media such as tris-acetate-phosphate (TAP) or BG-11. Microalgae are often used to polish secondary effluents coming out of activated sludge chamber. Activated sludge process is conventionally used to remove organic load and nutrients but somehow fails to remove the entire nutrient load. For microalgal population, secondary effluents provide a minimal growth medium. The effluent collected during sludge dewatering process provides an excellent medium for microalgal growth. The dewatering effluent, often termed as centrate, contains high concentrations of nitrogen (130 mg/L) and phosphorus (200 mg/L). It has been reported that on average the growth of microalgae is three times higher in the dewatering effluent than in raw wastewater. The removal efficiency for each contaminant, NH_4-N, total nitrogen, and phosphate, was achieved as over 90% in the dewatering effluent.

Treatment of agricultural/dairy industry wastewater using microalgal species comes with several challenges. The wastewater from a typical concentrated animal feeding operation (CAFO) contains high concentrations of suspended solids and ammonia that does not support microalgal growth. Anaerobic digestion is mostly used in dairy industry to remove COD from manure, but the anaerobic process cannot remove nutrients. Therefore, the effluent from the digestion process contains huge load of nutrients (NH_4-N, NO_3^-, PO_4^{3-}). Although algal growth is stimulated in the presence of nutrients, high concentration of nutrients inhibits the growth of algae. In most cases, the effluent received from the digestion chamber is mixed with other wastewater streams to dilute the nutrient load enough to support the growth of algal species.

There are no specific universal treatment methods for industrial wastewater as they vary from one industry to another. Complex physicochemical characteristics and the presence of toxic substances in the industrial wastewater make it difficult for algal species to grow. *Scenedesmus* sp. has been studied by many researchers for their ability to survive in toxic environments. Recent studies have shown that microalgal species like *Scenedesmus* sp. not only can be used to remove nutrients from toxic wastewater but also can be harvested downstream for the production of biofuel.[18] Some of the microalgae species, e.g., *Chlorella, Ankistrodesmus*, have been reported as ideal candidates to treat wastewater generated from food and paper industries.[19] The use of algae as biosorptive agent has also gained attention from the heavy metal industries. Brown algae *Sargassum bevanom* has been studied to remove 85% Cr (VI) from aqueous media.[20]

4.4 EMERGING POLLUTANTS AND WASTEWATER REMEDIATION

The scientific and pharmacological advances in the modern era have helped expand our arsenal of therapeutics which can now be used to treat several human and animal diseases. With the increase in the global demand for therapeutics there has been a concomitant rise in the pharmaceutical-derived emerging pollutants which pose a long-term risk to human health and the environment. Typically, WWTPs lack the design required for the treatment of emerging pollutants. The PhCs end up penetrating through the filtration stages in the conventional WWTPs, thereby, the PhC concentration in the effluents remains similar to that of the influents. Of the PhCs present in the effluent water from WWTPs, the highest concentrations are those of antibiotics, antihypertensive drugs, beta-blockers, diuretics, and anti-inflammatory drugs. Literature studies reveal that species of microalgae like *Chlamydomonas pitschmannii, Chlorella vulgaris, Tribonema aequale*, and *Ourococcus multisporus* harbor the capacity to degrade PhCs like ibuprofen, diclofenac, acetaminophen (anti-inflammatory agents/analgesics), levofloxacin, enrofloxacin, metronidazole, sulfamethazine (antibiotics), and carbamazepine, diazepam (psychiatric drugs).[21,22] In fact, microalgae can remove PhCs ranging up to 100% depending on the species, growth conditions, dissolved oxygen, treatment design, etc. Table 4.2, adapted from Sharma et al., summarizes the potential of several microalgal species for the treatment of pharmaceutical wastewater.[23]

Fungi constitutes another group of microorganisms that harbor the potential of bioremediation of emerging pollutants found in pharmaceutical wastewater. Fungi and their enzymes play a pivotal role in biotransformation of PhCs via hydroxylation, dealkylation, oxidation, and sulfoxidation reactions. Indeed, fungi belonging to *Basidiomycota, Mucoromycotina*, and *Ascomycota* harbor the ability to remove and transform PhCs including antibiotics, anticonvulsants, psychiatric and anti-inflammatory drugs via the production of hydroxylated, oxidated, and conjugated metabolites.[24] White-rot fungi (WRF), mainly basidiomycetes, possess lignin modifying enzymes (LME) which can degrade contaminants such as antibiotics and PhCs. LMEs act via the oxidation process leading to the formation of radicals and ROS which leads to the oxidation of the PhCs. The hyphae of the WRF can also aid in purification of wastewater through biosorption of PhCs which adhere to the cell surface or gets internalized into the cells. Fungi secrete extracellular enzymes like laccases and peroxidases (lignin peroxidase, manganese-dependent peroxidase, versatile peroxidase) which are nonspecific glycoprotein enzymes capable of catalyzing the oxidation of aromatic compounds like phenols, a common constituent of most drugs. Furthermore, the cytochrome P450 complex of fungi consists of enzymes that

TABLE 4.2
Species of Microalgae and their Potential for the Treatment of Pharmaceutical Wastewater

Microalgae Species Used	Pharmaceutical Treated	Removal Rate (%)
Chlorella sorokiniana	Diclofenac	40–60
Chlorella sp. and *Scenedesmus* sp.	Carbamazepine	20
Nannochloris sp.	Carbamazepine, ciprofloxacin	0–100
Navicula sp.	Ibuprofen	20–60
Scenedesmus obliquus, Chlamydomonas mexicana, Chlorella vulgaris, Ourococcus multisporus, Micractinium resseri	Enrofloxacin (1 mg/L)	18–25
C. mexicana, Chlamydomonas pitschmannii, O. multisporus, C. vulgaris	Ciprofloxacin (2 mg/L)	0–13
S. obliquus, and *Chlorella pyrenoidosa*	Cefradine	30–60
Chlorella sp., *Chlamydomonas* sp., *Mychonastes* sp., *C. pyrenoidosa*	7-Aminocephalosporanic acid (40–100 mg/L)	96–100
S. obliquus	Diclofenac	99

Source: Adapted from Sharma et al. [23].

can catalyze hydroxylation, dehalogenation, deamination, and dealkalization of the PhCs, leading to compound alteration and mineralization (Figure 4.2).[25]

Trametes versicolor, *Phanerochaete chrysosporium*, and *Phlebia tremellosa* constitute the most common WRF, playing an important role in the bioremediation of contaminants. *T. versicolor*, for example, can efficiently degrade drugs like naproxen, carbamazepine, clofibric acid, and ibuprofen present in wastewater. Effluents from WWTPs containing antidepressants like citalopram and fluoxetine can be degraded by incubation with *Bjerkandera adusta* and *P. chrysosporium*.[25,26]

Of the ascomycetes, *Epicoccum nigrum*, *Aspergillus nidulans*, and *Bipolaris tetramera* can transform and degrade nonsteroidal anti-inflammatory drugs (NSAIDs) such as diclofenac. Representatives of ascomycetes have also been found to be able to degrade antibiotics like quinolones and fluoroquinolones. For example, *Penicillium notatum*, *Penicillium frequentans*, *Penicillium expansum*, *Aspergillus fumigatus*, and *Trichoderma viride* can degrade ciprofloxacin, whereas *Beauveria bassiana* can transform cinoxacin.[24]

Of the *Mucoromycotina* group of fungi, *Cunninghamella elegans* has been extensively studied for its ability to metabolize xenobiotics. *C. elegans* can transform a broad spectrum of compounds including fluoroquinolones (antibiotics), furosemide (diuretic), warfarin (anticoagulant), brompheniramine, chlorpheniramine, and pheniramine (antihistamines), chlorpromazine and methdilazine (antipsychotics), mirtazapine (antidepressant), carbamazepine (anticonvulsant), and cyclobenzaprine (muscle relaxer),

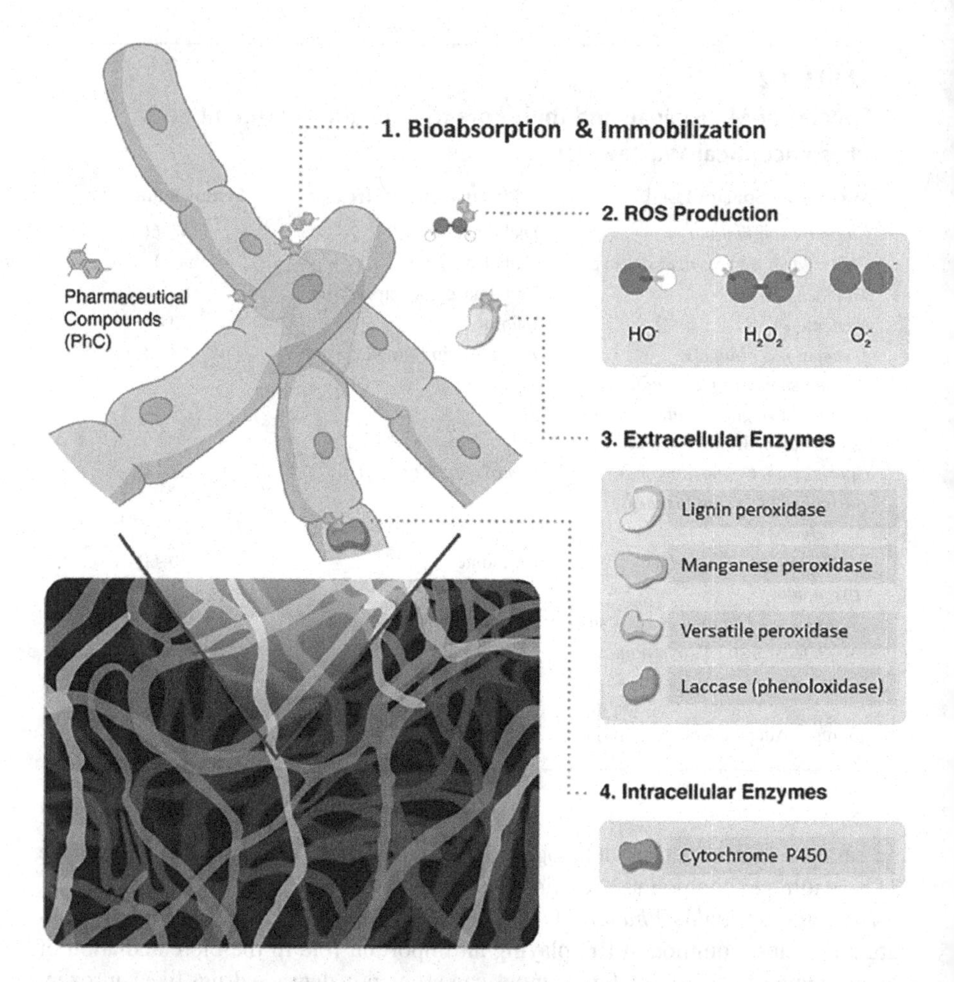

FIGURE 4.2 1. Bioabsorption and immobilization of PhCs by the fungal hyphae. 2. Production of ROS. 3. Extracellular enzymes produced by fungi (lignin peroxidase, manganese peroxidase, versatile peroxidase, laccase). 4. Intracellular enzymatic complex of fungi (cytochrome P450 complex). (Adapted from Akerman-Sanchez and Rojas-Jimenez [25].)

among others. Other representatives of the group such as *Umbelopsis ramanniana* and *Mucor rammanianus* have been investigated for their ability to transform anticonvulsant carbamazepine and fluoroquinolone antibiotics.[24]

In addition to whole cell fungal treatments, crude and purified extracts of enzymes such as laccases, versatile peroxidases, and manganese peroxidases can transform PhCs with high efficacy. Crude extract of *P. chrysosporium* containing manganese peroxidase can degrade tetracycline and oxytetracycline in 4 h. A purified enzymatic mix of laccase from *T. versicolor*, glucose oxidase from *Aspergillus niger*, and versatile peroxidase from *Bjerkandera adusta* proved to be stable and aided in the removal of PhCs like naproxen, acetaminophen, mefenamic acid, diclofenac, and indomethacin.[24]

4.5 MICROORGANISM-MEDIATED NUTRIENTS REMOVAL

The presence of nutrients in the wastewater effluents is one of the major problems because nutrients trigger eutrophication in surface water. WWTPs serve as a point source of nutrient pollution and affect the waterbodies located downstream. Nutrients play a major role in supporting the growth of microorganisms, but it plays even more favorable part providing an ideal growth platform for algae. Within days, nutrient-rich water triggers exponential proliferation of algal cells and eventually creates an anoxic environment, commonly termed as a dead zone. Microorganism-mediated removal of nutrients from wastewater is one of the best options but it often interferes with the overall retention time. Increased retention time also requires increased holding capacity of the treatment plant. The holding or flow capacity of a typical wastewater plant doesn't always depend on the average daily flow but is also affected by seasonal flow. Therefore, nutrient removal doesn't reach optimum level during high-flow season. Recent studies have indicated that a variety of fungal species are capable of performing denitrification in aqueous environment.[27,28] Typical wastewater nutrients are primarily comprised of nitrates (NO_3^-) and phosphates (PO_4^{3-}). Except the nitrates, other nitrogenous compounds in the municipal wastewater include nitrites, ammonium-nitrogen (NH_4-N), and organic nitrogen substances.[29] Techniques for capturing phosphate from wastewater are still limited, such as chemical precipitation, e.g., struvite precipitation. Some treatment plants are reliant on the biological assimilation mediated primarily by microorganisms.[30] In terms of cost and quality, biological treatments are always preferred over chemical ones. Use of filamentous fungi for the removal of phosphorus from wastewater has been found to be both financially and environmentally friendly.[31] Harvesting microorganisms for recovering resources from aqueous matrix often poses a challenge from the perspective of feasibility and cost. In this regard, filament-forming microbes provide a better rate of harvesting because of their structural integrity. Some filamentous fungi exhibit higher tolerance to toxic substances present in wastewater than bacteria.[32,33]

As a eukaryotic organism, fungal metabolism is more complex and diverse than prokaryotes. For decades, scientific evidence backed the fact that fungal organisms are one of the ideal candidates to remove and detoxify contaminants of emerging concern (CECs) from various environmental matrices. Fungal enzymes are capable of catalyzing diverse biochemical reactions to break intermolecular bonds, e.g., demethylation, hydroxylation, opening of cyclic structures. Majority of the scientific studies with a focus on removing micropollutants from wastewater used pure culture of laboratory-grown microorganisms. In spite of showing promising outcome in a controlled research environment, majority of these microorganisms fail to prove their worth at pilot stage with real wastewater. Typical wastewater influent stream carries a plethora of microorganisms. In case of specific microorganism-mediated remediation treatment, achieving dominance over the rest of the microbial population is a critical factor. Therefore, environmental and biological scientists have expressed their interest toward utilizing natural microbial isolates for wastewater remediation. Generally, microorganisms isolated from natural habitat exhibit higher survival potential due to their adaptation toward the physical and chemical variables involved in the treatment operation.[27,32] Fungal organisms such as *Trametes versicolor* and *Aspergillus luchuensis* have been studied extensively to investigate their potential to degrade diverse organic material in wastewater including bulky, heterogeneous, and recalcitrant polymers. These organisms also release nonspecific enzymes in the aqueous media and trigger extracellular degradation of contaminants. As most of the fungal organisms prefer acidic environment for their growth and proliferation, adjusting pH of the medium helps them to achieve dominance. Manipulation of pH also results in the growth inhibition of other microorganisms, specifically bacterial population. Microorganisms play a major role in removing nutrients in a symbiotic setting with higher organisms. The ideal example of such kind of treatment is vegetated wetland treatment where microorganisms in association with plants initiate expedited removal of nutrients and organic materials. Microbes colonized in root zone create a microenvironment to convert the complex species of nitrate and phosphate into readily consumable compounds and, in return, receive nutrition from plants. These natural treatment technologies provide a green, sustainable alternative to the conventional WWTPs.

4.6 CHALLENGES TO THE PROCESSES

Almost all the novel processes developed for the sustainable treatment of wastewater from various sectors come with their own challenges. One of the major challenges includes the necessity of infrastructure development that often does not fulfill the criteria of optimum cost-benefit ratio. Even after demonstrating promising outcome in microcosm scale, many sophisticated wastewater remediation techniques fail to provide expected quality due to diverse environmental conditions. In terms of overcoming the obstacle posed by the infrastructure development, many environmental scientists and engineers are focusing on the possibility of retrofitting new technologies in the current and conventional infrastructures. The concept of retrofitting is specifically useful for urban wastewater treatment facilities where availability of space itself is a challenge. Not only the infrastructure development raises a financial concern when it comes to implementing novel techniques in wastewater treatment, but also the operation and management cost plays a major role. In case of developed nations, the personnel cost covers a significant portion of the financial allocation for any operation. Therefore, there is a growing interest for developing techniques that require minimal manual effort and meet the criteria for self-sustenance. Alongside developing physical environment for sustainable and novel wastewater remediation processes, scientific investigators are emphasizing toward automation of these processes and even they are inclining toward the implementation of artificial intelligence which is one of the revolutionary technologies in the field of next-generation computational advances.

As the global population index is on a steady rise over the past few decades, the wastewater generation especially in the urban areas has significantly increased. For decades, the field of wastewater treatment has not observed any significant improvement and therefore, along with the financial inflation the treatment cost of wastewater management has reached a tipping point. One of the major costs of conventional wastewater treatment is associated with the energy consumption. A rough investigation estimates the energy consumption comprises at least 3% of global electricity consumption.[34] Recent scientific investigations are trying to overcome these major challenges by incorporating interdisciplinary technologies, e.g., microbial fuel cells. Depending on the serving sectors, wastewater may contain a number of valuable elements such as nutrients and heavy metals. Majority of these elements get compartmentalized in the sludge zone after secondary treatment. It has been found the disposal cost of sludge generated from wastewater treatment comprises 50% of the total cost of treatment. Appropriate treatment techniques focusing on maximizing the resource recovery may reduce the cost of wastewater treatment operations and management. Recent studies indicate that microbe-mediated electrochemical operation can help recover valuable products like silver (Ag), chromium (Cr), ammonia (NH_3), and nutrients like nitrate and phosphate.[35,36] The application of biosolids generated from wastewater sludge in the agricultural fields also supports the circular economy as well as sustainability of the whole system. Microbial bioaugmentation to remove wastewater pollutants has shown promising outcomes in many studies. However, one of the studies investigated to project the cost of fungal treatment to remove conventional contaminants from wastewater such as phosphorus, ammonium-nitrogen (NH_4-N), and total organic carbon (TOC). According to the study, the unit cost of fungal-mediated nutrient removal was determined to be $0.86–8.6/gal which is significantly (almost 200 times) higher than conventional treatment methods (Figure 4.3).[37]

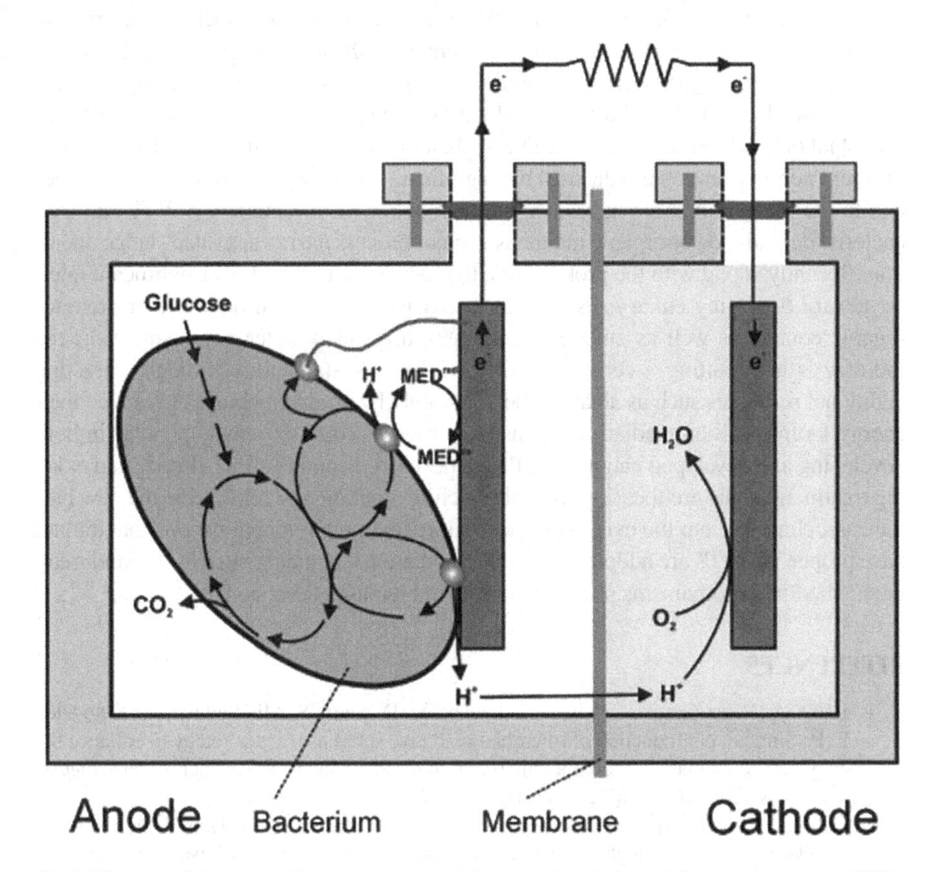

FIGURE 4.3 Schematic diagram of microbial fuel cells. (Adapted from Logan et al. [38].)

4.7 CONCLUSION

For over a century, we have heavily relied upon fossil fuel-based energy production. Most wastewater treatment industries have been a massive consumer of energy and therefore, significantly contributed toward both onsite and offsite greenhouse gas emissions. Environmental issues like global warming and climate change as well as geopolitical conflict due to depleting fossil fuel reserves have gradually created a pressure to shift our reliance from non-renewable resources. The modern treatment industries are eventually changing their perspectives and instead of considering the wastewater treatment processes as an energy burden, there is a constant effort to make the whole process energy neutral. It is high time to consider the influent sewage load not just a waste but as a resource. As we are looking for alternate fuel to meet the global energy demand, the perspective of solid and liquid waste needs a thorough reconsideration. Urban sewer system carries a huge load of organically rich biomass every day. This biomass has a significant potential to be converted into several products such as fuel resource for energy. The biomass also carries a plethora of emerging contaminants which upon releasing into the environment may trigger bigger problems such as antibiotics-induced antimicrobial resistance. We are in a situation where we have no other choices but to implement wastewater recycling and resource recovery because we need to meet the energy demand as well as prevent environmental degradation caused by wastewater pollutants such as nutrients and CECs. With the application of appropriate technique, the biomass present in the wastewater can be either recovered as a resource or be utilized as an alternative energy source. There is no doubt that the best way of biomass conversion is achieved either through microbial or biochemical processes. The application of microorganisms and enzymes in different non-conventional industries has significantly increased in the past few decades due to the recent advances in science and technology. Conventional WWTPs utilized bacterial degradation of organic matter as a major process during activated sludge operation. Recently, along with the prokaryotes, the industry has also shown significant interest toward deploying eukaryotes such as microalgae and fungal organisms to remove organic content as well as emerging pollutants from wastewater. Simultaneously, the industry is also putting a constant effort to achieve sustainability through extracting additional resources such as algal biofuel, or simply by producing biogas to reduce their energy footprint. Bioremediation of wastewater has become extremely popular in both developing and developed nations. While the developed countries that already had existing treatment plants are focusing toward switching operation or retrofitting the new biological techniques into the existing infrastructure, many developing nations that did not have proper WWTPs are adopting natural remediation techniques such as wetland treatment where microorganisms play a major role in biomass conversion.

REFERENCES

1. Bains, A.; Perez-Garcia, O.; Lear, G.; Greenwood, D.; Swift, S.; Middleditch, M.; Kolodziej, E. P.; Singhal, N., Induction of microbial oxidative stress as a new strategy to enhance the enzymatic degradation of organic micropollutants in synthetic wastewater. *Environmental Science & Technology* **2019,** *53*, (16), 9553–9563.
2. Galus, M.; Jeyaranjaan, J.; Smith, E.; Li, H.; Metcalfe, C.; Wilson, J. Y., Chronic effects of exposure to a pharmaceutical mixture and municipal wastewater in zebrafish. *Aquatic Toxicology* **2013,** *132*, 212–222.

3. Oh, S.; Logan, B. E., Hydrogen and electricity production from a food processing wastewater using fermentation and microbial fuel cell technologies. *Water Research* **2005,** *39*, (19), 4673–4682.
4. Munoz-Cupa, C.; Hu, Y.; Xu, C.; Bassi, A., An overview of microbial fuel cell usage in wastewater treatment, resource recovery and energy production. *Science of the Total Environment* **2021,** *754*, 142429.
5. Feng, Y.; Wang, X.; Logan, B. E.; Lee, H., Brewery wastewater treatment using air-cathode microbial fuel cells. *Applied Microbiology and Biotechnology* **2008,** *78*, (5), 873–880.
6. Wen, Q.; Wu, Y.; Zhao, L.; Sun, Q., Production of electricity from the treatment of continuous brewery wastewater using a microbial fuel cell. *Fuel* **2010,** *89*, (7), 1381–1385.
7. Gude, V. G., Wastewater treatment in microbial fuel cells-an overview. *Journal of Cleaner Production* **2016,** *122*, 287–307.
8. Marassi, R. J.; Queiroz, L. G.; Silva, D. C. V.; da Silva, F. T.; Silva, G. C.; de Paiva, T. C. B., Performance and toxicity assessment of an up-flow tubular microbial fuel cell during long-term operation with high-strength dairy wastewater. *Journal of Cleaner Production* **2020,** *259*, 120882.
9. Torres, E.; Bustos-Jaimes, I.; Le Borgne, S., Potential use of oxidative enzymes for the detoxification of organic pollutants. *Applied Catalysis B: Environmental* **2003,** *46*, (1), 1–15.
10. Falås, P.; Wick, A.; Castronovo, S.; Habermacher, J.; Ternes, T. A.; Joss, A., Tracing the limits of organic micropollutant removal in biological wastewater treatment. *Water Research* **2016,** *95*, 240–249.
11. Imlay, J. A., The molecular mechanisms and physiological consequences of oxidative stress: lessons from a model bacterium. *Nature Reviews Microbiology* **2013,** *11*, (7), 443–454.
12. Rao, M.; Scelza, R.; Acevedo, F.; Diez, M.; Gianfreda, L., Enzymes as useful tools for environmental purposes. *Chemosphere* **2014,** *107*, 145–162.
13. Gianfreda, L.; Xu, F.; Bollag, J.-M., Laccases: a useful group of oxidoreductive enzymes. *Bioremediation Journal* **1999,** *3*, (1), 1–26.
14. Li, K.; Liu, Q.; Fang, F.; Luo, R.; Lu, Q.; Zhou, W.; Huo, S.; Cheng, P.; Liu, J.; Addy, M., Microalgae-based wastewater treatment for nutrients recovery: a review. *Bioresource Technology* **2019,** *291*, 121934.
15. Kumar, R.; Pal, P., Assessing the feasibility of N and P recovery by struvite precipitation from nutrient-rich wastewater: a review. *Environmental Science and Pollution Research* **2015,** *22*, (22), 17453–17464.
16. Jin, L.; Zhang, G.; Tian, H., Current state of sewage treatment in China. *Water Research* **2014,** *66*, 85–98.
17. Chorus, I.; Welker, M., *Toxic cyanobacteria in water: a guide to their public health consequences, monitoring and management*. Taylor & Francis: 2021.
18. Abraham, J.; Lin, Y.; RoyChowdhury, A.; Christodoulatos, C.; Conway, M.; Smolinski, B.; Braida, W., Algae toxicological assessment and valorization of energetic-laden wastewater streams using Scenedesmus obliquus. *Journal of Cleaner Production* **2018,** *202*, 838–845.
19. Rawat, I.; Kumar, R. R.; Mutanda, T.; Bux, F., Dual role of microalgae: phycoremediation of domestic wastewater and biomass production for sustainable biofuels production. *Applied Energy* **2011,** *88*, (10), 3411–3424.
20. Javadian, H.; Ahmadi, M.; Ghiasvand, M.; Kahrizi, S.; Katal, R., Removal of Cr (VI) by modified brown algae Sargassum bevanom from aqueous solution and industrial wastewater. *Journal of the Taiwan Institute of Chemical Engineers* **2013,** *44*, (6), 977–989.
21. Leng, L.; Wei, L.; Xiong, Q.; Xu, S.; Li, W.; Lv, S.; Lu, Q.; Wan, L.; Wen, Z.; Zhou, W., Use of microalgae based technology for the removal of antibiotics from wastewater: a review. *Chemosphere* **2020,** *238*, 124680.
22. Garcia-Galan, M. J.; Arashiro, L.; Santos, L.; Insa, S.; Rodriguez-Mozaz, S.; Barcelo, D.; Ferrer, I.; Garfi, M., Fate of priority pharmaceuticals and their main metabolites and transformation products in microalgae-based wastewater treatment systems. *Journal of Hazardous Materials* **2020,** *390*, 121771.

23. Sharma, R.; Mishra, A.; Pant, D.; Malaviya, P., Recent advances in microalgae-based remediation of industrial and non-industrial wastewaters with simultaneous recovery of value-added products. *Bioresource Technology* **2022,** *344,* (Pt B), 126129.
24. Olicon-Hernandez, D. R.; Gonzalez-Lopez, J.; Aranda, E., Overview on the biochemical potential of filamentous fungi to degrade pharmaceutical compounds. *Frontiers in Microbiology* **2017,** *8,* 1792.
25. Akerman-Sanchez, G.; Rojas-Jimenez, K., Fungi for the bioremediation of pharmaceutical-derived pollutants: a bioengineering approach to water treatment. *Environmental Advances* **2021,** 100071.
26. Marco-Urrea, E.; Perez-Trujillo, M.; Vicent, T.; Caminal, G., Ability of white-rot fungi to remove selected pharmaceuticals and identification of degradation products of ibuprofen by Trametes versicolor. *Chemosphere* **2009,** *74,* (6), 765–772.
27. Dalecka, B.; Strods, M.; Juhna, T.; Rajarao, G. K., Removal of total phosphorus, ammonia nitrogen and organic carbon from non-sterile municipal wastewater with Trametes versicolor and Aspergillus luchuensis. *Microbiological Research* **2020,** *241,* 126586.
28. Shoun, H.; Kim, D.-H.; Uchiyama, H.; Sugiyama, J., Denitrification by fungi. *FEMS Microbiology Letters* **1992,** *94,* (3), 277–281.
29. Sankaran, S.; Khanal, S. K.; Jasti, N.; Jin, B.; Pometto III, A. L.; Van Leeuwen, J. H., Use of filamentous fungi for wastewater treatment and production of high value fungal byproducts: a review. *Critical Reviews in Environmental Science and Technology* **2010,** *40,* (5), 400–449.
30. Ye, Y.; Gan, J.; Hu, B., Screening of phosphorus-accumulating fungi and their potential for phosphorus removal from waste streams. *Applied Biochemistry and Biotechnology* **2015,** *177,* (5), 1127–1136.
31. He, Q.; Rajendran, A.; Gan, J.; Lin, H.; Felt, C. A.; Hu, B., Phosphorus recovery from dairy manure wastewater by fungal biomass treatment. *Water and Environment Journal* **2019,** *33,* (4), 508–517.
32. Guest, R.; Smith, D., Isolation and screening of fungi to determine potential for ammonia nitrogen treatment in wastewater. *Journal of Environmental Engineering and Science* **2007,** *6,* (2), 209–217.
33. Millán, B.; Lucas, R.; Robles, A.; García, T.; de Cienfuegos, G. A.; Gálvez, A., A study on the microbiota from olive-mill wastewater (OMW) disposal lagoons, with emphasis on filamentous fungi and their biodegradative potential. *Microbiological Research* **2000,** *155,* (3), 143–147.
34. Saba, B.; Christy, A. D.; Yu, Z.; Co, A. C., Sustainable power generation from bacterio-algal microbial fuel cells (MFCs): an overview. *Renewable and Sustainable Energy Reviews* **2017,** *73,* 75–84.
35. Ali, J.; Wang, L.; Waseem, H.; Sharif, H. M. A.; Djellabi, R.; Zhang, C.; Pan, G., Bioelectrochemical recovery of silver from wastewater with sustainable power generation and its reuse for biofouling mitigation. *Journal of Cleaner Production* **2019,** *235,* 1425–1437.
36. Jadhav, D. A.; Ray, S. G.; Ghangrekar, M. M., Third generation in bio-electrochemical system research-A systematic review on mechanisms for recovery of valuable by-products from wastewater. *Renewable and Sustainable Energy Reviews* **2017,** *76,* 1022–1031.
37. Hansen, R.; Thøgersen, T.; Rogalla, F., Comparing cost and process performance of activated sludge (AS) and biological aerated filters (BAF) over ten years of full sale operation. *Water Science and Technology* **2007,** *55,* (8–9), 99–106.
38. Logan, B. E.; Hamelers, B.; Rozendal, R.; Schröder, U.; Keller, J.; Freguia, S.; Aelterman, P.; Verstraete, W.; Rabaey, K., Microbial fuel cells: methodology and technology. *Environmental Science & Technology* **2006,** *40,* (17), 5181–5192.

5 Sustainable Wastewater Treatment Strategies

Toward a Circular Economy

Arindam Mitra and Nikhi Verma

5.1 INTRODUCTION

Sustainable Development Goals (SDGs) are seventeen goals that members of the United Nations, comprising 193 nations, adopted in the United Nations General Assembly in 2015. These goals aim to reduce poverty and improve health, education, and worldwide prosperity by 2030 (https://sdgs.un.org/goals). These goals are blueprints to strengthen equity, reduce disparities worldwide, and require global collaborations. Water is of essence to life, health, sustainable environment, and essential to the economy. However, the access or distribution to water in all regions of the earth is not uniform due to anthropogenic activities, urbanization, climate change, and increasing population. The consumption of water due to agriculture and industrial use is one of the highest. A severe water shortfall is expected by 2030 if water is not correctly used, managed, recycled, or reused. Scarcity of water is a global challenge, more so in arid regions, and the availability of water is also dependent on many factors, including pollution of water from agriculture and industry, rendering the water unusable. More than half of the world's population do not have safe sanitation services or access to basic sanitation infrastructure. Globally, approximately one-third of the world's population does not have toilets.

Several factors require wastewater treatment (WWT) to be sustainable, including water shortage, environmental and aquatic pollution, depletion of fossil fuels, and reuse of water. In many parts of the world, wastewater is not treated or adequately treated before being released into the river, which burdens health, the environment, and the economy. As of 2018, about 80% of the world's wastewater from human activities is not treated to remove pollutants before being released into water bodies such as groundwater, lakes, rivers, and oceans. Even where WWT is in place, most WWT is neither energy nor cost-efficient or environment friendly or creates a lot of residuals. WWT should ideally be cost-effective, environment friendly, ethical, and require reduced resource consumption. One SDG aims to treat wastewater of 2.3 billion people living in urban environments by 2030. When performed sustainably, WWT can accomplish as many as one-half of the SDGs directly or indirectly to bring sustainable solutions (Figure 5.1).

Several SDGs are in alignment with sustainable WWT. WWT is a sustainable solution for water reuse and can address water scarcity in many parts of the world

DOI: 10.1201/9781003517238-5

FIGURE 5.1 Sustainable WWT in alignment with SDGs.

(Tortajada 2020). According to United Nations Environment Programme, it is predicted that by 2025, two-thirds of the world population will face water scarcity. Recycling and reusing water is a sustainable solution for this planet, which can be achieved by proper treatment of water disposed of. SDG6 calls for accessibility and sustainable management of water and sanitation throughout the world (Guppy, Mehta, and Qadir 2019). Goal 6 aims for water access and sanitation for all in alignment with sustainable WWT. Sanitation includes drinking water services related to sewage and waste disposal, and it impacts public health, biodiversity, environment, education, and daily life of citizens. By ensuring proper treatment of wastewater, many diseases related to public health may be reduced, which aligns with Goal 3, which aims for good health and well-being. The health of aquatic animals is improved due to WWT, and hence it is also aligned with Goal 14, life below water. Similarly, WWT, when sustainable and does not require the use of fossil fuels, can offer affordable and clean energy, which is Goal 7. For this reason, it is also in alignment with Goal 11 for sustainable cities and communities. By offering employment to individuals in WWT processes, Goal 1, which aims to reduce poverty, can be covered by WWT, and Goal 8 aims to achieve decent work and economic growth.

On the other hand, untreated or improperly treated wastewater can burden public health and the economy when released in the aquatic environment. These costs include hospitalization expenses, and physicians' visits, loss of productive hours, and loss of human lives. The release of nitrogen and phosphorus and organic compounds into the river increases growth of cyanobacteria and results in algal blooms leading to eutrophication, affecting the aquatic environment. Waterborne diseases such as cholera, diarrhea, amebiasis, and various viral diseases are a significant burden to a community, mainly where adequate healthcare, including appropriate diagnostics, are not available. Children below five are most vulnerable to waterborne diseases. A study about drinking water quality indicated that natural water sources had fecal contamination, whereas treated municipal water did not detect any fecal contaminants (Edokpayi et al. 2018). Similarly, a lack of awareness in specific communities

TABLE 5.1
Strengths and Limitations of Some Sustainable WWT

Approaches	Strengths	Limitations
Microbial fuel cells	Clean energy generation of electricity	Can be expensive
Anaerobic remediation of wastewater	Biogas generation, environment friendly	Incomplete degradation of organic compounds
Photocatalysis/ozonization	Advanced oxidation process	Expensive process
Constructed wetlands	Cost-efficient, green technology	Slow process
Nanomaterials	Improved catalysis, strong adsorption property for pollutants	Toxicity, difficulty in disposing
Biochar/layered double hydroxide (LDH) composites	Clean process, low cost	May not be efficient in the removal of critical pollutants
Phytoremediation	Environment friendly and cost-efficient	Does not remove heavy toxic metals
Cold plasma technology	Energy efficient	Lack of awareness

regarding water quality for bathing and drinking can be challenging. In many countries worldwide, sewage and other industrial wastes are often discharged into river or surface water without proper or adequate treatment.

The challenges of WWT in resource-constrained regions are many, including high energy and time consumption, disposal of sludge, and emission of greenhouse gases such as carbon dioxide and nitrous oxide. Other challenges include weak governance at regional or county levels to implement or enforce adequate or sustainable WWT, lack of regulations or monitoring of wastewater, low priority to public health, lack of funding or workforce, or lack of other resources. However, despite these challenges, there are many approaches by which WWT can be made sustainable. These strategies are listed in Table 5.1.

5.2 APPROACHES OF SUSTAINABLE WWT

Several conventional technologies are available for WWT and this includes filtration, sedimentation, degasification, reverse osmosis (RO), flocculation, adsorption, and precipitation (Saravanan et al. 2021). Some of the significant sustainable approaches for WWT are summarized next (Bousquet et al. 2017, Do et al. 2018, Gherghel, Teodosiu, and De Gisi 2019, He et al. 2017, Lv et al. 2007, Kwon et al. 2012, Sarpong and Gude 2020, Shen et al. 2018, Tyagi and Lo 2013).

5.2.1 Microbial Fuel Cells

Microbial fuel cells (MFCs) are a cost-efficient, rapid approach that can be used to remove pollutants such as organic carbon from wastewater with electricity generation, making them an excellent alternative, sustainable renewable energy generation technology. They can be used as a biosensor for monitoring essential parameters. This technology uses a bioelectrochemical system (BES) that converts chemical energy in organic compounds such as glucose or similar compounds present in wastewater into electrical power (Santoro et al. 2017). Heavy metals such as mercury, cadmium, chromium, silver, thallium, and copper from wastewater are efficiently removed by MFC. This technology does not require fossil fuels and is considered cleaner. However, the efficiency and operation costs are considered higher in MFC. An iron-air fuel cell was tested to remove and reuse phosphorous and energy generation from wastewater (Wang et al. 2021). However, there are challenges such as relatively low electricity generation levels, instability of currents, cost of materials, and internal resistance. Investigations improve performance while reducing operating and construction costs (Do et al. 2018).

5.2.2 Microbial Remediation of Wastewater

Microbes in biofilms are often used for the remediation of wastewater. The biofilm reactors provide a stable matrix on which microbes are immobilized to degrade pollutants from water. A wide variety of reactors are used to remove carbon, phosphorus, bromates, perchlorates, and other organic matter. Bacteria isolated from untreated wastewater capable of degrading various pollutants can be immobilized in bioreactors to remove harsh chemicals such as 4-chlorophenol in wastewater (Patel et al. 2022). A green alga, *Klebsormidium nitens*-based bioreactor, has been shown to remove phosphorous from wastewater while being less energy intensive, having less land area requirement, and resistant to culture contamination (Valchev et al. 2021). Another study demonstrated that a low-cost reusable electrode based on TiO_2/Co-WO_3/SiC could be used to treat saline wastewater.

5.2.3 Nanomaterials

Nanomaterials are another option for sustainable WWT. Nanomaterials are smaller than 100 μm and possess excellent adsorption capacity for various pollutants due to the high surface area to volume ratio. They offer chemical stability, efficient catalysis, redox-active media, and ease of functionality. Nanomaterials used in WWT include carbon, nanotubes of carbon, nanoparticles of metal and metal oxides, and covalent organic frameworks. Many pollutants, including dyes, heavy metals, solvents, and various pathogens and toxins, have been successfully removed by nanomaterials (Bhat et al. 2022). The presence of antibiotics in drinking water can have health consequences, including hormonal disorders and toxicity, besides ecological disturbances. Nanomaterials are an excellent choice for removing antibiotics discharged from hospital wastes, farms, or aquaculture settings (Singh et al. 2021).

5.2.4 Biochar/Layered Double Hydroxide Composites

Biochar is a carbon-based material generated by pyrolysis of various industrial and solid wastes and others in limited oxygen supply. Biochar has a high surface area with reliable adsorbent capacity and is helpful in water purification. Besides, biochar reduces greenhouse gases and is not cost-intensive, unlike conventional technologies. To improve biochar's surface area and adsorption characteristics, layered double hydroxides (LDHs) in combination with biochar are also used. Such composites are sustainable, cost-effective alternatives to remove harmful pollutants compared to conventional technologies (Zubair et al. 2021).

5.2.5 Constructed Wetland

Wastewater from tanneries carries inorganic and organic pollutants, including chromium, with high biochemical oxygen demand (BOD) and COD with strong color and pH. When discharged in an aquatic environment without proper treatment, chromium-containing wastewater can significantly affect public and animal health as chromium VI is carcinogenic and teratogenic. This is also an issue when Cr-containing wastewater is used in irrigation and human consumption. Constructed wetland (CW) is a sustainable WWT strategy as costs involved in construction, operation, and maintenance are lower than other technologies (Masi, Rizzo, and Regelsberger 2018). CW is an environment-friendly technology applicable in resource-constrained regions (Younas et al. 2022).

5.2.6 Photocatalysis

Photocatalysts play an essential role in WWT. A recent study evaluated the role of sandwich-like composite catalysts that exhibit excellent degradation to 2-nitroaniline and 4-nitrophenol and a high utilization rate (Yin et al. 2021). Photocatalysts are typically immobilized metal oxides or semiconductors on fixed solid support to degrade pollutants. Photocatalysts are sustainable as they use naturally available resources such as solar energy to degrade pharmaceuticals and personal care products (Oluwole, Omotola, and Olatunji 2020). Various nanocomposites have been used to synthesize photocatalysts to mineralize organic pollutants via the advanced oxidation process. Advanced oxidation method using solar photocatalysis is an efficient and cost-effective process for sustainable WWT and use (Pandey et al. 2021). Combining photocatalysis with ozonization enhances the advanced oxidation process by generating higher free oxygen radicals and oxidizes the volatile organic compounds, industrial surfactants, present in the aqueous system (da Costa Filho et al. 2019, Zsilak et al. 2014, Parrino et al. 2014).

5.2.7 Anaerobic Treatment of Sludge

Activated sludge, treatment of municipal water in the developed world, requires aeration to remove organic carbon and nitrogen, which consumes energy. Consumption of energy depends on various factors such as the treatment process, the capacity of wastewater plants, geographical location, and others. Anaerobic treatment, in contrast, can bypass the requirement of oxygen and reduce energy consumption. Anammox bacteria convert nitrate and ammonium to nitrogen in anaerobic conditions and facilitate the removal of nitrogen with reduced energy consumption and greenhouse gas emissions compared to other biological processes. Anammox oxidizing bacteria with anammox bacteria convert ammonium to nitrogen via nitrite formation in a limited oxygen environment. This combination of nitritation-anammox systems is used to remove nitrogen in saline wastewater (Guo et al. 2020, Ge et al. 2019). A sustainable and optimized A–B process of nitritation and denitration shows nitrogen removal by 80%, besides sludge reduction and potential energy recovery from municipal wastewater (Gu, Yang, and Liu 2018).

Sludge is a pollutant and needs to be appropriately disposed of. Sludge management and its disposal are sustainable strategies in alignment with SDGs (Spinosa and Doshi 2021, Kacprzak et al. 2017). Sludges can also produce biogas, directly used as sustainable clean energy for electricity, heat, and transportation fuel. The anaerobic digestion of sludge produces biogas in the presence of microorganisms. Sludges can also be used to extract and recover nitrogen and phosphorous (Spinosa and Doshi 2021). The water treatment residual consisting of nontoxic chemical mixed with trace contaminants can be used for landfill in alignment with SDG12 (Consumption and production) and 15 (Terrestrial ecosystems).

5.2.8 Phytoremediation

Phytoremediation uses green plants to degrade pollutants from the environment, including wastewater. The emerging phytoremediation technology is eco-friendly, cost-efficient, and an alternative to conventional remediation strategies. Another advantage is that it can be applied in large areas and removed. Plants chosen for phytoremediation should absorb organic and inorganic pollutants, grow quickly in wastewater, and be controllable. Water hyacinth (*Eichhornia crassipes*), an aquatic plant, was used worldwide to reduce BOD, chemical oxygen demand (COD), suspended solids, nitrogen and phosphorus, and other nutrients. This plant also removes herbicides, is capable of biological control, and grows in nutrient-rich water. Heavy metals are often released in the wastewater due to urbanization and industrialization which can be toxic to aquatic animals and human health. Many plants such as hyacinth, sunflower, and duckweed can absorb heavy metals from wastewater in their roots. Approaches such as phytostabilization, phytoextraction via the application of recombinant DNA technology, and microbe-assisted remediation are effective (Wei et al. 2021, Yan et al. 2020). *Salvinia molesta* has been used to reduce the color, pH, BOD, COD of domestic wastewater (Mustafa and Hayder 2021).

5.3 CIRCULAR ECONOMY

A significant sustainability issue is the continuous production and consumption of natural resources, leading to depletion of natural resources, increased water stress, pollution, loss of biodiversity, and climate changes. The traditional take-make-waste model of natural resources is in sharp contrast to circular economy (CE), a process or framework that aims to reduce or minimize or even attempt to eliminate waste of natural resources. The concept is to keep raw materials and products to remain in the economy as long as possible without the need to waste, which can be used as secondary raw materials for recycling and reusing purposes. CE attempts to produce or consume resources sustainably and designs products that can be recycled or reused. The maintenance, recovery, and reuse of waste materials improve the life cycle of products and reduce waste generation. It attempts to design products that can

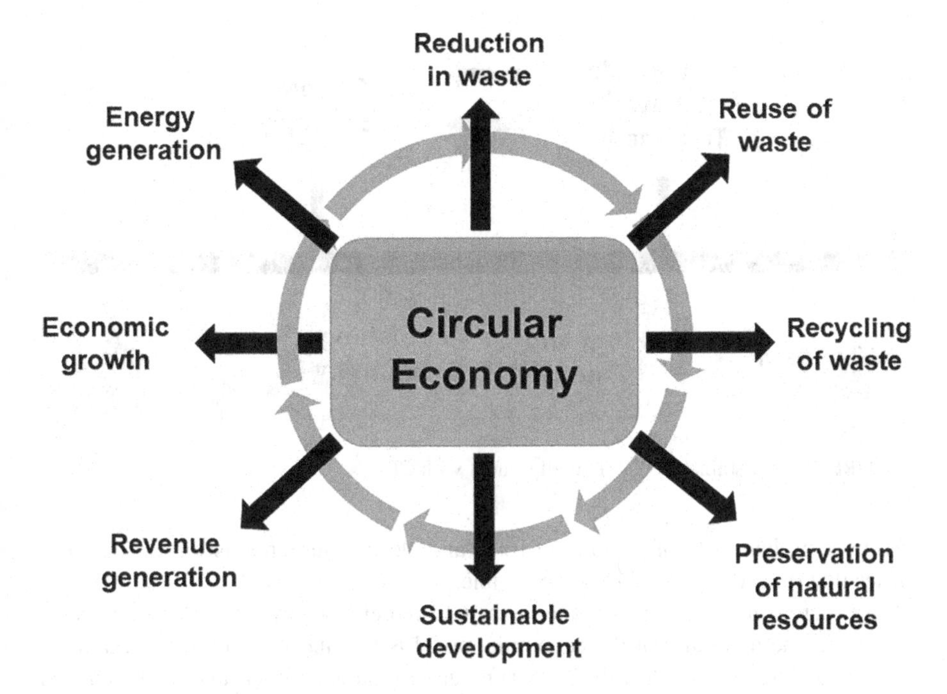

FIGURE 5.2 Diverse advantages of a CE.

be recovered and reused so that the need for new raw material is minimized, if not eliminated. The benefits of using a CE are illustrated in Figure 5.2.

The CE concept starts from the hypothetical step of an organization, a product, or amenities to prolong a product's life cycle. The CE is a concept where the raw materials and products shall persist within the economy for the longest possible duration. After that, waste generated should be considered the raw materials (secondary) for recycling leading to the production of other products that can be reused (Ghisellini, Cialani, and Ulgiati 2016). The CE is different from the linear economy, which is based on the 'take-make-waste' system wherein the waste is usually the last by-product and disposed of (Bocken et al. 2016). The CE advances sustainable energy and materials management by reducing waste production and reuse (Neczaj and Grosser 2018). Other reasons for implementing the CE approach include limited availability of starting materials, valuable or expensive products, barrier to import of raw materials, and competitiveness.

5.4 A MODEL OF CE IN THE CONTEXT OF WWT

WWT also brings circularity in the economy as the process is less dependent on external resources and relies on reusing existing resources or wastes or by-products. This strategic use of resources improves the economic impact. It could boost tourism and improve the health of aquatic animals and humans, water sports, and the fishing industry, among others. The primary target of SDG is to reduce the quantity of wastewater by untreated half and thus accelerate recycling. Secure reuse entrusts

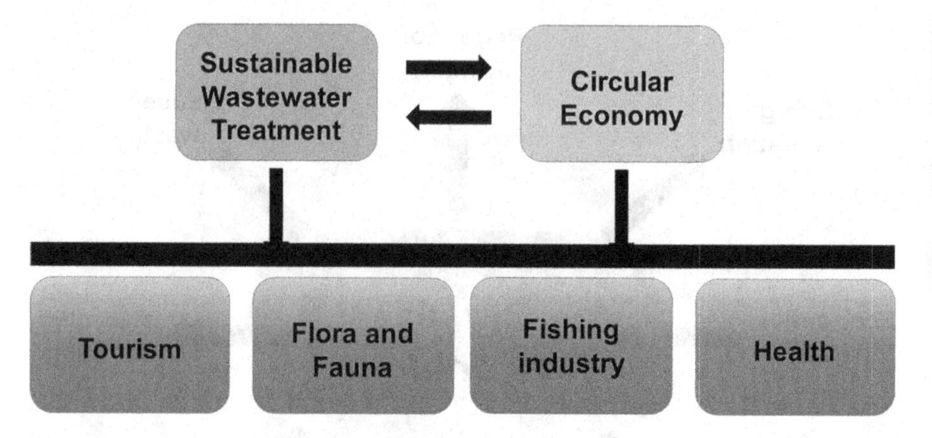

FIGURE 5.3 Sustainable WWT in alignment with CE.

different probabilities of generating fresh and clear water streams for various purposes that otherwise wouldn't be accessible.

An increasing curiosity is found in the CE model as many industries show commitment. The main aim of the economic model is to remove waste in an organized manner during the complete life cycles and utilization of products and their elements (Ghimire, Sarpong, and Gude 2021). According to Ellen MacArthur Foundation, it is recognized as a model that aids in producing long-lasting products, accelerates remodeling, and sees shifts in products or services wherever possible. Furthermore, the CE creates a value for municipalities and local bodies (communities) to arrange material, energy recovery, and local recirculation for water.

The CE principle is based on zero waste and grounded on three rules:

i. All durables that are products with prolonged or boundless life span should keep their value and must be reused, however not discarded or sloping downward. In other words, it means to break down into fragments and remodel into products that are new but with lesser value.
ii. All consumables that are products having a small life span must be utilized as frequently as possible before they return harmlessly into the biosphere.
iii. Further in 2015, it was added that natural resources might be utilized to a range only if they could be revived.

Figure 5.3 represents the benefits of having CE in alignment with SDG.

5.5 CE FOR RECOVERY OF VALUABLE PRODUCTS AND ENERGY FROM WASTEWATER

As of now, most of the thermal energy, chemical energy, nutrients, and water are not recovered from wastewater. However, if the wastewater is treated and recovered, the waste could be converted into resources and used for heating, cooling, biogas, power, and recovery of valuable chemicals. Circular approaches should be considered to recover and reuse valuable products and energy from wastewater. The goal is to keep these products in the economy as long as possible. Simultaneously it reduces the adverse effects of industrialization on the environment as far as possible. Further, energy efficiency would be promising if there is the recovery of waste materials and the policy of its reuse. Recovery of valuable resources from wastewater also generates revenues for the WWT plants and carbon credits. These revenues could then provide financial support for the WWT itself, making the process circular and sustainable, particularly in the context of the increasing global population stronger regulations for discharge in the future. Production of energy could surpass the energy required in WWT, leading to a positive energy balance that is sustainable.

The reuse of water as a resource can solve the issue of water scarcity; however, the contaminant level in recycled water should be regulated and monitored while maintaining a CE approach (Voulvoulis 2018). Recycling nutrients reduces the demand for fossil-based fertilizer, thus reducing water and energy consumption. The nitrogen and phosphorus can be recovered from the wastewater and reused as fertilizers. The reuse of struvite-based fertilizer and other nutrients from wastewater reduces the import from different regions and emissions while facilitating a self-sustainable CE (Gowd, Ramakrishna, and Rajendran 2021). Similarly, municipal sewage sludge can be a source of feedstock for biodiesel production in a cost-effective manner (Srivastava et al. 2018). It is believed that a linear economy doesn't have high compatibility with SDG. There is a need to manage or retrieve valuable raw materials or energy in this particular situation. For example, most ammonium is converted to nitrogen gas by the present biological nitrogen removal processes in municipal wastewater. Another study suggested using integrated anaerobic bioreactor and RO processes for ammonium recovery from municipal wastewater leading to a CE (Zhang and Liu 2021). Some wastewater plants utilize another anaerobic digester to improve the efficiency of wastewater materials by co-design. These plants recover fats, oils, grease, food waste, and primary and secondary sludge, and other technologies such as turbines and flow cells may be employed to recover energy efficiently. Rare earth elements such as lanthanum and cerium can be recovered and reused from industrial wastewater using zeolite and bacterial biofilms (Barros et al. 2019).

The CE model can be depicted as a structured framework for analyzing regional wastewater management in the environment. It minimizes waste and also processes the water for reuse, recovery, and recycling as far as possible. It involves all the actions and feasible methods in WWT for the CE principles of implementation, in alignment with the environmental, societal, organizational, and technological needs (Smol, Adam, and Preisner 2020). Treatment of swine wastewater in different hydraulic times enabled the cultivation of microalgae and bacteria and recovery of nutrients and biomass that can be used in swine feed, leading to CE in pig farming (Younas et al. 2022). Important indicators suggested for integration in the WWT framework and policy are nutrient removal efficiency, environmental aspects, sewage sludge processes, and biobased fertilizer production within the wastewater sector to reach CE objectives (Preisner et al. 2021). The capture of carbon dioxide from wastewater can be utilized in several technologies such as biochar production from sludge, CW application, and microbial electrochemical processes related to WWT. Such approaches are sustainable, environment friendly, and cost-effective with the possibility of internal rate of return and revenue generation in line with CE (Pahunang et al. 2021). The nutrients of wastewater can be used for the growth of aquatic plants and plant biomass, which can be a source for bioenergy, biogas, biofuels, biochar, animal feed, fertilizer, adsorbents, and other valuable products that aid in the CE approaches toward WWT (Kurniawan et al. 2021). A combination of MFC with a chemical precipitation reaction system harvests mineral struvite from mineral-rich wastewater along with generation of electricity and significant removal of carbon, nitrogen, and phosphorus from wastewater, facilitating a robust CE in the field of agriculture (Kim et al. 2021). Many countries are now moving into zero waste and toward waste to products policy in a biobased economy, particularly regarding sludge management for the production of biogas and other value-added products toward a CE (Kaszycki, Glodniok, and Petryszak 2021). To determine optimum operations parameters for sustainable WWT, the performance of an anaerobic membrane bioreactor treatment of domestic wastewater for one year period has been assessed by removal efficiency of COD and BOD and biogas production (Ji et al. 2021). For assessing sustainability in arid regions, a continuous performance indicator framework has been developed to benchmark WWT plants. Such framework determines action plan for long-term sustainability, such as reusing treated effluents in the current and future scenario, regulation of chlorination, and enhancement of removal of total solids, among others (Haider et al. 2021). An approach termed Energy and Raw Materials Factory (ERMF) aims to solve the water scarcity problem and recover resources such as bioplastics, cellulose from the aerobic granular sludge, and biomass. This concept proposes to have high financial return on investment, which is environmentally friendly while bringing a circularity to the economy (van Leeuwen et al. 2018).

5.6 ADVANCES IN WWT TOWARD SDGS

The scarcity of water is both an artificial and a natural problem. Water scarcity lies in its uneven distribution, wastage, contamination, and lack of regulation. WWT combats this issue by transforming wastewater into discharge (effluent), reusing, or releasing it into the environment (Voulvoulis 2018). The treatment of sewage by employing conventional technologies involves the primary treatment based on the addition of flocculants or chemicals. Secondary treatment involves biological methods and an adsorption process where activated carbon performs the polishing treatment. The technologies for conventional WWT are not sustainable due to elevated consumption levels for chemicals, discarding of sludge, excessive energy, and space demands. Furthermore, the efficient elimination of complex organic compounds, restriction of controlling wastewater compared to the specified design proportion, and shortage of proficient personnel are also major operational concerns in such technologies.

On account of all the above technological and operational hurdles in technologies employed in the conventional WWT, researchers are making an effort to come out with new types of advanced technologies for WWT to overcome the preceding problems. The technologies needed to be incorporated for advanced WWT are:

i. Advanced process of oxidation;
ii. Less quantity of bioflocculants or consumption of chemicals;

iii. Membrane technology;
iv. Generation of sludge in small amount and in case sludge is generated then how to use the sludge alternatively in place of disposing at site of landfill;
v. Economical adsorption materials;
vi. Nanomaterials of a new class for WWT.

The implementation of conventional technologies in combination with advanced wastewater technologies might result in the successful treatment of wastewater and thus escalate the policy of recycling and reusing treated water. Innovative solutions are required to generate usable and safe water, significant recovery of nutrients, energy, and other resources from wastewater in alignment with SDGs. Some of these solutions are decentralized approaches, low or no flush toilets, waste flow separation, and converting sludge into energy and others. These are listed in Table 5.2.

Optimization of parameters of WWT has been studied using multi-agent deep reinforcement learning to optimize parameters such as chemical dosage and dissolved oxygen in WWT plants (Chen et al. 2021). Besides, sludge-based adsorbents can also be used to adsorb various pollutants such as heavy metals, antibiotics, pharmaceuticals, dyes, and others. WWT sludge has shown the potential to remove emerging contaminants such as hormones from wastewater (Dias et al. 2021). Another study highlighted low pathogen risk for the use of water treatment residual in the amendment of nutrient-poor soils (Stone et al. 2021). A combination of electrochemical techniques with activated carbon reduces micropollutants from wastewater by adsorption as a part of advanced WWT (Sher et al. 2021).

Ecological engineering mimics natural ecosystems and may involve using macrophytic plants such as water hyacinths and other environmental conditions to remove nitrates, phosphates, and sulfates from wastewater. This strategy could effectively restore natural lakes and control the ecological flow sustainably (Selvaraj and Velvizhi 2021). Contrastingly, another study reports the use of willow in removing wastewater pollutants as these trees are tolerant to wastewater contaminants, remove the high percentage of nitrogen, and enable significant biomass improvements to produce biofuels (Sas et al. 2021). A strain of cyanobacteria, *Synechococcus elongatus*, has been shown as a sustainable method for removing nitrates from industrial

TABLE 5.2
Challenges, Causes, and Solutions Toward Sustainable WWT

Challenges	Causes	Proposed Solution
Lack of WWT plant or drainage system	Increasing urbanization, lack of funding or regulations for wastewater disposal	Decentralized solutions, regulatory monitoring, and guidelines of wastewater
Lack of access to toilets	Lack of public awareness, lack of proper design of housing	Economical and sustainable, low or no flush toilets
Poor governance or lack of sustainable wastewater management framework	Absence or low public awareness, education, public health priority	Implement policies and laws related to WWTs, decentralized solutions

wastewater and saline dairy wastewater. The use of cyanobacteria in sustainable industrial WWT can fulfill the goals of recovery of resources, high-value products generation, carbon fixation, and mitigating climate change (Samiotis, Stamatakis, and Amanatidou 2021).

Cut flowers and floral waste (CFW) can be reused as an excellent adsorbent capable of adsorbing antibiotics such as levofloxacin and lead ions from water significantly (Sabri et al. 2021). Countries worldwide also need to take appropriate WWT to contain SARS-CoV-2 as the virus might spread via sewage to attain SDG, including health and sanitation (Rahimi et al. 2021). A three-dimensional MFC based on graphite granules removed sodium dodecyl sulfate (SDS), COD, and oil and grease from carwash shampoo water (Radeef and Ismail 2021). Another study reported the degradation of organic pollutants by using reactive oxygen species generated by a combination of mechanical energy and a low-powered ultrasonic system. This study reported a novel insight on advanced oxidation strategies of pollutants by ultrasound and mechanical energy combination (Nie et al. 2021). Wastewater from sugar industry can be remediated by a variety of microbial strains and wastes can be converted into succinic acid, L-arabinose, and other value-added products toward CE (Nawaz et al. 2021). Advanced biological oxidation process has been found to be effective in removing the toxic chemicals and reduce sludge quantity from paint industry wastewater (Nair, Manu, and Azhoni 2021). Fruit and vegetable wastes can be reused as eco-friendly adsorbents for various inorganic and organic pollutants in wastewater (Matei et al. 2021). Total organic carbon and total nitrogen can be removed significantly in winery wastewater by green microalga, *Chlorella vulgaris*, and when combined with photocatalytic process, the level of COD in the treated water can go less than 150 mg/L, which then allows the treated water to be released in the surface water (Marchao et al. 2021). Similarly, various microalgal species can treat wastewater from the pork industry and remove a high quantity of nitrogen and phosphorus, minimizing eutrophication, generating valuable products such as pigment and polysaccharides and products of commercial value biogas, biohydrogen, etc. Such use of microalgae is sustainable and results in a CE (Lopez-Pacheco et al. 2021). A consortium of bacteria and microalgae in a photo-sequencing batch reactor efficiently removes COD and total Kjeldahl nitrogen (TKN) without aeration in a cost-efficient manner (Petrini, Foladori, and Andreottola 2018). Tetraselmis spp., a robust strain of algae, has been suggested to reduce nutrient and heavy metal levels from municipal and industrial wastewater. At the same time, carbohydrates and lipids of algal biomass can be used as feedstock.

Furthermore, photobioreactor use improves biomass yield and efficiency of removal of pollutants (Goswami et al. 2021). Nitrifying bacteria encapsulated in calcium alginate and $CaCO_3$ colloidosomes for a sustainable ammonium nitrogen removal of wastewater have been proposed (Liu et al. 2021). Silica-based nanomaterial adsorbents and biohybrids are efficient in removing uranium from nuclear wastewater which can then be reused by various processes including sonochemical leaching. These processes are sustainable as it recovers a limitedly available, valuable fissile material while minimizing environmental hazards (Lahiri et al. 2021). Various framework of assessment of water systems including WWT have been undertaken by various countries around the world (Garcia-Sanchez and Guereca 2019, Carvalho et al. 2019). These advances in WWT toward SDG are highlighted in Figure 5.4.

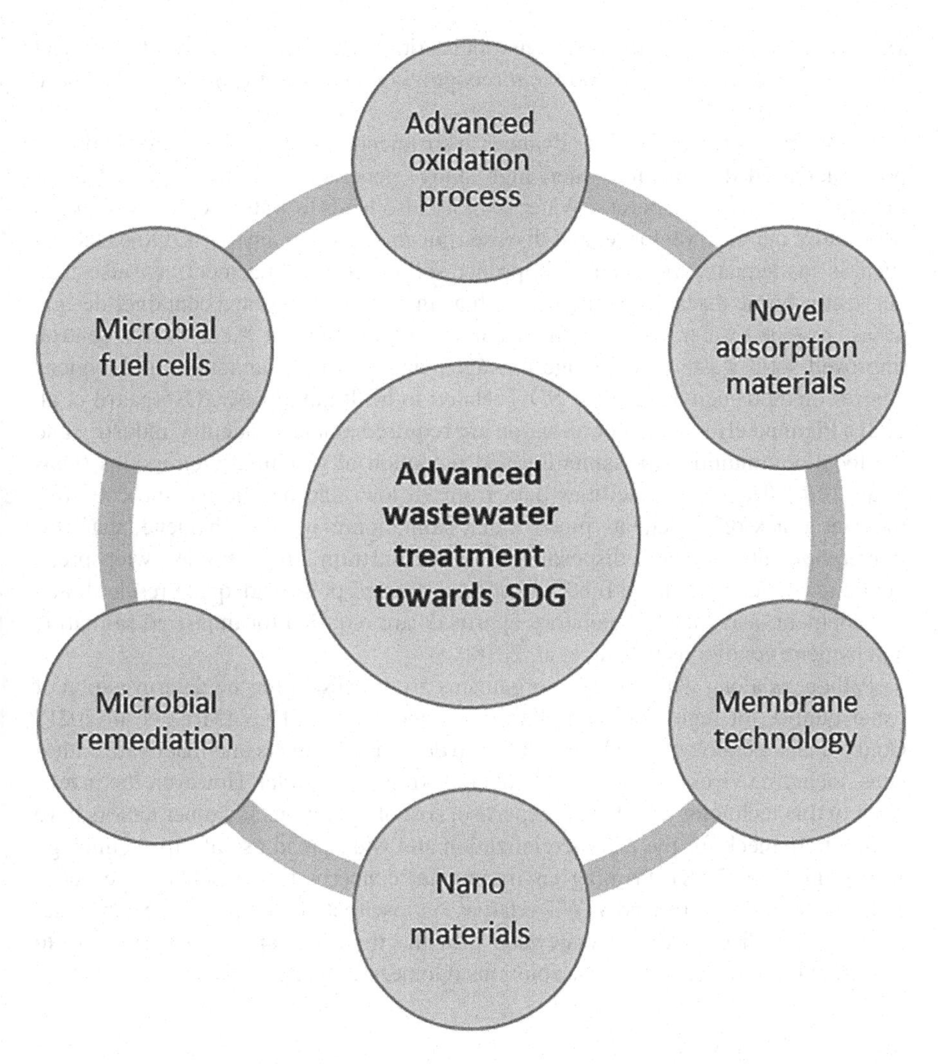

FIGURE 5.4 Advances in WWT toward SDGs.

5.7 CHALLENGES OF CURRENT WWT TOWARD ACHIEVING SUSTAINABILITY

One of the most difficult challenges in the 21st century is the convenience of adequate clean water setup without contaminants, particularly in resource-constrained regions (Johnson, Berg, and Sabatini 2014, Edokpayi et al. 2018). The treatment of wastewater is an indispensable part of SDGs. Despite all the efforts and developments, WWT presents many challenges worldwide. In addition, the average proportion for WWT advances around 70% in wealthy countries while in countries with middle income and lower middle income it is just 38% and 28%, respectively. Furthermore, worldwide water reuse is less than 5%. The challenges for WWT differ, thus hanging on various characteristics and conditions (legislations and socioeconomic) for effluent control. Therefore, there is a problem recognizing a common challenge relevant to all circumstances.

Moreover, there is a lack of advanced instruments, qualified staff, financial support, increased demand for water, high sludge generation, high energy consumption, and scarcity of resources. Water scarcity also leads to water storage, which can potentially increase vector-related disease transmission in many parts. Other factors such as inadequate sanitation or improper solid waste management can also trigger vector-borne diseases. Strategies such as innovation in storage container design, access to piped water, avoiding the accumulation of stagnant water, vector control, improved solid waste and garbage management, improved sanitation, and hygiene control methods can all achieve SDG related to health and water (Overgaard et al. 2021). Planned efforts and coordination are required among academia, industry, and the local communities for sanitation and reduction of greenhouse emissions (Dias et al. 2018). Healthcare facilities data from 78 low- and middle-income countries indicate a lack of inadequate piped water, sanitation conditions, hygiene, sterilization equipment, and waste disposal (Cronk and Bartram 2018). Besides, widespread public awareness, adequate funding support, aligned policy, adequate research and development, and speedy regulatory approvals are required for improved sanitation and hygiene conditions (Cheng et al. 2018).

Pathogens along with indicator organisms are suggested for the determination of water quality for regulatory compliance and meeting of SDGs (Mraz et al. 2021). Cold plasma technology can be employed to degrade pollutants and inactivate pathogens, including viruses such as coronaviruses from wastewater. However, the investment in this technology, technical expertise, cost of operation, and other factors have been a bottleneck for the commercialization and widespread use of this technology (Gururani et al. 2021). Priming environmental concerns to the public have better acceptance of decentralized WWT relative to those that are not primed. Such social acceptance of decentralized wastewater provides the necessary support and will to implement and adopt sustainable solutions (Gomez-Roman et al. 2021).

5.8 CONCLUSIONS

Due to anthropogenic activities such as increasing global population, urbanization, agriculture, and global climate change, most of the water is not reusable. Managing wastewater is one of the keys to sustainable development in alignment with SDGs adopted globally by the members of the United Nations. Sustainable WWT can bring sustainable solutions in alignment with SDG6 and others, but it also can recover and reuse valuable resources and water in line with the CE. The notion of converting waste to help is sustainable in generating energy, power, nutrients, and reuse of water which can minimize global water scarcity. This chapter has discussed and highlighted the concept that sustainable WWT is a way forward toward many United Nations SDGs. It also brings a CE, which itself is sustainable. We have discussed some of the advances and challenges in achieving these environment-friendly and cost-efficient wastewater management solutions for a sustainable future.

REFERENCES

Barros, O., L. Costa, F. Costa, A. Lago, V. Rocha, Z. Vipotnik, B. Silva, and T. Tavares. 2019. "Recovery of rare earth elements from wastewater towards a circular economy." *Molecules* 24 (6). doi: 10.3390/molecules24061005.

Bhat, S. A., F. Sher, M. Hameed, O. Bashir, R. Kumar, D. N. Vo, P. Ahmad, and E. C. Lima. 2022. "Sustainable nanotechnology based wastewater treatment strategies: achievements, challenges and future perspectives." *Chemosphere* 288 (Pt 3):132606. doi: 10.1016/j.chemosphere.2021.132606.

Bocken, N. M. P., I. de Pauw, C. Bakker, and B. van der Grinten. 2016. "Product design and business model strategies for a circular economy." *JIPE* 33 (5):308–320. doi: 10.1080/21681015.2016.1172124.

Bousquet, C., I. Samora, P. Manso, L. Rossi, P. Heller, and A. J. Schleiss. 2017. "Assessment of hydropower potential in wastewater systems and application to Switzerland." *Renew Energy* 113:64–73. doi: 10.1016/j.renene.2017.05.062.

Carvalho, L., E. B. Mackay, A. C. Cardoso, A. Baattrup-Pedersen, S. Birk, K. L. Blackstock, G. Borics, A. Borja, C. K. Feld, M. T. Ferreira, L. Globevnik, B. Grizzetti, S. Hendry, D. Hering, M. Kelly, S. Langaas, K. Meissner, Y. Panagopoulos, E. Penning, J. Rouillard, S. Sabater, U. Schmedtje, B. M. Spears, M. Venohr, W. van de Bund, and A. L. Solheim. 2019. "Protecting and restoring Europe's waters: an analysis of the future development needs of the water framework directive." *Sci Total Environ* 658:1228–1238. doi: 10.1016/j.scitotenv.2018.12.255.

Chen, K., H. Wang, B. Valverde-Perez, S. Zhai, L. Vezzaro, and A. Wang. 2021. "Optimal control towards sustainable wastewater treatment plants based on multi-agent reinforcement learning." *Chemosphere* 279:130498. doi: 10.1016/j.chemosphere.2021.130498.

Cheng, S., Z. Li, S. M. N. Uddin, H. P. Mang, X. Zhou, J. Zhang, L. Zheng, and L. Zhang. 2018. "Toilet revolution in China." *J Environ Manage* 216:347–356. doi: 10.1016/j.jenvman.2017.09.043.

Cronk, R., and J. Bartram. 2018. "Environmental conditions in health care facilities in low- and middle-income countries: coverage and inequalities." *Int J Hyg Environ Health* 221 (3):409–422. doi: 10.1016/j.ijheh.2018.01.004.

da Costa Filho, B. M., G. V. Silva, R. A. R. Boaventura, M. M. Dias, J. C. B. Lopes, and V. J. P. Vilar. 2019. "Ozonation and ozone-enhanced photocatalysis for VOC removal from air streams: process optimization, synergy and mechanism assessment." *Sci Total Environ* 687:1357–1368. doi: 10.1016/j.scitotenv.2019.05.365.

Dias, C. M. M., L. P. Rosa, J. M. A. Gomez, and A. D'Avignon. 2018. "Achieving the Sustainable Development Goal 06 in Brazil: the universal access to sanitation as a possible mission." *An Acad Bras Cienc* 90 (2):1337–1367. doi: 10.1590/0001-3765201820170590.

Dias, R., D. Sousa, M. Bernardo, I. Matos, I. Fonseca, V. Vale Cardoso, R. Neves Carneiro, S. Silva, P. Fontes, M. A. Daam, and R. Mauricio. 2021. "Study of the potential of water treatment sludges in the removal of emerging pollutants." *Molecules* 26 (4). doi: 10.3390/molecules26041010.

Do, M. H., H. H. Ngo, W. S. Guo, Y. Liu, S. W. Chang, D. D. Nguyen, L. D. Nghiem, and B. J. Ni. 2018. "Challenges in the application of microbial fuel cells to wastewater treatment and energy production: a mini review." *Sci Total Environ* 639:910–920. doi: 10.1016/j.scitotenv.2018.05.136.

Edokpayi, J. N., E. T. Rogawski, D. M. Kahler, C. L. Hill, C. Reynolds, E. Nyathi, J. A. Smith, J. O. Odiyo, A. Samie, P. Bessong, and R. Dillingham. 2018. "Challenges to sustainable safe drinking water: a case study of water quality and use across seasons in rural communities in Limpopo Province, South Africa." *Water (Basel)* 10 (2). doi: 10.3390/w10020159.

Garcia-Sanchez, M., and L. P. Guereca. 2019. "Environmental and social life cycle assessment of urban water systems: the case of Mexico City." *Sci Total Environ* 693:133464. doi: 10.1016/j.scitotenv.2019.07.270.

Ge, C. H., Y. Dong, H. Li, Q. Li, S. Q. Ni, B. Gao, S. Xu, Z. Qiao, and S. Ding. 2019. "Nitritation-anammox process – a realizable and satisfactory way to remove nitrogen from high saline wastewater." *Bioresour Technol* 275:86–93. doi: 10.1016/j.biortech.2018.12.032.

Gherghel, A., C. Teodosiu, and S. De Gisi. 2019. "A review on wastewater sludge valorisation and its challenges in the context of circular economy." *J Clean Prod* 228:244–263. doi: 10.1016/j.jclepro.2019.04.240.

Ghimire, U., G. Sarpong, and V. G. Gude. 2021. "Transitioning wastewater treatment plants toward circular economy and energy sustainability." *ACS Omega* 6 (18):11794–11803. doi: 10.1021/acsomega.0c05827.

Ghisellini, P., C. Cialani, and S. Ulgiati. 2016. "A review on circular economy: the expected transition to a balanced interplay of environmental and economic systems." *J Clean Prod* 114:11–32. doi: 10.1016/j.jclepro.2015.09.007.

Gomez-Roman, C., J. M. Sabucedo, M. Alzate, and B. Medina. 2021. "Environmental concern priming and social acceptance of sustainable technologies: the case of decentralized wastewater treatment systems." *Front Psychol* 12:647406. doi: 10.3389/fpsyg.2021.647406.

Goswami, R. K., K. Agrawal, S. Mehariya, and P. Verma. 2021. "Current perspective on wastewater treatment using photobioreactor for Tetraselmis sp.: an emerging and foreseeable sustainable approach." *Environ Sci Pollut Res Int*. doi: 10.1007/s11356-021-16860-5.

Gowd, S. C., S. Ramakrishna, and K. Rajendran. 2021. "Wastewater in India: an untapped and under-tapped resource for nutrient recovery towards attaining a sustainable circular economy." *Chemosphere*:132753. doi: 10.1016/j.chemosphere.2021.132753.

Gu, J., Q. Yang, and Y. Liu. 2018. "A novel strategy towards sustainable and stable nitritation-denitritation in an A-B process for mainstream municipal wastewater treatment." *Chemosphere* 193:921–927. doi: 10.1016/j.chemosphere.2017.11.038.

Guo, Y., T. Sugano, Y. Song, C. Xie, Y. Chen, Y. Xue, and Y. Y. Li. 2020. "The performance of freshwater one-stage partial nitritation/anammox process with the increase of salinity up to 3.0." *Bioresour Technol* 311:123489. doi: 10.1016/j.biortech.2020.123489.

Guppy, L., P. Mehta, and M. Qadir. 2019. "Sustainable development goal 6: two gaps in the race for indicators." *Sustain Sci* 14 (2):501–513. doi: 10.1007/s11625-018-0649-z.

Gururani, P., P. Bhatnagar, B. Bisht, V. Kumar, N. C. Joshi, M. S. Tomar, and B. Pathak. 2021. "Cold plasma technology: advanced and sustainable approach for wastewater treatment." *Environ Sci Pollut Res Int* 28 (46):65062–65082. doi: 10.1007/s11356-021-16741-x.

Haider, H., M. AlHetari, A. R. Ghumman, I. S. Al-Salamah, H. Thabit, and M. Shafiquzzaman. 2021. "Continuous performance improvement framework for sustainable wastewater treatment facilities in arid regions: case of Wadi Rumah in Qassim, Saudi Arabia." *Int J Environ Res Public Health* 18 (13). doi: 10.3390/ijerph18136857.

He, L., P. Du, Y. Chen, H. Lu, X. Cheng, B. Chang, and Z. Wang. 2017. "Advances in microbial fuel cells for wastewater treatment." *Renew Sustain Energy Rev* 71:388–403. doi: 10.1016/j.rser.2016.12.069.

Ji, J., Y. Chen, Y. Hu, A. Ohtsu, J. Ni, Y. Li, S. Sakuma, T. Hojo, R. Chen, and Y. Y. Li. 2021. "One-year operation of a 20-L submerged anaerobic membrane bioreactor for real domestic wastewater treatment at room temperature: pursuing the optimal HRT and sustainable flux." *Sci Total Environ* 775:145799. doi: 10.1016/j.scitotenv.2021.145799.

Johnson, C. A., M. Berg, and D. Sabatini. 2014. "Towards sustainable safe drinking water supply in low- and middle-income countries: the challenges of geogenic contaminants and mitigation measures." *Sci Total Environ* 488-489:475–6. doi: 10.1016/j.scitotenv.2014.01.131.

Kacprzak, M., E. Neczaj, K. Fijalkowski, A. Grobelak, A. Grosser, M. Worwag, A. Rorat, H. Brattebo, A. Almas, and B. R. Singh. 2017. "Sewage sludge disposal strategies for sustainable development." *Environ Res* 156:39–46. doi: 10.1016/j.envres.2017.03.010.

Kaszycki, P., M. Glodniok, and P. Petryszak. 2021. "Towards a bio-based circular economy in organic waste management and wastewater treatment – the Polish perspective." *N Biotechnol* 61:80–89. doi: 10.1016/j.nbt.2020.11.005.

Kim, B., N. Jang, M. Lee, J. K. Jang, and I. S. Chang. 2021. "Microbial fuel cell driven mineral rich wastewater treatment process for circular economy by creating virtuous cycles." *Bioresour Technol* 320 (Pt A):124254. doi: 10.1016/j.biortech.2020.124254.

Kurniawan, S. B., A. Ahmad, N. S. M. Said, M. F. Imron, S. R. S. Abdullah, A. R. Othman, I. F. Purwanti, and H. A. Hasan. 2021. "Macrophytes as wastewater treatment agents: nutrient uptake and potential of produced biomass utilization toward circular economy initiatives." *Sci Total Environ* 790:148219. doi: 10.1016/j.scitotenv.2021.148219.

Kwon, E. E., S. Kim, Y. J. Jeon, and H. Yi. 2012. "Biodiesel production from sewage sludge: new paradigm for mining energy from municipal hazardous material." *Environ Sci Technol* 46 (18):10222–8. doi: 10.1021/es3019435.

Lahiri, S., A. Mishra, D. Mandal, R. L. Bhardwaj, and P. R. Gogate. 2021. "Sonochemical recovery of uranium from nanosilica-based sorbent and its biohybrid." *Ultrason Sonochem* 76:105667. doi: 10.1016/j.ultsonch.2021.105667.

Liu, G., T. Du, J. Chen, X. Hao, F. Yang, H. He, T. Meng, and Y. Wang. 2021. "Microfluidic aqueous two-phase system-based nitrifying bacteria encapsulated colloidosomes for green and sustainable ammonium-nitrogen wastewater treatment." *Bioresour Technol* 342:126019. doi: 10.1016/j.biortech.2021.126019.

Lopez-Pacheco, I. Y., A. Silva-Nunez, J. S. Garcia-Perez, D. Carrillo-Nieves, C. Salinas-Salazar, C. Castillo-Zacarias, S. Afewerki, D. Barcelo, H. N. M. Iqbal, and R. Parra-Saldivar. 2021. "Phyco-remediation of swine wastewater as a sustainable model based on circular economy." *J Environ Manage* 278 (Pt 2):111534. doi: 10.1016/j.jenvman.2020.111534.

Lv, Pengmei, Zhenhong Yuan, Chuangzhi Wu, Longlong Ma, Yong Chen, and Noritatsu Tsubaki. 2007. "Bio-syngas production from biomass catalytic gasification." *Energy Convers Manage* 48 (4):1132–1139. doi: 10.1016/j.enconman.2006.10.014.

Marchao, L., J. R. Fernandes, A. Sampaio, J. A. Peres, P. B. Tavares, and M. S. Lucas. 2021. "Microalgae and immobilized TiO2/UV-A LEDs as a sustainable alternative for winery wastewater treatment." *Water Res* 203:117464. doi: 10.1016/j.watres.2021.117464.

Masi, F., A. Rizzo, and M. Regelsberger. 2018. "The role of constructed wetlands in a new circular economy, resource oriented, and ecosystem services paradigm." *J Environ Manage* 216:275–284. doi: 10.1016/j.jenvman.2017.11.086.

Matei, E., M. Rapa, A. M. Predescu, A. A. Turcanu, R. Vidu, C. Predescu, C. Bobirica, L. Bobirica, and C. Orbeci. 2021. "Valorization of agri-food wastes as sustainable eco-materials for wastewater treatment: current state and new perspectives." *Materials (Basel)* 14 (16). doi: 10.3390/ma14164581.

Mraz, A. L., I. K. Tumwebaze, S. R. McLoughlin, M. E. McCarthy, M. E. Verbyla, N. Hofstra, J. B. Rose, and H. M. Murphy. 2021. "Why pathogens matter for meeting the united nations' sustainable development goal 6 on safely managed water and sanitation." *Water Res* 189:116591. doi: 10.1016/j.watres.2020.116591.

Mustafa, H. M., and G. Hayder. 2021. "Performance of Salvinia molesta plants in tertiary treatment of domestic wastewater." *Heliyon* 7 (1):e06040. doi: 10.1016/j.heliyon.2021.e06040.

Nair, K. S., B. Manu, and A. Azhoni. 2021. "Sustainable treatment of paint industry wastewater: current techniques and challenges." *J Environ Manage* 296:113105. doi: 10.1016/j.jenvman.2021.113105.

Nawaz, M. Z., M. Bilal, A. Tariq, H. M. N. Iqbal, H. A. Alghamdi, and H. Cheng. 2021. "Bio-purification of sugar industry wastewater and production of high-value industrial products with a zero-waste concept." *Crit Rev Food Sci Nutr* 61 (21):3537–3554. doi: 10.1080/10408398.2020.1802696.

Neczaj, Ewa, and Anna Grosser. 2018. "Circular economy in wastewater treatment plant-challenges and barriers." *Proceedings* 2 (11):614.

Nie, G., K. Hu, W. Ren, P. Zhou, X. Duan, L. Xiao, and S. Wang. 2021. "Mechanical agitation accelerated ultrasonication for wastewater treatment: sustainable production of hydroxyl radicals." *Water Res* 198:117124. doi: 10.1016/j.watres.2021.117124.

Oluwole, A. O., E. O. Omotola, and O. S. Olatunji. 2020. "Pharmaceuticals and personal care products in water and wastewater: a review of treatment processes and use of photocatalyst immobilized on functionalized carbon in AOP degradation." *BMC Chem* 14 (1):62. doi: 10.1186/s13065-020-00714-1.

Overgaard, H. J., N. Dada, A. Lenhart, T. A. B. Stenstrom, and N. Alexander. 2021. "Integrated disease management: arboviral infections and waterborne diarrhoea." *Bull World Health Organ* 99 (8):583–592. doi: 10.2471/BLT.20.269985.

Pahunang, R. R., A. Buonerba, V. Senatore, G. Oliva, M. Ouda, T. Zarra, R. Munoz, S. Puig, F. C. Ballesteros, Jr., C. W. Li, S. W. Hasan, V. Belgiorno, and V. Naddeo. 2021. "Advances in technological control of greenhouse gas emissions from wastewater in the context of circular economy." *Sci Total Environ* 792:148479. doi: 10.1016/j.scitotenv.2021.148479.

Pandey, A. K., R. Reji Kumar, K. B, I. A. Laghari, M. Samykano, R. Kothari, A. M. Abusorrah, K. Sharma, and V. V. Tyagi. 2021. "Utilization of solar energy for wastewater treatment: challenges and progressive research trends." *J Environ Manage* 297:113300. doi: 10.1016/j.jenvman.2021.113300.

Parrino, F., G. Camera-Roda, V. Loddo, G. Palmisano, and V. Augugliaro. 2014. "Combination of ozonation and photocatalysis for purification of aqueous effluents containing formic acid as probe pollutant and bromide ion." *Water Res* 50:189–99. doi: 10.1016/j.watres.2013.12.001.

Patel, N., S. Shahane, B. Bhunia, U. Mishra, V. K. Chaudhary, and A. L. Srivastav. 2022. "Biodegradation of 4-chlorophenol in batch and continuous packed bed reactor by isolated Bacillus subtilis." *J Environ Manage* 301:113851. doi: 10.1016/j.jenvman.2021.113851.

Petrini, S., P. Foladori, and G. Andreottola. 2018. "Laboratory-scale investigation on the role of microalgae towards a sustainable treatment of real municipal wastewater." *Water Sci Technol* 78 (8):1726–1732. doi: 10.2166/wst.2018.453.

Preisner, M., M. Smol, M. Horttanainen, I. Deviatkin, J. Havukainen, M. Klavins, R. Ozola-Davidane, J. Kruopiene, B. Szatkowska, L. Appels, S. Houtmeyers, and K. Roosalu. 2021. "Indicators for resource recovery monitoring within the circular economy model implementation in the wastewater sector." *J Environ Manage* 304:114261. doi: 10.1016/j.jenvman.2021.114261.

Radeef, A. Y., and Z. Z. Ismail. 2021. "Bioelectrochemical treatment of actual carwash wastewater associated with sustainable energy generation in three-dimensional microbial fuel cell." *Bioelectrochemistry* 142:107925. doi: 10.1016/j.bioelechem.2021.107925.

Rahimi, N. R., R. Fouladi-Fard, R. Aali, A. Shahryari, M. Rezaali, Y. Ghafouri, M. R. Ghalhari, M. Asadi-Ghalhari, B. Farzinnia, O. Conti Gea, and M. Fiore. 2021. "Bidirectional association between COVID-19 and the environment: a systematic review." *Environ Res* 194:110692. doi: 10.1016/j.envres.2020.110692.

Sabri, M. A., T. H. Ibrahim, M. I. Khamis, A. Ludwick, and P. Nancarrow. 2021. "Sustainable management of cut flowers waste by activation and its application in wastewater treatment technology." *Environ Sci Pollut Res Int* 28 (24):31803–31813. doi: 10.1007/s11356-021-13002-9.

Samiotis, G., K. Stamatakis, and E. Amanatidou. 2021. "Assessment of Synechococcus elongatus PCC 7942 as an option for sustainable wastewater treatment." *Water Sci Technol* 84 (6):1438–1451. doi: 10.2166/wst.2021.319.

Santoro, C., C. Arbizzani, B. Erable, and I. Ieropoulos. 2017. "Microbial fuel cells: from fundamentals to applications. A review." *J Power Sources* 356:225–244. doi: 10.1016/j.jpowsour.2017.03.109.

Saravanan, A., P. Senthil Kumar, S. Jeevanantham, S. Karishma, B. Tajsabreen, P. R. Yaashikaa, and B. Reshma. 2021. "Effective water/wastewater treatment methodologies for toxic pollutants removal: processes and applications towards sustainable development." *Chemosphere* 280:130595. doi: 10.1016/j.chemosphere.2021.130595.

Sarpong, Gideon, and Veera Gnaneswar Gude. 2020. "Near future energy self-sufficient wastewater treatment schemes." *IJER* 14 (4):479–488. doi: 10.1007/s41742-020-00262-5.

Sas, E., L. M. Hennequin, A. Fremont, A. Jerbi, N. Legault, J. Lamontagne, N. Fagoaga, M. Sarrazin, J. P. Hallett, P. S. Fennell, S. Barnabe, M. Labrecque, N. J. B. Brereton, and F. E. Pitre. 2021. "Biorefinery potential of sustainable municipal wastewater treatment using fast-growing willow." *Sci Total Environ* 792:148146. doi: 10.1016/j.scitotenv.2021.148146.

Selvaraj, D., and G. Velvizhi. 2021. "Sustainable ecological engineering systems for the treatment of domestic wastewater using emerging, floating and submerged macrophytes." *J Environ Manage* 286:112253. doi: 10.1016/j.jenvman.2021.112253.

Shen, Chao, Zhuoyu Lei, Yuan Wang, Chenghu Zhang, and Yang Yao. 2018. "A review on the current research and application of wastewater source heat pumps in China." *TSEP* 6:140–156. doi: 10.1016/j.tsep.2018.03.007.

Sher, F., S. Z. Iqbal, T. Rasheed, K. Hanif, J. Sulejmanovic, F. Zafar, and E. C. Lima. 2021. "Coupling of electrocoagulation and powder activated carbon for the treatment of sustainable wastewater." *Environ Sci Pollut Res Int* 28 (35):48505–48516. doi: 10.1007/s11356-021-14129-5.

Singh, S., V. Kumar, A. G. Anil, D. Kapoor, S. Khasnabis, S. Shekar, N. Pavithra, J. Samuel, S. Subramanian, J. Singh, and P. C. Ramamurthy. 2021. "Adsorption and detoxification of pharmaceutical compounds from wastewater using nanomaterials: a review on mechanism, kinetics, valorization and circular economy." *J Environ Manage* 300:113569. doi: 10.1016/j.jenvman.2021.113569.

Smol, Marzena, C. Adam, and M. Preisner. 2020. "Circular economy model framework in the European water and wastewater sector." *J Mater Cycles Waste Manag* 22 (3):682–697. doi: 10.1007/s10163-019-00960-z.

Spinosa, L., and P. Doshi. 2021. "Re-thinking sludge management within the sustainable development goal 6.2." *J Environ Manage* 287:112338. doi: 10.1016/j.jenvman.2021.112338.

Srivastava, N., M. Srivastava, V. K. Gupta, A. Manikanta, K. Mishra, S. Singh, S. Singh, P. W. Ramteke, and P. K. Mishra. 2018. "Recent development on sustainable biodiesel production using sewage sludge." *3 Biotech* 8 (5):245. doi: 10.1007/s13205-018-1264-5.

Stone, W., N. S. Lukashe, L. I. Blake, T. Gwandu, A. G. Hardie, J. Quinton, K. Johnson, and C. E. Clarke. 2021. "The microbiology of rebuilding soils with water treatment residual co-amendments: risks and benefits." *J Environ Qual.* doi: 10.1002/jeq2.20286.

Tortajada, Cecilia. 2020. "Contributions of recycled wastewater to clean water and sanitation sustainable development goals." *npj Clean Water* 3 (1):22. doi: 10.1038/s41545-020-0069-3.

Tyagi, V. K., and S.-L. Lo. 2013. "Sludge: a waste or renewable source for energy and resources recovery?" *Renew Sustain Energy Rev* 25:708–728. doi: 10.1016/j.rser.2013.05.029.

Valchev, D., I. Ribarova, B. Uzunov, and M. Stoyneva-Gartner. 2021. "Photo-sequencing batch reactor with Klebsormidium nitens: a promising microalgal biotechnology for sustainable phosphorus management in wastewater treatment plants." *Water Sci Technol* 83 (10):2463–2476. doi: 10.2166/wst.2021.149.

van Leeuwen, K., E. de Vries, S. Koop, and K. Roest. 2018. "The energy & raw materials factory: role and potential contribution to the circular economy of the Netherlands." *Environ Manage* 61 (5):786–795. doi: 10.1007/s00267-018-0995-8.

Voulvoulis, Nikolaos. 2018. "Water reuse from a circular economy perspective and potential risks from an unregulated approach." *Curr Opin Environ Sci Health* 2:32–45. doi: 10.1016/j.coesh.2018.01.005.

Wang, R., M. Y. Liu, M. Zhang, A. Ghulam, and L. J. Yuan. 2021. "An iron-air fuel cell system towards concurrent phosphorus removal and resource recovery in the form of vivianite and energy generation in wastewater treatment: a sustainable technology regarding phosphorus." *Sci Total Environ* 791:148213. doi: 10.1016/j.scitotenv.2021.148213.

Wei, Z., Q. Van Le, W. Peng, Y. Yang, H. Yang, H. Gu, S. S. Lam, and C. Sonne. 2021. "A review on phytoremediation of contaminants in air, water and soil." *J Hazard Mater* 403:123658. doi: 10.1016/j.jhazmat.2020.123658.

Yan, A., Y. Wang, S. N. Tan, M. L. Mohd Yusof, S. Ghosh, and Z. Chen. 2020. "Phytoremediation: a promising approach for revegetation of heavy metal-polluted land." *Front Plant Sci* 11:359. doi: 10.3389/fpls.2020.00359.

Yin, J., B. Ge, T. Jiao, Z. Qin, M. Yu, L. Zhang, Q. Zhang, and Q. Peng. 2021. "Self-Assembled sandwich-like MXene-derived composites as highly efficient and sustainable catalysts for wastewater treatment." *Langmuir* 37 (3):1267–1278. doi: 10.1021/acs.langmuir.0c03297.

Younas, F., N. K. Niazi, I. Bibi, M. Afzal, K. Hussain, M. Shahid, Z. Aslam, S. Bashir, M. M. Hussain, and J. Bundschuh. 2022. "Constructed wetlands as a sustainable technology for wastewater treatment with emphasis on chromium-rich tannery wastewater." *J Hazard Mater* 422:126926. doi: 10.1016/j.jhazmat.2021.126926.

Zhang, X., and Y. Liu. 2021. "Circular economy-driven ammonium recovery from municipal wastewater: state of the art, challenges and solutions forward." *Bioresour Technol* 334:125231. doi: 10.1016/j.biortech.2021.125231.

Zsilak, Z., O. Fonagy, E. Szabo-Bardos, O. Horvath, K. Horvath, and P. Hajos. 2014. "Degradation of industrial surfactants by photocatalysis combined with ozonation." *Environ Sci Pollut Res Int* 21 (19):11126–34. doi: 10.1007/s11356-014-2527-2.

Zubair, M., I. Ihsanullah, H. Abdul Aziz, M. Azmier Ahmad, and M. A. Al-Harthi. 2021. "Sustainable wastewater treatment by biochar/layered double hydroxide composites: progress, challenges, and outlook." *Bioresour Technol* 319:124128. doi: 10.1016/j.biortech.2020.124128.

6 Advanced Enzymatic Systems for Wastewater Treatment

Future Prospects

Saba Ghattavi and Ahmad Homaei

6.1 INTRODUCTION

Rapid industrialization and development have increased pollution discharge into the environment (Das and Dey, 2020; Shindhal et al., 2021; Sivasubramaniam and Franks, 2016). Water is the most essential element for all life on this planet and a critical source for human society (Morrison et al., 2020; Rubilar et al., 2020). Every day, billions of gallons of wastewater are created in various industries and discharged into the sea, posing a threat to the environment (Karthik et al., 2021; Surendhiran et al., 2017). In recent years, water contamination has become a global environmental concern (Acharya et al., 2020; Aleya et al., 2020; Bilal et al., 2020a; Husain and Ulber, 2011; Lee et al., 2003; Parra-Saldivar et al., 2020; Ponce-Rodríguez et al., 2020) because of the rapid increase in population, urban growth and industrialization (Husain and Ulber, 2011). Industrial pollutants are one of the most serious environmental issues because of its impact on surface and groundwater resources, as well as on human health (Anas et al., 2016; de Vidales et al., 2019; Sivasubramaniam and Franks, 2016).

One of the primary sources of water pollution is the discharge of enormous volumes of industrial waste (Figure 6.1) into rivers and lakes without sufficient waste treatment facilities (Husain and Ulber, 2011; Jun et al., 2019). A significant number of organic contaminants, such as dyes, pesticides, medicines, oestrogens, phenols and personal care products, are being identified in global water sources and wastewater (Aleya et al., 2020; Bilal et al., 2020a; Parra-Saldivar et al., 2020). Their presence in uncontrolled and unregulated doses adversely affects water quality and poses a major risk to humans and aquatic species (Bilal et al., 2020b; Kallel et al., 2020). The World Health Organization and environmental protection agencies have announced that the concentration of toxic pollutants in water is higher than global standards, and about a billion people do not have access to healthy drinking water in the next two decades (Edokpayi et al., 2020; Jiang et al., 2019), and also waterborne illnesses claim the lives of almost 10–20 million people each year (Leonard et al., 2003). The majority of toxic pollutants in water are also known as environmental contaminants, and the World Health Organization classifies them as carcinogens in Group 1 owing to their

DOI: 10.1201/9781003517238-6

potentially carcinogenic and mutagenic nature (Bilal and Iqbal, 2019). This problem is exacerbated by a lack of effective garbage disposal, management and recycling technology. Furthermore, the majority of these compounds are resistant to classical treatment methods and are not adequately eliminated in modern wastewater treatment facilities (Ho et al., 2021; Zdarta et al., 2021). Most recently developed pollutant removal technologies suffer from low efficiency and the generation of many wastes and by-products (Kadam et al., 2020; Wang et al., 2020). As a result, advanced technology for wastewater treatment and drinkable water supply from water resources are needed (Sahu et al., 2019). To this end, the enzymatic system can be a good candidate for wastewater treatment.

Enzyme engineering has progressed to an extent that we are always developing new approaches to discover effective answers to the majority of the challenges in each industry such as pharmaceutical industry (Apetrei et al., 2013), biofuel industry (Selvakumar et al., 2021; Selvakumar et al., 2018), food industry (Ismail and Nielsen, 2010), textile industry and diagnostics (Srivastava et al., 2020). Enzymatic system has opened up a new world of use for wastewater treatment and in terms of high

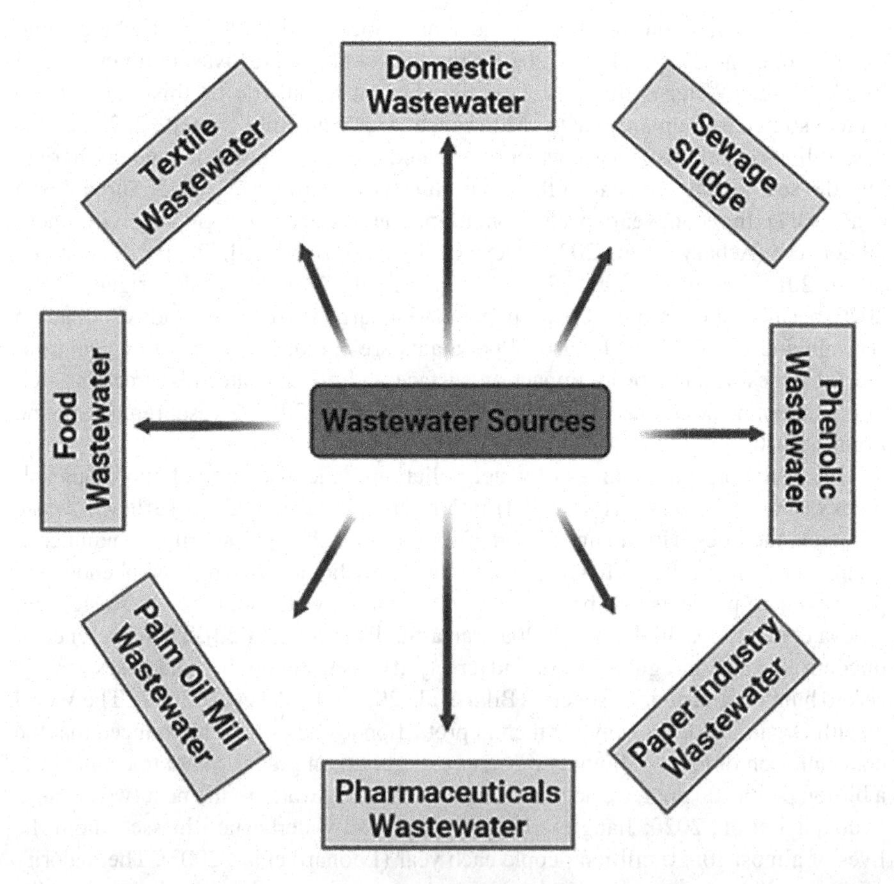

FIGURE 6.1 Various types of wastewater. (Reproduced with permission from Bhatia et al., 2020).

precision and selectivity, environmentally friendly and biodegradability, the enzymatic system outperforms conventional methods (Gholami-Borujeni et al., 2011b). Enzymes, as opposed to organic catalysts, have high efficiencies in transforming complex chemical structures under moderate reaction conditions (Jun et al., 2019; Kılıç et al., 2016; Sanchez and Demain, 2011). Enzymes, as a biocatalyst, accelerate various chemical and biological processes (Apetrei et al., 2013).

The enzymatic system has the ability to regulate enzymes, inhibit enzyme inactivation, increase enzymatic operation, minimize vulnerability to microbial infection, as well as resolve several limitations (Jun et al., 2019). Furthermore, immobilized enzymes have a significant advantage over free enzymes. These benefits include the ability of immobilized enzymes to be reused after recovery from the reaction mixture and the preservation of enzymatic activity for a longer period of time, and also free enzymes are more unstable and have certain drawbacks in terms of industrial uses (Karthik et al., 2021; Tonini and Astrup, 2012). Many studies have confirmed the treatment of wastewater by using a broad range of enzymes (Ba et al., 2013; Dambmann and Aunstrup, 1981; Jamie et al., 2016; Todorova et al., 2019). In addition, recent developments in nanotechnology research have improved the efficiency of enzymes in wastewater treatment applications (Jun et al., 2019) by which they can also be combined with other types of water treatment techniques (Ang et al., 2015; Karri et al., 2017; Lingamdinne et al., 2017; Sillanpää et al., 2018). In this chapter, we try to provide an overview of the application of enzymatic systems in wastewater treatment and discuss the advantages of enzymatic systems and also the importance of enzymatic systems for wastewater treatment.

6.2 ENZYMATIC SYSTEMS AND THEIR IMPORTANCE IN WASTEWATER TREATMENT

Due to water scarcity in the world, wastewater treatment and restoring consumed water to the cycle have become increasingly important. Today, wastewater is one of the valuable and essential sources of water supply. Untreated wastewater may contain organic trash that has degraded. If left for an extended period of time, it can release a large number of gases with a foul odour. It also contains large amounts of toxic compounds that are potentially damaging to the environment. Wastewater treatment methods are important because of the lack of water and contamination of the environment due to abandonment in nature. The removal of toxic products from water is a much-needed technique in today's globe, as pollution levels increase with rapid urbanization and industrialization (Karthik et al., 2021). Several methods are used for pollutant degradation such as substrates (organic, inorganic and hybrid substrates), functionalized materials, biodegradation and enzymatic systems (Kołodziejczak-Radzimska et al., 2021) (Figure 6.2).

In recent years, enzymes are being used extensively in wastewater treatment and have important roles in it. Enzymes, particularly those from the oxidoreductase group, such as peroxidases, tyrosinases and laccases, have emerged as a viable option for biological water and wastewater treatment. These biomolecules may catalyse the oxidation processes of a wide range of organic compounds, primarily phenolic and non-phenolic aromatic molecules (Barber et al., 2020; Bilal et al., 2021;

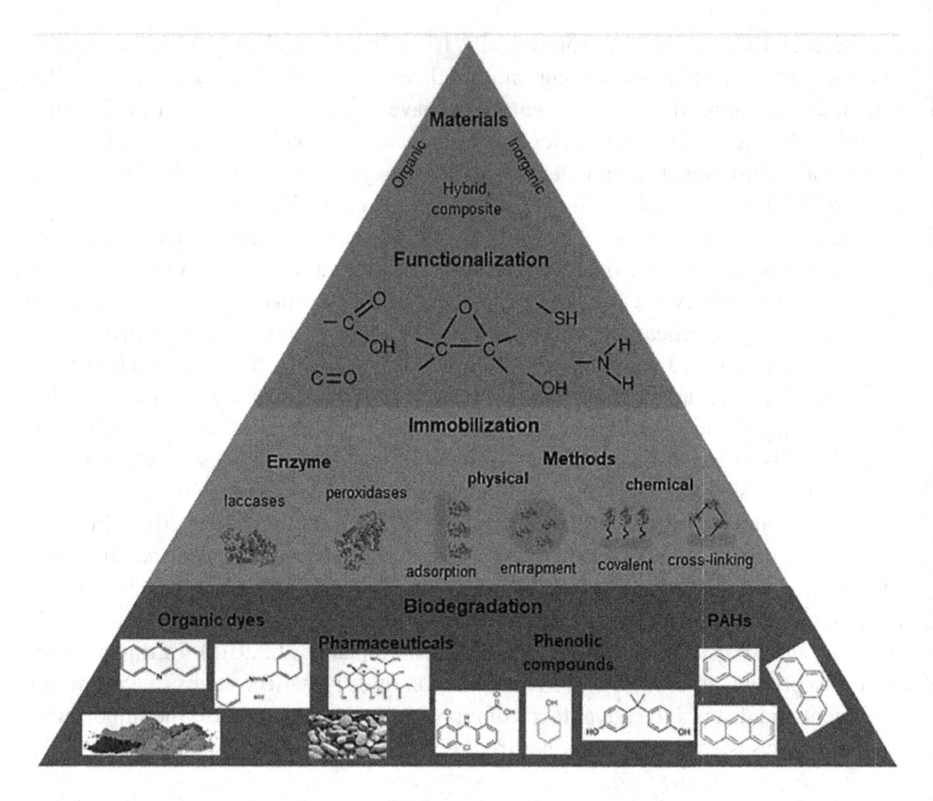

FIGURE 6.2 Several methods that are used for pollutant degradation. (Reproduced with permission from Kołodziejczak-Radzimska et al., 2021).

Kaushal et al., 2018; Morsi et al., 2020). The enzymatic system effectively breaks down organic materials in the wastewater. Enzyme-based water treatment is fast, sensitive and long-lasting, and several tests have shown that it outperforms current technologies (Sharma et al., 2018). In total, enzymes are important biological macromolecules that serve as a catalyst for a variety of biochemical reactions that occur in contaminant degradation pathways (Kalogerakis et al., 2017). The catalytic capabilities of such enzymes make them appealing for the transformation of a variety of contaminants (Alneyadi et al., 2018; Bilal et al., 2019). They have the ability to selectively eliminate a target pollutant without harming the other constituents of the effluent. As a result, enzymatic treatment is appropriate for effluents containing comparatively high levels of recalcitrant goal contaminants in contrast to others (Adam et al., 1999). Enzymatic conversion produces less toxic compounds that can be easily removed from the post-reaction mixture, in addition to a high removal rate (Liu and Smith, 2020). Bioremediation of contamination in a specific environment depends on the catalytic activity of the enzyme secreted by microbes, not the growth of a specific microorganism in an infected environment (Ruggaber and Talley, 2006). Enzymatic systems do not produce by-products during the treatment and removal of contaminants (Gianfreda and Bollag, 2002) and also enzymatic systems are very important in wastewater treatment because they can be used several times to remove contaminants at a higher rate (Sharma et al., 2018) and also because of their biodegradability, enzymes are more environmentally friendly (Adam et al., 1999).

6.3 ECO-FRIENDLY TECHNOLOGIES FOR WASTEWATER TREATMENT

Environmental contamination (organic and inorganic) has emerged as one of the most important challenges facing the world today. As a result, renewable technology and eco-friendly strategies are in desperate need of acceptance in order to preserve a healthy human community and environment (Premkumar et al., 2013).

Toxic and dangerous chemicals are being removed from polluted areas by using variety of methods. Many advanced remediation techniques, such as physical, chemical and biological, are in operation today, but they are inadequate to protect the environment (Sharma et al., 2018). Traditional physical and chemical methods of pollution removal, including ozonation, sedimentation, advanced chemical oxidation and adsorption, are commonly utilized. Unfortunately, because of their disadvantages such as production of toxic by-products, usage of hazardous material, the need for specialized equipment, high overall costs or even incompatibility with the elimination of all phenolic compounds, more research has been dedicated to the treatment of these substances by biological technique (Ontanon et al., 2017). The use of physicochemical methods for the treatment and maintenance of toxic environments is popular, but these approaches are environmentally unfriendly, due to the production of secondary contaminants. In general, for wastewater treatment, physical and chemical methods are used, which have many disadvantages. The treatment of wastewater by these methods is that without treatment they can cause neurological and cancerous diseases in humans and create many problems for human health and can also create unknown deleterious effects in the environment (Dearfield et al., 1988; Mallevialle et al., 1984; Özacar and Şengil, 2003). It is because they may successfully remove phenolic pollutants in line with green chemistry standards, under moderate process conditions, without the use of extra hazardous chemicals, and the creation of toxic by-products and sludge may be avoided (Ontanon et al., 2017; Zdarta et al., 2021).

Environmentally safe methods, such as bioremediation techniques, may be a long-term alternative for the prevention and disposal of polluted environment (Bharagava, 2020; Mallevialle et al., 1984). Biological pollution removal methods that are used in laboratory and large-scale applications can be split into two categories:

- treatment with microorganisms and
- treatment with enzymes (both in immobilized and free form) (Meena et al., 2021) (Table 6.1)

Living creatures have been regarded as the primary source of physiologically active enzymes. More than half of the enzymes used in industry are derived from fungi and yeast, more than a third from bacteria and the remainder from animal and plant sources.

TABLE 6.1
Treatment Technologies for Wastewater Pollution (Russo et al., 2020).

Advanced oxidation processes	Phase changing technologies	Biological treatments
Advanced oxidation processes	Phase-changing technologies	Biological treatments
UV	Adsorption using activated carbon	Microorganism treatment
Ozone	Adsorption using biochar	Enzymatic treatment
Ozone/H_2O_2	Adsorption by clay mineral	
Fenton process	Membrane technology	
Photo process	Other sorbent	
Sono chemical	Adsorption in carbon nanotube	
UV/Ozone		
UV/H_2O_2		

Microbial sources of enzymes are chosen over nonmicrobial sources for a variety of industrial applications because their production costs are lower, raw materials are readily available and they are of stable composition as needed, with fast microbial growth and cellular metabolism. Nonmicrobial sources, including animal and plant tissues, contain or produce more potentially hazardous chemicals than bacteria (Anbu et al., 2015; Kołodziejczak-Radzimska et al., 2021).

In wastewater treatment with microorganisms because of the wide diversity of microbial species involved, the process can be performed in aerobic, anaerobic or mixed aerobic–anaerobic environments. There is oxygen in aerobic treatments, but there is no oxygen in anaerobic treatment, which results in the creation of very toxic aromatic amines (Ali et al., 2017; Zhu et al., 2020). On the other hand, aerobic conversion produces less harmful compounds but requires a longer processing period. Bacterial microorganisms produce and activate sludge that can remove contaminants through both conversion and adsorption. Many microorganisms are widely utilized in the removal of contaminants from wastewaters. *Pseudomonas* and *Streptomyces* are the most commonly utilized bacteria for wastewater treatment, whereas white rot fungal species such as *Trametes versicolor* and *Pleurotus ostreatus* are the most commonly used fungi (Ali et al., 2017). Microbiological treatments are successful in removing the majority of contaminants found in wastewater, but they are ineffective in removing phenolic compounds such as medicines, dyes and phenols and pharmaceuticals. These bacterial microorganisms produce activated sludge capable of removing contaminants not only via conversion but also through adsorption of hazardous contaminants (Thwaites et al., 2018). This is because of the compounds' resistance to microbiological treatment, insufficiency of enzymes in microbial cells and diffusional constraints in substrate transport (Thwaites et al., 2018; Zdarta et al., 2021).

Enzyme-mediated processes are fast gaining popularity in a variety of sectors (Choi et al., 2015; Li et al., 2012). Enzymatic treatment has created new possibilities for wastewater treatment. In terms of biodegradability, environmentally friendly, and high precision and selectivity, the enzymatic solution outperforms traditional approaches (Gholami-Borujeni et al., 2011b). Recently, there has been a surge of interest in the production of immobilized enzyme technologies, which are more cost-effective, efficient and environmentally sustainable (Jun et al., 2019). Enzymatic systems can save energy (Villegas et al., 2016), and they are one of the environmentally friendly methods for wastewater treatment; in contrast to inorganic catalysts, enzymes have some significant advantages (Hermes et al., 1987). Enzymes can work under moderate conditions such as neutral pH, ambient temperature and atmospheric pressure, create less waste and induce fewer lateral reactions; as a result, they can be an effective and sustainable alternative to environmental chemicals and inorganic catalysts (Liu and Smith, 2020). Enzyme-based processes also have a number of additional advantages, which are low toxicity, the capacity to function in moderate circumstances, a lower volume of sludge and lower by-product creation (Ivanov et al., 2019; Jäger and Croft, 2018). In general, enzymatic system is superior to other methods. Unfortunately, using free enzymes for such applications on an industrial scale has significant challenges, including low biocatalyst stability under severe process conditions and highly restricted reusability of the free enzymes (Jesionowski et al., 2014; Zdarta et al., 2019). Immobilized enzymes are often utilized and researched in enzymatic systems (Chang et al., 2015; Fan et al., 2018; Quintanilla-Guerrero et al., 2008).

The immobilization of enzymes leads to the creation of heterogeneous catalysts that are more stable under severe process conditions. Furthermore, when enzymes are immobilized, their storage stability and operability improve dramatically. Finally, the most significant benefit of the immobilized enzyme is improved enzyme recycling and reusability, which boosts biocatalytic productivity, reduces further pollution and lowers overall costs (Bilal et al., 2019; Chang et al., 2015; Fan et al., 2018; Li et al., 2019; Quintanilla-Guerrero et al., 2008; Zdarta et al., 2021). It is worth noting that in these systems, immobilization techniques and support materials have a significant impact on the activity of enzymes, treatment efficiency, enzymatic kinetics and cost of enzymatic operations (Quintanilla-Guerrero et al., 2008).

6.4 THE ADVANTAGES OF ENZYMATIC SYSTEMS OVER OTHER METHODS IN WASTEWATER TREATMENT

Enzymes are biological reagents that operate as catalytic agents in the conversion of compounds into products, therefore lowering the activation energy of the process (Rao et al., 2014). The ever-increasing use of enzymes is driving up demand for biocatalysts with better or novel properties (De Carvalho, 2011). Enzymes may eliminate recalcitrant contaminants through precipitation or turn them into other materials by operating on them directly. Enzymatic systems have a distinct advantage over conventional chemical or physical approaches in that they are considered safe and environmentally friendly (Feng et al., 2013; Liu and Smith, 2020; Nicell et al., 2003). The advantages of different physicochemical treatments include temperature flexibility and ease of control, but there are also significant drawbacks that include high energy demand and chemical costs (Kurniawan et al., 2006). In general, various factors have developed many effective environmental methods, such as high cost of chemical treatment to remove pollutants, producing solid waste through other methods, and also government rules that set strict limits on the amount of pollution that should be allowed in the environment. Enzymes are catalysts with a high specificity and performance (Nelson and Cox, 2004) and they are more environmentally friendly due to their biodegradability (Funabashi et al., 2017; Tischer and Wedekind, 1999). Enzymatic treatment, inspired by biological treatment, employs a biocatalyst, which results in their isolation from water. This procedure has a huge advantage in that it is much less expensive than conventional biological and chemical treatments (Steevensz et al., 2014). This method can be used as a stand-alone treatment or in conjunction with a biological method. Unlike biological treatment, this technique entails polymerization of the target molecules, rather than degradation, before the materials exceed their solubility limits and precipitate (Villegas et al., 2016). In total, the enzymatic advantage with other methods is that they can reduce its negative environmental effects (Funabashi et al., 2017), be ready for the next catalytic cycle (Pandey et al., 2017) and it's a cost-effective and safe approach (Karigar and Rao, 2011). Another advantage that enzymes have is the recovery of useful and advantageous substances from industrial wastewater (Al-Mutwalli et al., 2021). As a result, it is an environmentally friendly strategy for wastewater treatment. Figure 6.3 and Table 6.2 show the advantages of enzymatic systems over other methods in wastewater treatment.

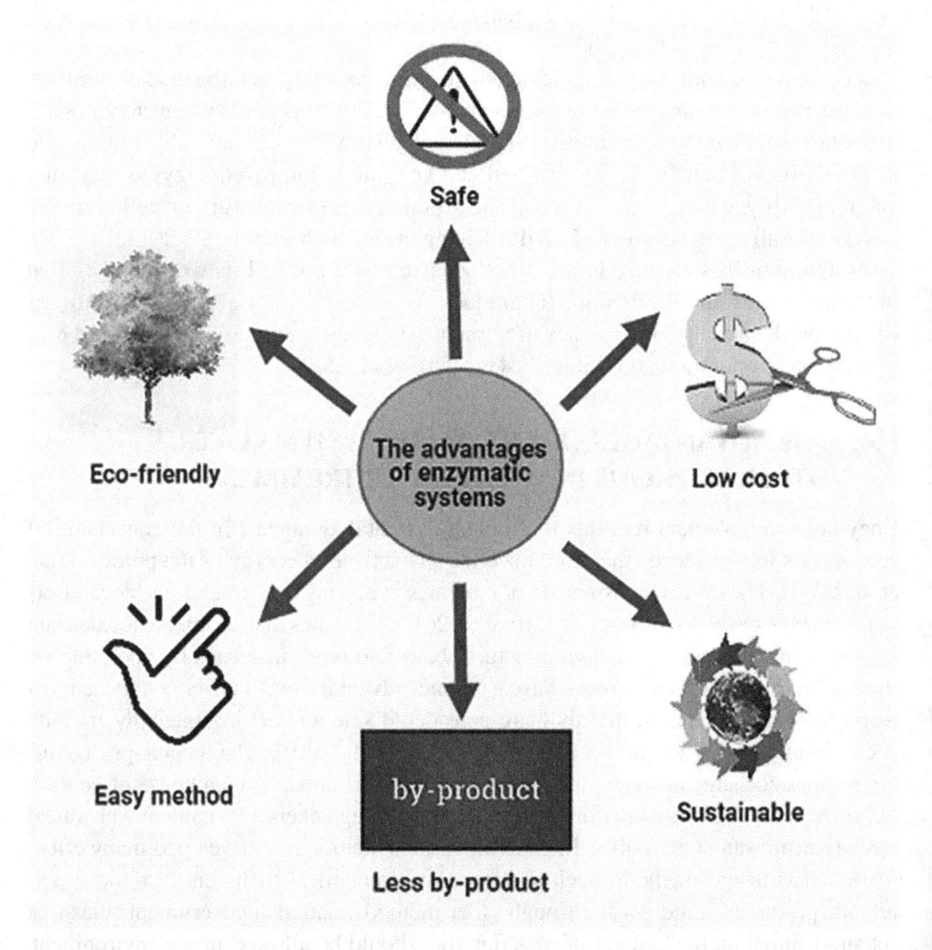

FIGURE 6.3 Some advantages of enzymatic systems over other methods in wastewater treatment.

TABLE 6.2
Advantages and Disadvantages of Wastewater Treatment Methods

Methods	Advantages	Disadvantages
Physical	– High performance for removing pollutants – Used one kind of waste to treat another type of waste – Excellent adsorbent	– High dissolved oxygen demand – Colourants that are light resistant are inefficient – Disposal of waste after treatment
Chemical	– Excellent removal systems – Low operating costs	– Publishing aromatic – High cost – Incompatible for scatter dyes – Not eco-friendly
Biological	– Colour elimination in conjunction with biological oxygen requirement – Removing the chemical demand for oxygen	– Long maintenance period – Poor removal ability for recalcitrant contaminants
Enzymatic system	– Reduce negative environmental effect – Eco-friendly – Safe approach – High performance – Does not produce by-products – Be used several times – Operation over a wide range of pH – Activity at high and low contaminant levels – Operation over a broad temperature and salinity – Controllability and simplicity of the operation – Reduction of sludge volume without the generation of biomass	– Cost of enzymes – Inhibition of the substance

6.5 APPLICATION OF ENZYMATIC SYSTEMS FOR WASTEWATER TREATMENT

Enzymes are incredibly effective and natural catalysts that can accelerate the convention of target compounds while causing no harm to other molecules (Al-Mutwalli et al., 2021; Yao et al., 2020). In comparison to other chemical catalysts, enzymes have several desirable qualities such as high efficiency, high selectivity and functioning in moderate conditions (Gholami-Borujeni et al., 2011a; Torres et al., 2017). Enzymes have a low environmental impact since they are biodegradable, owing to their natural origins (Mojsov et al., 2016). Enzymes might lose their functioning if the environment becomes inappropriate because enzymes are highly sensitive to changes in the local environment, such as pH and temperature (Jaiswal et al., 2016).

In recent years, enzymes have gained a significant reputation for their potential to treat or eliminate various contaminants, especially in wastewater treatment (Arca-Ramos et al., 2015) and they have different advantages in wastewater treatment than microorganisms. Relevant enzymatic systems can be constructed with the aim of achieving long-term enzymatic degradation in order to produce a promising wastewater treatment (Al-Mutwalli et al., 2021). Different enzymes play an important role in wastewater treatment (Pant and Adholeya, 2007). Enzymatic treatment, like other wastewater treatment methods, can be suitable in some cases (Aitken, 1993). Enzymatic treatment is suitable for wastewater treatment because of particular compounds; it is efficient, with a shorter expected touch time and shock loadings have no effect on it (Premkumar et al., 2018). The ability of bacteria and fungi to biologically convert and destroy xenobiotic compounds in sludge, waste, soil or water is used in bioremediation purification processes where enzymes catalyse intracellular and extracellular metabolic reactions (Alcalde et al., 2006; Dua et al., 2002; Gavrilescu et al., 2015). Since enzymes are the biocatalysts in these reactions, the use of separate enzyme preparations rather than whole organism structures tends to be a more creative and systematic approach (Stadlmair et al., 2018). Various enzymes have different applications in wastewater treatment, including treatment of oily wastewater (Jamie et al., 2016), removing detergent (Jayasekara and Ratnayake, 2019), degradation (Bilal et al., 2018), decolorization (Chiong et al., 2016), treatment of saline wastewater (Sivaprakasam et al., 2011) and removing micropollutants (Zhou et al., 2020). Lipase is one of the enzymes that play a very important role in wastewater treatment. Lipase is a very ubiquitous enzyme that is produced from a variety of natural sources, including microorganisms, plants and animals (Bharathiraja et al., 2014). Microbial lipases are one of the most commonly utilized because of their unique qualities, which include high stability in a variety of physicochemical environments, selectivity, substrate specificity and ease of manufacturing and separation (Szymczak et al., 2021). Lipase can break down triglycerol into a glycerol and fatty acids, in wastewater treatment (Nimkande and Bafana, 2022). Cellulases are inducible enzymes produced by a wide range of microorganisms during their development on cellulosic materials, including fungus, bacteria, and actinomycetes where they can break down complex cellulosic products into simple sugars (Khan et al., 2016; Kubicek, 1993; Lee et al., 2001; Pei et al., 2010). In addition, laccases are generated by a wide range of organisms (Gasser et al., 2014; Giardina et al., 2010); they are multicopper oxidases with at least one copper atom (Ba and Vinoth Kumar, 2017; Gasser et al., 2014). Laccase enzymes are potentially useful for decreasing phenol contamination in wastewater treatment (Lante et al., 2000; Madhavi and Lele, 2009). Furthermore, oxygenases are important for the breakdown of aromatic substances in wastewater (Jadeja et al., 2014). Oxidase plays a critical part in the activated sludge process since most wastewater treatment plants use the aerobic route of degradation (Jadeja et al., 2014; Phale et al., 2007). Monooxygenases (enzymes that catalyse the insertion of a single oxygenase atom) and dioxygenases (enzymes that catalyse the insertion of both oxygenase atoms) are the two forms of oxygenases (Kapley et al., 2001; Kulakov et al., 1998; Selvakumaran et al., 2011; Sharma et al., 2012). The first step in degradation is to oxidize the aromatic ring, rendering it more vulnerable to cleavage by ring-cleaving dioxygenases (Eltis and Bolin, 1996; Jadeja et al., 2014; Phale et al., 2007). In Figure 6.4, some enzymes and their functions are presented in wastewater treatment (Sharma et al., 2018).

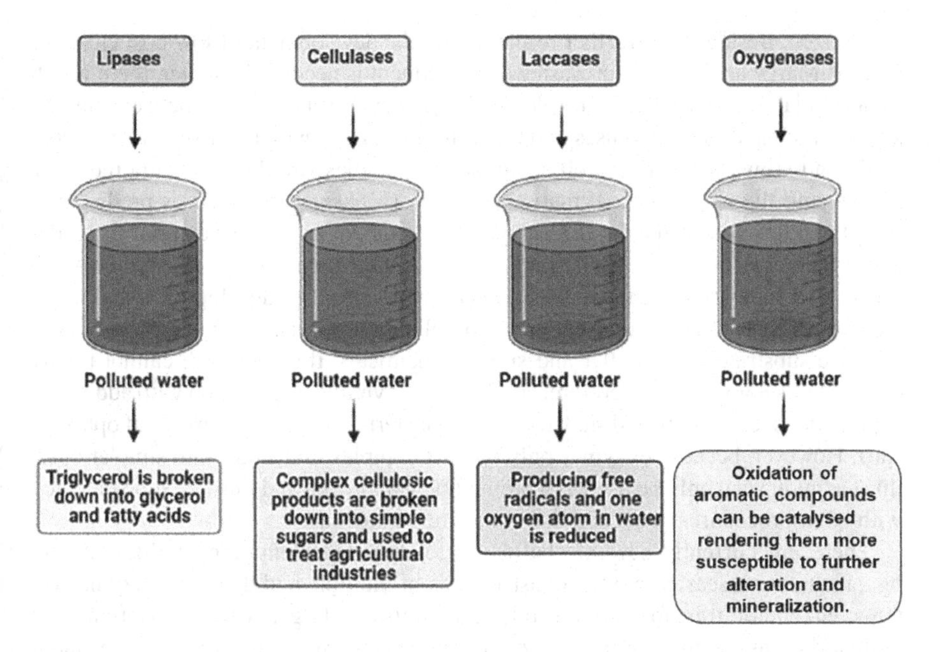

FIGURE 6.4 Some enzymes and their functions in wastewater treatment.

6.6 CONCLUSION AND FUTURE PERSPECTIVES

Pollutant concentration in the earth has reached an unprecedented degree in recent years as a result of urban development, population growth and industrial production. During the past few years, bioremediation methods, in which biological agents are used to remove residual toxins, have gained popularity. During the past year, the majority of wastewater treatment systems have been found to be limited by their inherent constraints. Using enzymes is a good and cost-effective choice for one of the eco-friendly solutions to this issue. As previously stated, enzymatic methods have a lot of possibilities or excellent potential in wastewater treatment. Enzyme engineering is closely related to three key and essential disciplines of chemical biotechnology that include single-step biocatalysis, metabolic engineering and enzymatic cascades. Enzyme types and resources are important in enzymatic reactions. As a result, additional research is needed to investigate the effectiveness of enzymes on particular wastewater. Enzyme-based wastewater treatment has demonstrated its efficacy in the removal of resistant contaminants. Relevant recalcitrant contaminants may be removed or converted to less toxic items by enzymes. Enzymes, for their ability to withstand extreme environments, green and clean technologies and their biodegradability, are well-known. The use of an enzymatic approach to treat wastewater will be more practical and environmentally sustainable. To date, most research on the use of enzymes has been carefully controlled and the use of enzymes in wastewater treatment has a bright future. There is also a need to explore new enzymes and their roles in environmental improvement because many microflora in the environment have not yet been recognized. Besides that, further research into the development of low-cost enzymes and their large applications in wastewater treatment is needed. However, there is still a substantial difference between field and laboratory testing, and also between university research and industrial uses and there must be coordination between them. More study on real wastewater, as well as larger-scale studies, are thus critically required. As a result, the use of an enzymatic system in wastewater treatment is predicted to be a breakthrough in the future. Enzymatic processes have the potential to biodegrade refractory pollutants such as insecticides, medications, personal care items, oil, grease and industrial chemicals. As a result, it is critical to develop immobilization techniques that allow several enzymes to collaborate. Because enzymes only convert one substrate into smaller and simpler chemicals, this approach cannot totally eliminate contaminants. Therefore, combining enzymatic degradation with additional approaches such as activated sludge or anaerobic fermentation is an excellent option to start. However, because enzymes only convert complex molecules into simpler ones, this technique can only be used as a pretreatment approach and must be supplemented with other procedures to accomplish complete treatment.

There are currently several challenges to the implementation of this technology, including research into the constructions of multiple and different enzyme systems, enzymatic transformation products monitoring, larger-scale applications and real-world wastewater, and also enzymes' ability augmentation. In addition, more improved approaches and technologies will be useful in the future for researching and inventing novel enzymes with desirable properties derived from natural resources, as well as modifying the current repertoire of enzymes for desired uses. As a result, in the future, a smart combination of sophisticated experimental and analytical techniques of advanced enzyme systems, and related disciplines will eventually aid in the development of designer enzymes with stable structures, broad substrate uniqueness, innovative and multipurpose actions for desired wastewater treatment. It is better to reduce the cost of enzyme production by creating new technologies or optimize existing methods, improve process in order to develop extremely active and long-term stable biocatalytic systems, and use innovative, custom-made support materials to increase enzyme activity and utilization. Enzymatic reactors and enzymatic membrane bioreactors with immobilized biocatalysts like catalytic beads are used for long-term treatment of models and actual water treatments. Evaluation of innovative and promising technologies is needed for more effective wastewater treatment, including enzymatic conversion, biodegradation and simultaneous adsorption by photocatalytic activity. Further research into the uses of enzymes extracted from marine organisms in wastewater treatment is needed in the future. Consequently, a deeper knowledge on the chemical structure of marine enzymes will be required.

REFERENCES

Acharya, K., Blackburn, A., Mohammed, J., Haile, A.T., Hiruy, A.M., Werner, D., 2020. Metagenomic water quality monitoring with a portable laboratory. *Water research* 184, 116112.

Adam, W., Lazarus, M., Saha-Möller, C.R., Weichold, O., Hoch, U., Häring, D., Schreier, P., 1999. Biotransformations with peroxidases. *Biotransformations* 63, 73–108.

Aitken, M.D., 1993. Waste treatment applications of enzymes: opportunities and obstacles. *The Chemical Engineering Journal* 52, B49–B58.

Al-Mutwalli, S.A., Korkut, S., Kilic, M.S., Imer, D.Y., 2021. *Enzymatic Degradation of Industrial Wastewater Pollutants, Removal of Emerging Contaminants through Microbial Processes*. Springer, pp. 373–398.

Alcalde, M., Ferrer, M., Plou, F.J., Ballesteros, A., 2006. Environmental biocatalysis: from remediation with enzymes to novel green processes. *TRENDS in Biotechnology* 24, 281–287.

Aleya, L., Uddin, M., 2020. *Environmental Pollutants and the Risk of Neurological Disorders*. Springer, pp. 44657–44658.

Ali, S., Rizwan, M., Ibrahim, M., Nafees, M., Waseem, M., 2017. Role of bioremediation agents (bacteria, fungi, and algae) in alleviating heavy metal toxicity, *Probiotics in Agroecosystem*. Springer, pp. 517–537.

Alneyadi, A.H., Rauf, M.A., Ashraf, S.S., 2018. Oxidoreductases for the remediation of organic pollutants in water-a critical review. *Critical Reviews in Biotechnology* 38, 971–988.

Anas, M., Han, D.S., Mahmoud, K., Park, H., Abdel-Wahab, A.., 2016. Photocatalytic degradation of organic dye using titanium dioxide modified with metal and non-metal deposition. *Materials Science in Semiconductor Processing* 41, 209–218.

Anbu, P., Gopinath, S.C., Chaulagain, B.P., Tang, T.-H., Citartan, M., 2015. *Microbial Enzymes and their Applications in Industries and Medicine 2014*. BioMed Research International. Hindawi.

Ang, W.L., Mohammad, A.W., Hilal, N., Leo, C.P., 2015. A review on the applicability of integrated/hybrid membrane processes in water treatment and desalination plants. *Desalination* 363, 2–18.

Apetrei, I., Rodriguez-Mendez, M., Apetrei, C., De Saja, J.J.S., Chemical, A.B., 2013. Enzyme sensor based on carbon nanotubes/cobalt (II) phthalocyanine and tyrosinase used in pharmaceutical analysis. *Sensors and Actuators B: Chemical* 177, 138–144.

Arca-Ramos, A., Eibes, G., Feijoo, G., Lema, J., Moreira, M., 2015. Potentiality of a ceramic membrane reactor for the laccase-catalyzed removal of bisphenol A from secondary effluents. *Applied Microbiology and Biotechnology* 99, 9299–9308.

Ba, S., Arsenault, A., Hassani, T., Jones, J.P., Cabana, H., 2013. Laccase immobilization and insolubilization: from fundamentals to applications for the elimination of emerging contaminants in wastewater treatment. *Critical Reviews in Biotechnology* 33, 404–418.

Ba, S., Vinoth Kumar, V., 2017. Recent developments in the use of tyrosinase and laccase in environmental applications. Current Research in Biotechnology 37, 819–832.

Barber, E.A., Liu, Z., Smith, S.R., 2020. Organic contaminant biodegradation by oxidoreductase enzymes in wastewater treatment. Journal of Microbiology 8, 122.

Bharagava, R.N., 2020. *Emerging Eco-friendly Green Technologies for Wastewater Treatment*. Berlin, Heidelberg, Germany: Springer.

Bharathiraja, B., Chakravarthy, M., Kumar, R.R., Yuvaraj, D., Jayamuthunagai, J., Kumar, R.P., Palani, S., 2014. Biodiesel production using chemical and biological methods-A review of process, catalyst, acyl acceptor, source and process variables. Renewable and Sustainable Energy Reviews 38, 368–382.

Bhatia, R.K., Sakhuja, D., Mundhe, S., Walia, A.., 2020. Renewable energy products through bioremediation of wastewater. *Sustainability* 12, 7501.

Bilal, M., Iqbal, H.M., 2019. Persistence and impact of steroidal estrogens on the environment and their laccase-assisted removal. *Science of the Total Environment* 690, 447–459.

Bilal, M., Rasheed, T., Iqbal, H.M., Hu, H., Wang, W., Zhang, X., 2018. Horseradish peroxidase immobilization by copolymerization into cross-linked polyacrylamide gel and its dye degradation and detoxification potential. *International Journal of Biological Macromolecules* 113, 983–990.

Bilal, M., Adeel, M., Rasheed, T., Zhao, Y., Iqbal, H.M., 2019. Emerging contaminants of high concern and their enzyme-assisted biodegradation-a review. *Environmental International* 124, 336–353.

Bilal, M., Ashraf, S.S., Cui, J., Lou, W.-Y., Franco, M., Mulla, S.I., Iqbal, H.M.., 2021. Harnessing the biocatalytic attributes and applied perspectives of nanoengineered laccases-a review. *International Journal of Biological Macromolecules* 166, 352–373.

Bilal, M., Iqbal, H., Barceló, D., 2020a. *Perspectives on the Feasibility of using Enzymes for Pharmaceutical Removal in Wastewater, Removal and Degradation of Pharmaceutically Active Compounds in Wastewater Treatment.* Springer, Berlin, pp. 119–143.

Bilal, M., Mehmood, S., Rasheed, T., Iqbal, H.M.J., 2020b. Antibiotics traces in the aquatic environment: persistence and adverse environmental impact. Current Opinion in Environmental Science & Health 13, 68–74.

Chang, Q., Jiang, G., Tang, H., Li, N., Huang, J., Wu, L.J., 2015. Enzymatic removal of chlorophenols using horseradish peroxidase immobilized on superparamagnetic Fe3O4/graphene oxide nanocomposite. *Chinese Journal of Catalysis* 36, 961–968.

Chiong, T., Lau, S.Y., Lek, Z.H., Koh, B.Y., Danquah, M.K., 2016. Enzymatic treatment of methyl orange dye in synthetic wastewater by plant-based peroxidase enzymes. *Journal of Environmental Chemical Engineering* 4, 2500–2509.

Choi, J.-M., Han, S.-S., Kim, H.-S., 2015. Industrial applications of enzyme biocatalysis: current status and future aspects. *Biotechnology Advances* 33, 1443–1454.

Dambmann, C., Aunstrup, K., 1981. *The Variety of Serine Proteases and Their Industrial Significance, Proteinases and Their Inhibitors.* Elsevier, Pergamon, pp. 231–244.

Das, A., Dey, A., 2020. P-Nitrophenol-Bioremediation using potent Pseudomonas strain from the textile dye industry effluent. *Journal of Environmental Chemical Engineering* 8, 103830.

De Carvalho, C.C., 2011. Enzymatic and whole cell catalysis: finding new strategies for old processes. *Biotechnology Advances* 29, 75–83.

de Vidales, M.J.M., Nieto-Márquez, A., Morcuende, D., Atanes, E., Blaya, F., Soriano, E., Fernández-Martínez, F.J., 2019. 3D printed floating photocatalysts for wastewater treatment. *Chemical Engineering Journal* 328, 157–163.

Dearfield, K.L., Abernathy, C.O., Ottley, M.S., Brantner, J.H., Hayes, P.F., 1988. Acrylamide: its metabolism, developmental and reproductive effects, genotoxicity, and carcinogenicity. *Mutation Research/Reviews in Genetic Toxicology* 195, 45–77.

Dua, M., Singh, A., Sethunathan, N., Johri, A., 2002. Biotechnology and bioremediation: successes and limitations. *Applied Microbiology and Biotechnology* 59, 143–152.

Edokpayi, J.N., Makungo, R., Mathivha, F., Rivers, N., Volenzo, T., Odiyo, J.O., 2020. Influence of global climate change on water resources in South Africa: toward an adaptive management approach. In: *Water Conservation and Wastewater Treatment in BRICS Nations.* Elsevier, Amsterdam, pp. 83–115.

Eltis, L.D., Bolin, J.T., 1996. Evolutionary relationships among extradiol dioxygenases. *Journal of Bacteriology* 178, 5930–5937.

Fan, X., Hu, M., Li, S., Zhai, Q., Wang, F., Jiang, Y.J., 2018. Charge controlled immobilization of chloroperoxidase on both inner/outer wall of NHT: improved stability and catalytic performance in the degradation of pesticide. *Applied Catalysis B: Environmental* 163, 92–99.

Feng, W., Taylor, K.E., Biswas, N., Bewtra, J.K., 2013. Soybean peroxidase trapped in product precipitate during phenol polymerization retains activity and may be recycled. *Journal of Chemical Technology & Biotechnology* 88, 1429–1435.

Funabashi, H., Takeuchi, S., Tsujimura, S., 2017. Hierarchical meso/macro-porous carbon fabricated from dual MgO templates for direct electron transfer enzymatic electrodes. *Scientific reports* 7, 1–9.

Gasser, C.A., Ammann, E.M., Shahgaldian, P., Corvini, P.F.-X., 2014. Laccases to take on the challenge of emerging organic contaminants in wastewater. *Applied Microbiology and Biotechnology* 98, 9931–9952.

Gavrilescu, M., Demnerová, K., Aamand, J., Agathos, S., Fava, F., 2015. Emerging pollutants in the environment: present and future challenges in biomonitoring, ecological risks and bioremediation. *New Biotechnology* 32, 147–156.

Gholami-Borujeni, F., Mahvi, A.H., Naseri, S., Faramarzi, M.A., Nabizadeh, R., Alimohammadi, M., 2011a. Application of immobilized horseradish peroxidase for removal and detoxification of azo dye from aqueous solution. *Journal of Chemical Engineering* 15, 217–222.

Gholami-Borujeni, F., Mahvi, A.H., Nasseri, S., Faramarzi, M.A., Nabizadeh, R., Alimohammadi, M., 2011b. Enzymatic treatment and detoxification of acid orange 7 from textile wastewater. *Applied Biochemistry and Biotechnology* 165, 1274–1284.

Gianfreda, L., Bollag, J., 2002. Isolated enzymes for the transformation and detoxification of organic pollutants. In: *Enzymes in the Environment: Activity, Ecology, and Applications*, Marcel Dekker, New York, pp. 495–538.

Giardina, P., Faraco, V., Pezzella, C., Piscitelli, A., Vanhulle, S., Sannia, G.J.C., Sciences, M.L., 2010. Laccases: a never-ending story. *Cellular and Molecular Life Sciences* 67, 369–385.

Hermes, J., Blacklow, S., Knowles, J., 1987. *The Development of Enzyme Catalytic Efficiency: An Experimental Approach, Cold Spring Harbor Symposia on Quantitative Biology*. Cold Spring Harbor Laboratory Press, pp. 597–602.

Ho, K.C., Teow, Y.H., Sum, J.Y., Ng, Z.J., Mohammad, A.W., 2021. Water pathways through the ages: integrated laundry wastewater treatment for pollution prevention. *Science of the Total Environment* 760, 143966.

Husain, Q., Ulber, R., 2011. Immobilized peroxidase as a valuable tool in the remediation of aromatic pollutants and xenobiotic compounds: a review. *Critical Reviews in Environmental Science and Technology* 41, 770–804.

Ismail, B., Nielsen, S.J., 2010. Invited review: plasmin protease in milk: current knowledge and relevance to dairy industry. *Journal of Dairy Science* 93, 4999–5009.

Ivanov, V., Stabnikov, V., Stabnikova, O., Kawasaki, S., 2019. Environmental safety and biosafety in construction biotechnology. *Journal of Waterway, Port, Coastal, and Ocean Engineering* 35, 1–11.

Jadeja, N.B., More, R.P., Purohit, H.J., Kapley, A., 2014. Metagenomic analysis of oxygenases from activated sludge. *Bioresource Technology* 165, 250–256.

Jäger, C.M., Croft, A.K., 2018. Anaerobic radical enzymes for biotechnology. *Chemical Reviews* 5, 143–162.

Jaiswal, N., Pandey, V.P., Dwivedi, U.N., 2016. Immobilization of papaya laccase in chitosan led to improved multipronged stability and dye discoloration. *International Journal of Biological Macromolecules* 86, 288–295.

Jamie, A., Alshami, A.S., Maliabari, Z.O., Ali Ateih, M., Al Hamouz, O.C.S., 2016. Immobilization and enhanced catalytic activity of lipase on modified MWCNT for oily wastewater treatment. *Environmental Progress & Sustainable Energy* 35, 1441–1449.

Jayasekara, S., Ratnayake, R., 2019. Microbial cellulases: an overview and applications. *Cellulose*. IntechOpen: London, UK 1-21

Jesionowski, T., Zdarta, J., Krajewska, B.J.A., 2014. Enzyme immobilization by adsorption: a review. *Adsorption* 20, 801–821.

Jiang, Y., Carrijo, D., Huang, S., Chen, J., Balaine, N., Zhang, W., van Groenigen, K.J., Linquist, B., 2019. Water management to mitigate the global warming potential of rice systems: a global meta-analysis. *Field Crops Research* 234, 47–54.

Jun, L.Y., Yon, L.S., Mubarak, N., Bing, C.H., Pan, S., Danquah, M.K., Abdullah, E., Khalid, M., 2019. An overview of immobilized enzyme technologies for dye and phenolic removal from wastewater. *Journal of Environmental Chemical Engineering* 7, 102961.

Kadam, R.L., Kim, Y., Gaikwad, S., Chang, M., Tarte, N.H., Han, S.J.C., 2020. Catalytic decolorization of Rhodamine B, Congo red, and crystal violet dyes, with a novel Niobium Oxide anchored Molybdenum (Nb-O-Mo). *Catalysts* 10, 491.

Kallel, A., Ksibi, M., Ben Dhia, H., Khélifi, N.J., 2020. *Pollutant Removal and the Health Effects of Environmental Pollution*. Springer, Tunisia, pp. 23375–23378.

Kalogerakis, N., Fava, F., Corvini, P., 2017. 6th European Bioremediation Conference, Chania, Crete, Greece, 29 June–2 July, 2015. *New Biotechnology* 38, 41–106.

Kapley, A., Purohit, H.J., 2001. Tracking of phenol degrading genotype. *Journal of Environmental Sciences* 8, 89–90.

Karigar, C.S., Rao, S.S., 2011. Role of microbial enzymes in the bioremediation of pollutants: a review. *Enzyme Research* 2011, 2011.

Karri, R.R., Sahu, J., Jayakumar, N., 2017. Optimal isotherm parameters for phenol adsorption from aqueous solutions onto coconut shell based activated carbon: error analysis of linear and non-linear methods. *Journal of the Taiwan Institute of Chemical Engineers* 80, 472–487.

Karthik, V., Senthil Kumar, P., Vo, D.-V.N., Selvakumar, P., Gokulakrishnan, M., Keerthana, P., Audilakshmi, V., Jeyanthi, J., 2021. Enzyme-loaded nanoparticles for the degradation of wastewater contaminants: a review. *Journal of Environmental Chemical Engineering* 19, 2331–2350.

Kaushal, J., Mehandia, S., Singh, G., Raina, A., Arya, S.K., 2018. Catalase enzyme: application in bioremediation and food industry. *Journal of Biotechnology and Applied Biochemistry* 16, 192–199.

Khan, M.N., Luna, I.Z., Islam, M.M., Sharmeen, S., Salem, K.S., Rashid, T.U., Zaman, A., Haque, P., Rahman, M.M., 2016. *Cellulase in Waste Management Applications, New and Future Developments in Microbial Biotechnology and Bioengineering*. Elsevier, Netherlands, pp. 237–256.

Kılıç, N., Nasiri, F., Cansaran-Duman, D., 2016. *Fungal Laccase Enzyme Applications in Bioremediation of Polluted Wastewater, Phytoremediation*. Springer, Cham, pp. 201–209.

Kołodziejczak-Radzimska, A., Nghiem, L.D., Jesionowski, T., 2021. Functionalized materials as a versatile platform for enzyme immobilization in wastewater treatment. *Chemical Papers* 7, 263–276.

Kubicek, C.Á., 1993. From cellulose to cellulase inducers: facts and fiction, *Proceedings of the Second TRICEL Symposium on Trichoderma reesei Cellulases and Other Hydrolytic Enzymes. Foundation of Biotechnical and Industrial Fermentation Research,* Espoo Finland, p. 188.

Kulakov, L.A., Delcroix, V.A., Larkin, M.J., Ksenzenko, V.N., Kulakova, A.N.J.M., 1998. Cloning of new Rhodococcus extradiol dioxygenase genes and study of their distribution in different Rhodococcus strains. *Journal of Microbiology* 144, 955–963.

Kurniawan, T.A., Chan, G.Y., Lo, W.-H., Babel, S., 2006. Physico-chemical treatment techniques for wastewater laden with heavy metals. *Chemical Engineering Journal* 118, 83–98.

Lante, A., Crapisi, A., Krastanov, A., Spettoli, P., 2000. Biodegradation of phenols by laccase immobilised in a membrane reactor. *Process Biochemistry* 36, 51–58.

Lee, B.C., Bae, J.T., Pyo, H.B., Choe, T.B., Kim, S.W., Hwang, H.J., Yun, J.W., 2003. Biological activities of the polysaccharides produced from submerged culture of the edible Basidiomycete Grifola frondosa. *Enzyme and Microbial Technology* 32, 574–581.

Lee, S.-M., Koo, Y.-M., 2001. Pilot-scale production of cellulase using Trichoderma Reesei rut C-30 fed-batch mode. *Journal of Microbiology and Biotechnology* 11, 229–233.

Leonard, P., Hearty, S., Brennan, J., Dunne, L., Quinn, J., Chakraborty, T., O'Kennedy, R., 2003. Advances in biosensors for detection of pathogens in food and water. *Enzyme and Microbial Technology* 32, 3–13.

Li, L.-J., Xia, W.-J., Ma, G.-P., Chen, Y.-L., Ma, Y.-Y., 2019. A study on the enzymatic properties and reuse of cellulase immobilized with carbon nanotubes and sodium alginate. *Journal of Analytical and Environmental Chemistry* 9, 1–8.

Li, S., Yang, X., Yang, S., Zhu, M., Wang, X., 2012. Technology prospecting on enzymes: application, marketing and engineering. *Journal of Biological Chemistry* 2, e201209017.

Lingamdinne, L.P., Kim, I.-S., Ha, J.-H., Chang, Y.-Y., Koduru, J.R., Yang, J.-K., 2017. Enhanced adsorption removal of Pb (II) and Cr (III) by using nickel ferrite-reduced graphene oxide nanocomposite. *Metals* 7, 225.

Liu, Z., Smith, S.R., 2020. Enzyme recovery from biological wastewater treatment. *Waste and Biomass Valorization*, 12, 1–27.

Madhavi, V., Lele, S., 2009. Laccase: properties and applications. *Bioresource Technology* 4, 1694–1717.

Mallevialle, J., Bruchet, A., Fiessinger, F., 1984. How safe are organic polymers in water treatment? *Journal-American Water Works Association* 76, 87–93.

Meena, M., Sonigra, P., Yadav, G.J., 2021. Biological-based methods for the removal of volatile organic compounds (VOCs) and heavy metals. *Environmental Science and Pollution Research* 28, 2485–2508.

Mojsov, K.D., Andronikov, D., Janevski, A., Kuzelov, A., Gaber, S.J., 2016. The application of enzymes for the removal of dyes from textile effluents. *Journal of Applied Technologies* 5, 81–86.

Morrison, E.S., Shields, M.R., Bianchi, T.S., Liu, Y., Newman, S., Tolic, N., Chu, R.K., 2020. Multiple biomarkers highlight the importance of water column processes in treatment wetland organic matter cycling. *Water Research* 168, 115153.

Morsi, R., Bilal, M., Iqbal, H.M., Ashraf, S.S., 2020. Laccases and peroxidases: the smart, greener and futuristic biocatalytic tools to mitigate recalcitrant emerging pollutants. *Science of the Total Environment* 714, 136572.

Nelson, D.L., Cox, M.M., 2004. *Lehninger Principles of Biochemistry Lecture Notebook.* Macmillan, New York.

Nicell, J.A., 2003. Enzymatic treatment of waters. *Chemical Engineering Journal* 423.

Nimkande, V.D., Bafana, A., 2022. A review on the utility of microbial lipases in wastewater treatment. *Journal of Water Process Engineering* 46, 102591.

Ontanon, O.M., González, P.S., Barros, G.G., Agostini, E., 2017. Improvement of simultaneous Cr (VI) and phenol removal by an immobilised bacterial consortium and characterisation of biodegradation products. *New Biotechnology* 37, 172–179.

Özacar, M., Şengil, İ.A., 2003. Evaluation of tannin biopolymer as a coagulant aid for coagulation of colloidal particles. *Colloids and Surfaces A: Physicochemical and Engineering Aspects* 229, 85–96.

Pandey, K., Singh, B., Pandey, A.K., Badruddin, I.J., Pandey, S., Mishra, V.K., Jain, P.A., 2017. Application of microbial enzymes in industrial waste water treatment. *International Journal of Current Microbiology and Applied Sciences* 6, 1243–1254.

Pant, D., Adholeya, A., 2007. Biological approaches for treatment of distillery wastewater: a review. *Bioresource Technology* 98, 2321–2334.

Parra-Saldivar, R., Castillo-Zacarías, C., Bilal, M., Iqbal, H., Barceló, D., 2020. *Sources of Pharmaceuticals in Water, Interaction and Fate of Pharmaceuticals in Soil-Crop Systems*. Springer, Cham, pp. 33–47.

Pei, H.-y., Hu, W.-r., Liu, Q.-H., 2010. Effect of protease and cellulase on the characteristic of activated sludge. *Journal of Hazardous Materials* 178, 397–403.

Phale, P.S., Basu, A., Majhi, P.D., Deveryshetty, J., Vamsee-Krishna, C., Shrivastava, R., 2007. Metabolic diversity in bacterial degradation of aromatic compounds. *Journal of Industrial Microbiology & Biotechnology* 11, 252–279.

Ponce-Rodríguez, H., Verdú-Andrés, J., Herráez-Hernández, R., Campíns-Falcó, P., 2020. Exploring hand-portable nano-liquid chromatography for in place water analysis: Determination of trimethylxanthines as a use case. *Science of The Total Environment* 747, 140966.

Premkumar, M.P., Kumar, V.V., Kumar, P.S., Baskaralingam, P., Sathyaselvabala, V., Vidhyadevi, T., Sivanesan, S., 2013. Kinetic and equilibrium studies on the biosorption of textile dyes onto Plantago ovata seeds. *Korean Journal of Chemical Engineering* 30, 1248–1256.

Premkumar, M.P., Thiruvengadaravi, K., Kumar, P.S., Nandagopal, J., Sivanesan, S., 2018. *Eco-Friendly Treatment Strategies for Wastewater Containing Dyes and Heavy Metals, Environmental Contaminants*. Springer, Singapore, pp. 317–360.

Quintanilla-Guerrero, F., Duarte-Vázquez, M., García-Almendarez, B., Tinoco, R., Vazquez-Duhalt, R., Regalado, C., 2008. Polyethylene glycol improves phenol removal by immobilized turnip peroxidase. *Journal of Biotechnology* 99, 8605–8611.

Rao, M., Scelza, R., Acevedo, F., Diez, M., Gianfreda, L.J.C., 2014. Enzymes as useful tools for environmental purposes. *Chemical Engineering Journal* 107, 145–162.

Rubilar, R., Hubbard, R., Emhart, V., Mardones, O., Quiroga, J.J., Medina, A., Valenzuela, H., Espinoza, J., Burgos, Y., Bozo, D., 2020. Climate and water availability impacts on early growth and growth efficiency of eucalyptus genotypes: the importance of GxE interactions. *Forest Ecology and Management* 458, 117763.

Ruggaber, T.P., Talley, J.W., 2006. Enhancing bioremediation with enzymatic processes: a review. *Practice Periodical of Hazardous, Toxic, and Radioactive Waste Management* 10, 73–85.

Russo, V., Hmoudah, M., Broccoli, F., Iesce, M.R., Jung, O.-S., Di Serio, M., 2020. Applications of metal organic frameworks in wastewater treatment: a review on adsorption and photodegradation. *Frontiers in Chemistry and Chemical Engineering* 2:581487.

Sahu, J., Karri, R.R., Zabed, H.M., Shams, S., Qi, X., 2019. Current perspectives and future prospects of nano-biotechnology in wastewater treatment. *Separation & Purification Reviews*, 1–20.

Sanchez, S., Demain, A.L., 2011. Enzymes and bioconversions of industrial, pharmaceutical, and biotechnological significance. *Organic Process Research & Development* 15, 224–230.

Selvakumar, P., Karthik, V., Kumar, P.S., Asaithambi, P., Kavitha, S., Sivashanmugam, P., 2021. Enhancement of ultrasound assisted aqueous extraction of polyphenols from waste fruit peel using dimethyl sulfoxide as surfactant: assessment of kinetic models. *Chemical Engineering Communications* 263, 128071.

Selvakumar, P., Sivashanmugam, P., 2018. Multi-hydrolytic biocatalyst from organic solid waste and its application in municipal waste activated sludge pre-treatment towards energy recovery. *Process Safety and Environmental Protection* 117, 1–10.

Selvakumaran, S., Kapley, A., Kashyap, S.M., Daginawala, H.F., Kalia, V.C., Purohit, H.J., 2011. Diversity of aromatic ring-hydroxylating dioxygenase gene in Citrobacter. *Journal of Biotechnology* 102, 4600–4609.

Sharma, B., Dangi, A.K., Shukla, P., 2018. Contemporary enzyme based technologies for bioremediation: a review. *Journal of environmental management* 210, 10–22.

Sharma, N., Tanksale, H., Kapley, A., Purohit, H.J., 2012. Mining the metagenome of activated biomass of an industrial wastewater treatment plant by a novel method. *Journal of Industrial Microbiology & Biotechnology* 52, 538–543.

Shindhal, T., Rakholiya, P., Varjani, S., Pandey, A., Ngo, H.H., Guo, W., Ng, H.Y., Taherzadeh, M.J., 2021. A critical review on advances in the practices and perspectives for the treatment of dye industry wastewater. *Journal of Bioscience and Bioengineering* 12, 70–87.

Sillanpää, M., Ncibi, M., Matilainen, A., Vepsäläinen, M., 2018. Removal of natural organic matter in drinking water treatment by coagulation: a comprehensive review. *Chemosphere*, v. 190..

Sivaprakasam, S., Dhandapani, B., Mahadevan, S., 2011. Optimization studies on production of a salt-tolerant protease from pseudomonas aeruginosa strain BC1 and its application on tannery saline wastewater treatment. *Brazilian Journal of Microbiology* 42, 1506–1515.

Sivasubramaniam, D., Franks, A.E., 2016. Bioengineering microbial communities: their potential to help, hinder and disgust. *Journal of Biological Engineering* 7, 137–144.

Srivastava, B., Singh, H., Khatri, M., Singh, G., Arya, S.K., 2020. Immobilization of keratinase on chitosan grafted-β-cyclodextrin for the improvement of the enzyme properties and application of free keratinase in the textile industry. *International Journal of Biological Macromolecules* 165, 1099–1110.

Stadlmair, L.F., Letzel, T., Drewes, J.E., Grassmann, J., 2018. Enzymes in removal of pharmaceuticals from wastewater: a critical review of challenges, applications and screening methods for their selection. *Chemosphere* 205, 649–661.

Steevensz, A., Cordova Villegas, L.G., Feng, W., Taylor, K.E., Bewtra, J.K., Biswas, N., 2014. Soybean peroxidase for industrial wastewater treatment: a mini review. *Journal of Environmental Engineering and Science* 9, 181–186.

Surendhiran, D., Sirajunnisa, A., Tamilselvam, K.J., 2017. Silver-magnetic nanocomposites for water purification. *Environmental Chemistry Letters* 15, 367–386.

Szymczak, T., Cybulska, J., Podleśny, M., Frąc, M.J., 2021. Various perspectives on microbial lipase production using agri-food waste and renewable products. *Agriculture* 11, 540.

Thwaites, B.J., Short, M.D., Stuetz, R.M., Reeve, P.J., Gaitan, J.-P.A., Dinesh, N., van den Akker, B., 2018. Comparing the performance of aerobic granular sludge versus conventional activated sludge for microbial log removal and effluent quality: implications for water reuse. *Water Research* 145, 442–452.

Tischer, W., Wedekind, F., 1999. Immobilized enzymes: methods and applications. *Biocatalysis-from Discovery to Application*, 200, 95–126.

Todorova, Y., Yotinov, I., Topalova, Y., Benova, E., Marinova, P., Tsonev, I., Bogdanov, T., 2019. Evaluation of the effect of cold atmospheric plasma on oxygenases' activities for application in water treatment technologies. *Environmental Technology* 40, 3783–3792.

Tonini, D., Astrup, T., 2012. Life-cycle assessment of a waste refinery process for enzymatic treatment of municipal solid waste. *Waste Management* 32, 165–176.

Torres, J., Nogueira, F., Silva, M., Lopes, J., Tavares, T., Ramalho, T., Corrêa, A., 2017. Novel eco-friendly biocatalyst: soybean peroxidase immobilized onto activated carbon obtained from agricultural waste. *RSC Advances* 7, 16460–16466.

Villegas, L.G.C., Mashhadi, N., Chen, M., Mukherjee, D., Taylor, K.E., Biswas, N., 2016. A short review of techniques for phenol removal from wastewater. *Current Pollution Reports* 2, 157–167.

Wang, Z., Zhang, Y., Li, K., Sun, Z., Wang, J., 2020. Enhanced mineralization of reactive brilliant red X-3B by UV driven photocatalytic membrane contact ozonation. *Journal of Hazardous Materials* 391, 122194.

Yao, Y., Wang, M., Liu, Y., Han, L., Liu, X., 2020. Insights into the improvement of the enzymatic hydrolysis of bovine bone protein using lipase pretreatment. *Food chemistry* 302, 125199.

Zdarta, J., Jankowska, K., Bachosz, K., Degórska, O., Kaźmierczak, K., Nguyen, L.N., Nghiem, L.D., Jesionowski, T., 2021. Enhanced wastewater treatment by immobilized enzymes. *Chemical Engineering Journal* 7, 167–179.

Zdarta, J., Meyer, A.S., Jesionowski, T., Pinelo, M., 2019. Multi-faceted strategy based on enzyme immobilization with reactant adsorption and membrane technology for biocatalytic removal of pollutants: a critical review. *Biotechnology Advances* 37, 107401.

Zhou, W., Zhang, W., Cai, Y., 2020. Laccase immobilization for water purification: a comprehensive review. *Chemical Engineering Journal*, 403, 126272.

Zhu, Y., Wang, W., Ni, J., Hu, B.J.C., 2020. Cultivation of granules containing anaerobic decolorization and aerobic degradation cultures for the complete mineralization of azo dyes in wastewater. *Chemosphere* 246, 125753.

7 Problems of Using Wastewater in Agriculture and Mitigating them through Microbial Bioremediation

Debraj Biswal

7.1 INTRODUCTION

Water crisis is an emerging problem that has its roots in two major causes – global climate change and population explosion. Annual rise in global temperature and alterations in precipitation patterns have already been recorded worldwide and they have been projected to become worse in the coming years (IPCC 2013). Temperature rise causes water scarcity by increasing the rates of evaporation which will hit the arid and semi-arid regions of the world harder than any other place (Piao et al. 2010; Hussain et al. 2019). The condition may trigger desertification in the greener parts of the globe too. Moreover, the rising sea levels may infiltrate the coastlines causing salinization of freshwater bodies (Nicholls et al. 2011). Thus, the net effect of global warming will be decreased availability of freshwater fit for consumption by humans, the livestock and the agricultural sector (Hussain et al. 2019). Unprecedented population growth will aggravate the conditions further due to increased demands of drinking water and food supply. The second problem has two primary aspects – (1) crop production and (2) the livestock industry, both of which require adequate freshwater supplies. The problems will worsen in the coming years due to global climate change.

It has been estimated that population growth has caused the reduction of total amount of available water to about 33×10^5L in 1960 and further to about 125×10^4L in 1995. According to experts, this will decrease further to approximately 65×10^4L in 2025 (Abdel-Raouf et al. 2012). The remaining water resources are threatened by increased production of wastewater due to the rising industrialization and urbanization along with population growth (Singh et al. 2018). Besides wastewater generation, the freshwater bodies have become increasingly vulnerable to pollution from industrial effluents, accidental oil spillage, mining, smelting of iron ores, agricultural and livestock by-products, etc. (Dixit et al. 2015). These, in turn, have jeopardized the availability of good quality water for irrigation. The problem is more severe in parts of the world with low annual rainfall. These include Iran, Southern Europe,

DOI: 10.1201/9781003517238-7

North Africa and even in sections of the United States and Australia (Aghalari et al. 2020). According to Mora-Ravelo et al. (2017), 77% of all available freshwater is used for agriculture but 15% of the aquifers have already been over-exploited. This indicates a future condition of water crisis that can be, to some extent, resolved by using treated wastewater for irrigation. Use of wastewater in agriculture has two main benefits – (1) it provides nutrients to the crop plants and (2) it reduces the production cost. Wastewater is usually high in nitrogen (N) and phosphorus (P) that nourishes the plants and automatically reduces the input of fertilizers, making it highly acceptable among the local farmers for food crop production (Plevich et al. 2012). According to Mora-Ravelo et al. (2017) the amount of N present in wastewater can nearly supplement the N in plant fertilizers.

Despite the available remedy, use of wastewater for irrigation is not free from risks and disadvantages. The presence of several hazardous contaminants such as heavy metals in wastewater makes it detrimental to anthropogenic health (Olmos and Alarcón-Herrera 2014). The most common heavy metals present in wastewater include Pb, Cd, Hg, Cu, As and Cr which have already been reported to cause health problems in living organisms (Kaur et al. 2018). The solubility and bioavailability of some of these metals facilitate their entry into the living systems, complicating the whole scenario (Barkat 2011). Additionally, the intensification of agriculture and industrial expansion have augmented the production of chemicals such as polyaromatic hydrocarbons (PAHs), azo dyes, pesticides, polychlorinated compounds, etc., which are not easily degradable. They can persist in the environment for a long period of time and become a source of threat to the ecosystem (Elekwachi et al. 2014). Many of these compounds are reportedly toxic, mutagenic, teratogenic and carcinogenic to humans as well as other animals (Liu et al. 2019). Unfortunately, they form a major part of the organic pollutants in the wastewater discharges. Thus, if the wastewater is not used properly many of these toxic substances can enter into the food web leading to physiological and even genetic complications in living organisms (Jan et al. 2015). In plants, heavy metals augment seed mortality and interfere with their growth (Singh et al. 2020). This necessitates the development of cost-effective, eco-friendly and simple wastewater treatment procedures to make it useful and save freshwater mainly for the purpose of consumption (Masciandaro et al. 2015).

The current chapter gives an overview of wastewater and the advantages and disadvantages of its application in agriculture. It also discusses about the various remediation techniques that have been developed over the years for rendering wastewater fit for irrigation. Specifically, it tries to elaborate the basic principle and processes of microbial bioremediation of wastewater. Last but not the least, it attempts to find the drawbacks of these remediation techniques and analyses their prospect of solving the projected future water crisis especially in the field of agriculture.

7.2 SOURCES, TYPES AND CHEMICAL COMPOSITION OF WASTEWATER

Wastewater is defined as the effluent water from various industrial, municipal, commercial, domestic, agricultural and/or livestock management centres having variable compositions of organic and inorganic substances some of which are toxic that need to be eliminated prior to use (Bhandari et al. 2021). Depending on their origin, the wastewaters have been broadly classified into industrial wastewater and urban wastewater. Urban wastewater mainly consists of domestic sewage and the by-products of water used for washing, making them rich in detergents. Thus, they have a more or less homogeneous composition. Industrial wastewater, on the other hand, has a heterogeneous composition that varies with the purpose it is being used for. This complicates the industrial wastewater treatment procedures (Mora-Ravelo et al. 2017). Figure 7.1 gives an overview of different types of wastewater mentioning their sources.

Wastewater mainly consists of organic and inorganic materials and various solid particles that can either be in the form of non-settleable colloids or immersed solids or as suspended solids that can easily be settled (Naidoo and Olaniran 2014). Proteins, lignin, soaps and detergents form the main organic components of wastewater while the inorganic substances primarily include metals such as Pb, Hg, Cu, Zn, etc. (Westerhoff et al. 2015). The cleaning and manufacturing processes in the industries generate industrial wastewaters rich in Ni, Zn, Cr, Fe, Ti and other toxic pollutants some of which are carcinogens (Bazrafshan et al. 2015). The composition varies with the industry. For instance, halogenated organic compounds used in the synthesis of fungicides, herbicides, insecticides, plasticizers, etc., are present in wastewater as contaminants (Fetzner and Lingens 1994). Similarly, the paper and pulp industry generates high amounts of lignin, terpenes, resin acids, catechols, bisphenol A and carboxylates while the pharmaceutical industry generates large amounts of diclofenac, metharbital, bromazepam, dexamethasone and antihistamine cimetidine among others (Dsikowitzky and Schwarzbauer 2013). It has been observed that some of these wastewaters, especially from the pharmaceutical and tannery industries are not easily degradable (Kavitha et al. 2012). Here lies the problem with it.

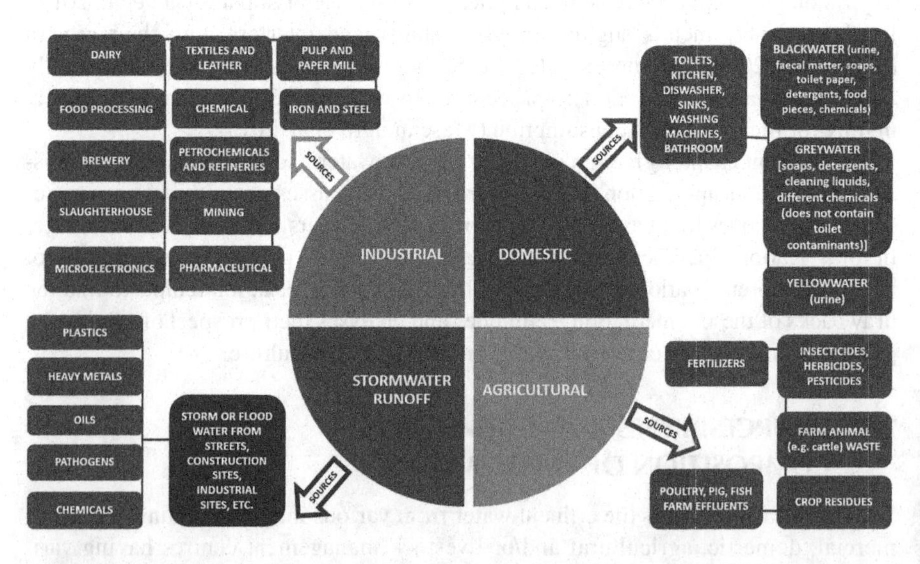

FIGURE 7.1 Broad classification of different types of wastewater along with their sources.

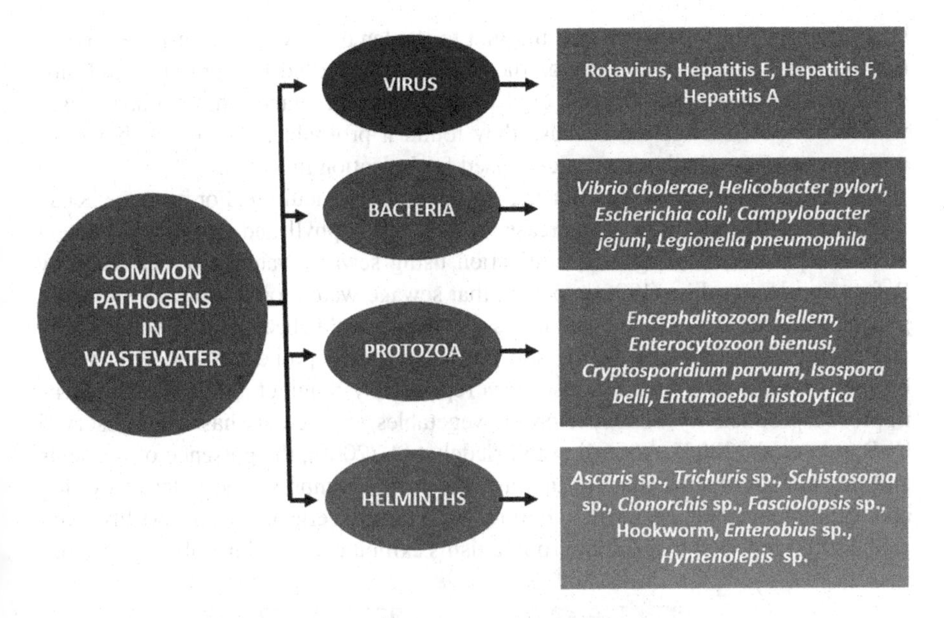

FIGURE 7.2 Some common pathogens present in wastewater.

Apart from the organic and inorganic contaminants, the wastewater also contains some biological contaminants – the pathogens and parasites which may cause health hazards for humans as well as animals (mainly the livestock) if not properly treated (Osuolale and Okoh 2017). Viruses, fungi, parasites (including their eggs and cysts), the faecal coliform bacteria, *Vibrio cholerae*, *Salmonella* sp., *Shigella* sp. and *Escherichia coli* are some of the most common biological pollutants present in wastewater (Aghalari et al. 2020). They can cause diseases such as typhoid, tuberculosis, cholera, hepatitis and dysentery (Jaromin-Gleń et al. 2017). Figure 7.2 groups some common pathogens present in wastewater based on reports from earlier studies. Thus, untreated wastewater when released to the environment directly may be a threat to the aquatic ecosystem as well as humans (Okeyo et al. 2018).

7.3 EFFECTS OF USING (UNTREATED) WASTEWATER IN AGRICULTURE

Records show that many parts of the world use untreated wastewater for irrigation. It either serves as a measure to cope up with water scarcity or to conserve freshwater for other uses. Published reports show its use in countries with poor freshwater supplies. They include Israel, Qatar, Saudi Arabia, United Arab Emirates, Cyprus, Kuwait, Bahrain, Malta and many others (Raschid-Sally and Jayakody 2008). It is also very popular in parts of Asia, Africa and Latin America (Balkhair and Ashraf 2016). However, using sewage water or wastewater for irrigating agricultural lands is not a recent practice. Literature shows that use of sewage water in irrigation was common in many countries of Western Europe even in the seventeenth century (Li et al. 2018). The oldest site of sewage irrigation has been recorded from Berlin in Germany

(Lottermoser 2012). America was the first to implement sewage water-based irrigation on a large scale in as early as the 1920s which called for some research and developments in the sector (Chen et al. 2000). Later on, many other countries used wastewater-based irrigation because they found it profitable. Figure 7.3 shows the different ways by which wastewater is used for irrigation in Korea.

There are many advantages of using wastewater in agriculture. For instance, Khan and Bano (2016) reported an increase in total chlorophyll and carotenoid content of maize plants following their cultivation using sewage water. Another study by Safary and Hajrasoliha (1995) reported that sewage water could improve the nitrogen, phosphorus and carbon content of soil while it could check rises in soil sodicity and salinity. Enhancement of the soil physicochemical properties and crop yields by sewage water irrigation has also been reported by Singh et al. (2011). Similarly, improved productivity of fodder crops, vegetables and cereals has been observed by Rattan et al. (2005). According to Friedel et al. (2000), the presence of nutrients and degradable organic matters in wastewater could improve the nutrient cycling and soil biological activity which, in turn, have positive effects on soil fertility. Meli et al. (2002) reported that soil microorganisms exhibit enhanced metabolic activities

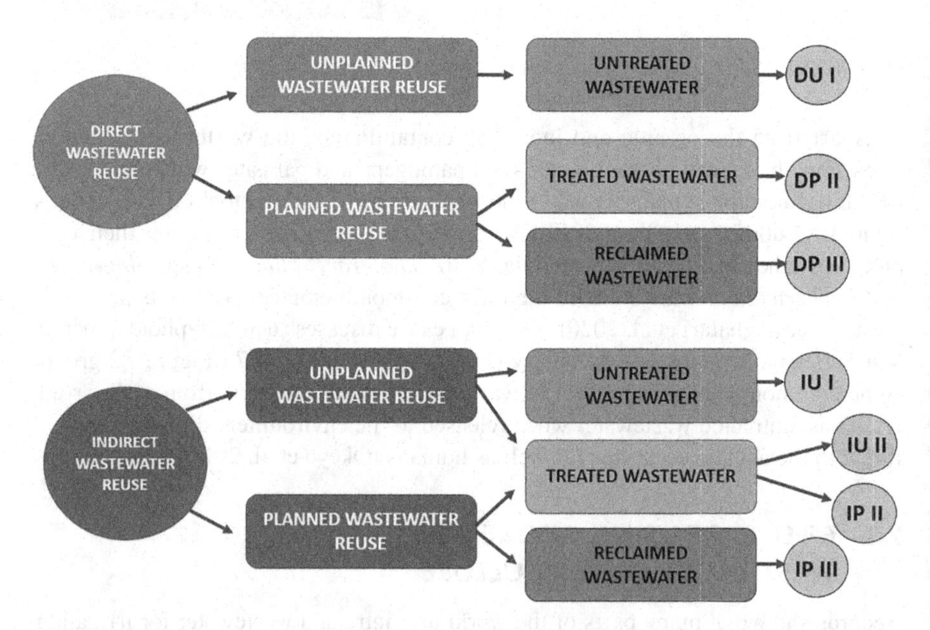

FIGURE 7.3 Different classes of wastewater reuse in Korea. (**DU I** – mainly involves the use of household water directly for irrigation without treatment, **DP II** – involves the use of the effluent directly from the wastewater treatment plants for irrigation, **DP III** – in this case the effluent from the wastewater treatment plants are treated further before use in irrigation, **IU I** – in this case the wastewater is first directly mixed with lake or stream water followed by using it for irrigation, **IU II** – this involves mixing of the effluent from the wastewater treatment plants with stream or lake water before being used for irrigation, **IP II** – here, the effluent from wastewater plants is diluted and mixed with freshwater before using it for irrigation, **IP III** – this involves treatment of the lake or stream water after it has been mixed with effluent from wastewater plants and then used for irrigation.)

under the influence of sewage irrigation water. Alghobar et al. (2014) concluded that sewage water could make the soil more fertile compared to normal water. According to Wang et al. (2012), the macro- and micronutrient contents of the soil could be enhanced by irrigating croplands with sewage water for a long time. It automatically improved the soil health. El-Nahhal et al. (2013) reported that irrigation with sewage water could yield crops with higher biomass compared to irrigation with freshwater. Similar observations were also made by Becerra-Castro et al. (2013). However, there are contradictory reports too.

Earlier publications show that application of wastewater for irrigating croplands has four immediate effects – (1) some of the pollutants may be detoxified by the natural remediation processes present in the soil, (2) some pollutants could be adsorbed and retained by the soil particles, (3) some pollutants and/or nutrients could be absorbed by the crops and accumulated in their fruits and/or foliage, and (4) some may percolate deeper and contaminate the groundwater (Li et al. 2018). Thus, long-term use of wastewater for irrigation affects the physicochemical as well as the biological properties of soil because some pollutants and/or nutrients are beyond the recommended levels of soil fertility and/or natural detoxification. For example, Wang and Lin (2003) reported that long-term use of sewage irrigation may cause hardening of the soil which directly affects the soil porosity, aeration and water holding capacity. It also has a negative effect on solute migration and plant root penetration (Hernández et al. 2015). Negative impacts of sewage water irrigation on formation of soil aggregates and increase in bulk density have been reported too (Li et al. 2018). Irrigation with sewage water increases the soil viscosity which also increases the field capacity or water holding capacity of the soil (Lan et al. 2000). This has impacts on plant growth.

Under normal conditions, soil often acts as a buffer to adjust the soil pH that facilitates plant growth (Masto et al. 2009). The stability is, however, destroyed when swage water is used for irrigation for a long period of time, resulting in the soil being either too acidic or too alkaline that directly affects the availability of nutrients to plants (Ma and Zhao 2010). The reason for alterations in soil pH mainly arise from accumulation and/or release of metal ions present in sewage water or alterations in nitrification and ammonification of soil organic matter or anaerobic decomposition of organic matter (Dheri et al. 2007).

The organic matter present in sewage water is often considered to enhance soil fertility through increases in organic composition but the effect varies with the soil type and also with the depth (Xue 2012). It has been observed that the organic enrichment due to sewage water occurs mostly on the surface layers which has important implications in global carbon cycle and, thereby, global warming (Rattan et al. 2005; Li et al. 2018). Sewage water reportedly causes nitrogen enrichment of the soil with NO_3^- ions percolating the deeper layers and polluting the aquifers (Liu and Lu 2002). It also causes excessive accumulation of phosphorus in the upper 5 cm of the soil which has direct impacts on plant growth (Li et al. 2018). The availability of potassium is also increased due to irrigation by sewage water which has a positive effect on soil fertility (Xu 2007). On the contrary, organic pollutants such as phenols, aromatic hydrocarbons, organic chlorines, plasticizers, sterols, etc., present in wastewater can contaminate the soil as well as the ground water deep below, with deleterious effects on human health (Du et al. 2010). Fang (2011) reported that rice grown on such polluted soils emits strong odours of oils and aromatic compounds.

The most serious effects are exerted by the heavy metals present abundantly as an important part of inorganic pollutants in the wastewater. Wastewater has been found to exhibit high concentrations of heavy metals such as Pb, Cr, Cd, As and Hg, often beyond permissible limits that serve to contaminate not only the soil but also the underground water (Lucho-Constantine et al. 2005; Flores-Magdaleno et al. 2011; Li et al. 2018). They can induce oxidative stress in living organisms by forming free radicals (Bhat and Khan 2011). Heavy metals can also disrupt pigments and enzymes by replacing the metals and interfere with their proper functioning (Kumar et al. 2014). Paradoxically, plants and animals require small amounts of these metals as essential micronutrients but their presence in concentrations beyond a certain limit is deleterious for them. Excessive heavy metals in soil can hinder plant growth and destroy the microbes, rendering the soil unfit for agricultural practices (Ayangbenro and Babalola 2017). The effects of metal contamination are very complicated because of the ease with which many of them can form complexes with impacts more severe than the heavy metals themselves. They can also be present as chlorides, sulphates and oxides (Wuana and Okieimen 2011). The effect, however, varies with the concentration of these heavy metals but are harmful both in elemental as well as in the soluble form (Chibuike and Obiora 2014). These metals may be adsorbed by the soil particles resulting in heavy metal pollution and allowing them to persist in nature (Wan et al. 2015). Heavy metals present in environment tends to bioaccumulate in animal tissues and often biomagnify, causing health deterioration that have been extensively studied and documented (Genuis and Kelln 2015). They can also enter human body through contaminated food and water leading to dire consequences (Singh et al. 2011). Thus, even trace amounts of these pollutants can have serious impacts either directly or indirectly by entering the food web (Wan et al. 2015).

Soil health is maintained by the soil microbes to a large extent. They mainly include the bacteria, fungi and the actinomycetes. Earlier studies show that irrigation with sewage water has negative impacts on the soil microbiota due to alterations in the soil microenvironments (Zhang et al. 2008). Usually, the population of actinomycetes and bacteria has been found to decline while the population of fungi has been found to increase slowly under these conditions (Ge et al. 2009). Nitrogen and phosphorus enrichment of soil due to sewage water has been found to increase the population of nitrogen-fixing bacteria and phosphorus bacteria, respectively (Zhang et al. 2007). This, however, has a positive effect on soil fertility.

Using wastewater for irrigation has also been found to affect the activity of soil enzymes. Soil enzymes include the enzymes produced by the plant roots, the soil microbes and the decomposing plant and animal matters in the soil (Cao et al. 2003). Some studies show that use of wastewater from the petroleum industry triggers the proliferation of heterotrophic soil bacteria and fungi. The activities of some soil enzymes such as catalase, dehydrogenase and polyphenol oxidase have been found to be stimulated due to petroleum wastewater while the activity of soil urease has been found to be negatively affected by the same (Li et al. 2018). Effects of heavy metal pollution on the action of soil enzymes have also been recorded. Liu (1996) reported that activities of soil peroxidase and urease were stimulated by increasing levels of Cd and Pb in the soil. Li (2001) reported that irrigation with sewage water often leads to secondary salinization that has negative impacts on the functioning of soil enzymes. This may lead to an overall decline of soil quality. Some studies reveal that the problems of using wastewater for irrigation are higher in the urban regions and their outskirts because of greater amount of wastewater generation in those regions (Sarabia-Meléndez et al. 2011). As a result, the soil samples collected from these regions often show high concentrations of salts and/or heavy metals (Chen et al. 2009).

Epidemiological impacts of using wastewater for agriculture and irrigation have also been investigated and the results are not at all encouraging. Experiments have revealed the presence of heavy loads of faecal coliform bacteria and helminth eggs in wastewater. They can have serious health impacts on the human population that depend on agricultural products obtained from farms irrigated with wastewater. High prevalence of gastrointestinal disorders caused by protozoan parasites such as *Giardia lamblia* and *Entamoeba histolytica* have already been recorded in such populations. In addition, high prevalence of helminthiasis caused by *Ascaris lumbricoides* has also been observed (Siebe and Cifuentes 1995).

7.4 TREATMENT OR REMEDIATION OF WASTEWATER

Gradually, when the pitfalls of using wastewater in irrigation of farmlands became more and more evident some standards were fixed by certain international bodies such as the World Health Organization (WHO) and the Food and Agriculture Organization (FAO). The first set of guidelines was published by the WHO in 1973 which were reset time and again, making the rules more and more stringent (Li et al. 2018). The wastewater to be used for agricultural purposes is required to conform to certain physicochemical and biological criteria so that the toxic pollutants present in it are not introduced into the food web causing damage to human health and to the ecosystem (Rooney et al. 2006). As a result, the wastewater treatment is a complicated procedure and involves different types of remediation techniques (Ferrera-Cerrato et al. 2006). There are several methods of wastewater treatments that depend on the nature of pollutants present in it and whether they are biodegradable or not. The methods used, in turn, determine the quality of the treated water and its subsequent application (Vassilev et al. 2004).

The conventional methods of removal of xenobiotics include their disposal into pits, incineration at high temperatures and decomposition by application of chemical or ultraviolet technologies (Bhandari et al. 2021). However, these processes have lost universal acceptance because they are considered unsafe with production of sludge that may generate harmful secondary pollutants if not processed properly prior to their disposal (Karigar and Rao 2011). Thus, the physical and chemical remediation techniques are expensive and generate sludge (consisting of wastewater solids) whose disposal is another issue because of their eco-toxic effects (Balakrishnan and Velu 2015). However, sewage sludge has been found to be beneficial because of high contents of nutrients such as carbon and nitrogen and absence of heavy metals. This enables its application as a fertilizer. On the other hand, industrial sludge contains a heavy load of toxic compounds together with heavy metals, making it very difficult to handle (Mtshali et al. 2014). The expense of the physical and chemical remediation techniques and their failure to address ecological issues forced the scientific community to devise cost-effective and eco-friendly strategies of wastewater management (Zheng et al. 2013). Bioremediation or remediation using living organisms and/or their products has been found to satisfy both these needs (Bansal et al. 2018). Additionally, bioremediation does not generate any toxic by-products (Mani and Kumar 2014).

The concept of bioremediation is based on the catabolic activities of some organisms that possess the natural ability to decompose organic substances for their own nutrition and energy (Pilon-Smits 2005). Bacteria, fungi and plants have been observed to be very efficient in detoxification of wastewater (Chowdhury et al. 2012). The efficiency, however, depends on the availability of pollutants to the bacteria and also on environmental factors (Chowdhury et al. 2012). Their ability can be exploited and/or augmented to breakdown, modify and/or eliminate the toxic organic substances from wastewater (Van Hamme et al. 2003). This serves two important functions – (1) disposal of untreated wastewater could have led to dire environmental consequences and (2) if the treatment procedure is appropriate, the purified wastewater can be used for irrigation purposes. Researchers are giving more importance to the second part because it enables a mechanism of recycling wastewater and address the problem of water scarcity.

Based on the biodegradable nature of the toxic pollutants present in wastewater, the process of purification varies. This is because biodegradable contaminants such as volatile organic compounds, polycyclic aromatic hydrocarbons, cis-dichloroethylene vinyl chloride, pentachlorophenol, polychlorinated biphenyls (PCBs), etc., can be easily degraded by biological treatments (Megharaj et al. 2011). However, remediation of non-biodegradable substances is complex and requires their conversion into biodegradable forms such as organic acids, ketones and aldehydes (Van Leeuwen et al. 2009). This can be achieved by chemical oxidation and/or ozonation of the activated sludge (Wang et al. 2018a).

Roughly, the whole purification process can be divided into three steps. The first step, commonly known as preliminary treatment, involves the removal of fatty contaminants and large solid residues such as bottles, rags, cans, etc. The second step, referred to as primary treatment, applies physical separation procedures such as filtration or gravity and/or chemical sedimentation techniques to discard the suspended solid particles and organic matters (Mora-Ravelo et al. 2017). The third step or the secondary treatment is rather complicated and involves the use of aerated lagoons, trickling filters, activated sludge and/or stabilization ponds to transform as much as 90% of the organic substances present in the effluent into a biological agglomeration consisting of organic matter, minerals and bacteria (Mora-Ravelo et al. 2017). This is often followed by tertiary treatment procedures that involve the use of aerobic, anaerobic and/or facultative techniques for complete degradation of the organic substances present in the wastewater and removal of suspended solid matters as well. The advantage of this step lies in the aerobic removal of pathogens, nutrients and heavy metals (Mora-Ravelo et al. 2017). Basically, the tertiary step exploits the enzymatic activities of the microbes such as bacteria and/or protozoans, either directly or fixed to a surface, to bring about the aerobic and/or anaerobic decomposition of the organic compounds encountered in the wastewater. There are more sophisticated techniques that employ bioengineered enzymes instead of using the organisms themselves to carry out the same processes. Organic substances, such as proteins, carbohydrates and fatty acids, are broken down by the microbes under aerobic conditions (Tchobanoglous and Burton 2005). In the presence of oxygen, the suspended solids and the microbes together form small clumps (by flocculation) followed by their precipitation in the form of sludge which can be subsequently removed by filtration techniques using biofilters or other sophisticated filtration apparatus. Figure 7.4 summarizes the different categories (or steps) of wastewater treatment for reuse.

The mechanism of action of microbes under conditions of limited oxygen supply (or anaerobic conditions) can be divided into two stages – the acidogenic and the acetogenic stages. In the first stage, the organic substances are degraded into simple organic acids while the second stage involves the decomposition of fatty acids into methane (Noyola 1999). Anaerobic reactors, lagoons or even anaerobic filters (on fixed surfaces) can be used for the anaerobic procedures (Mora-Ravelo et al. 2017). A simpler coagulation process involves addition of metal salts to the gel-like suspended substances in the wastewater that destabilizes the suspended matter and brings about the aggregation of dissolved organic compounds. Eventually, this facilitates the removal of the conglomerates of organic compounds and/or the microbes from the wastewater (Ramírez et al. 2004). Table 7.1 summarizes various engineered structures and their basic mechanisms of remediation of wastewater. Figure 7.5 represents the basic outlay of activated sludge wastewater treatment method.

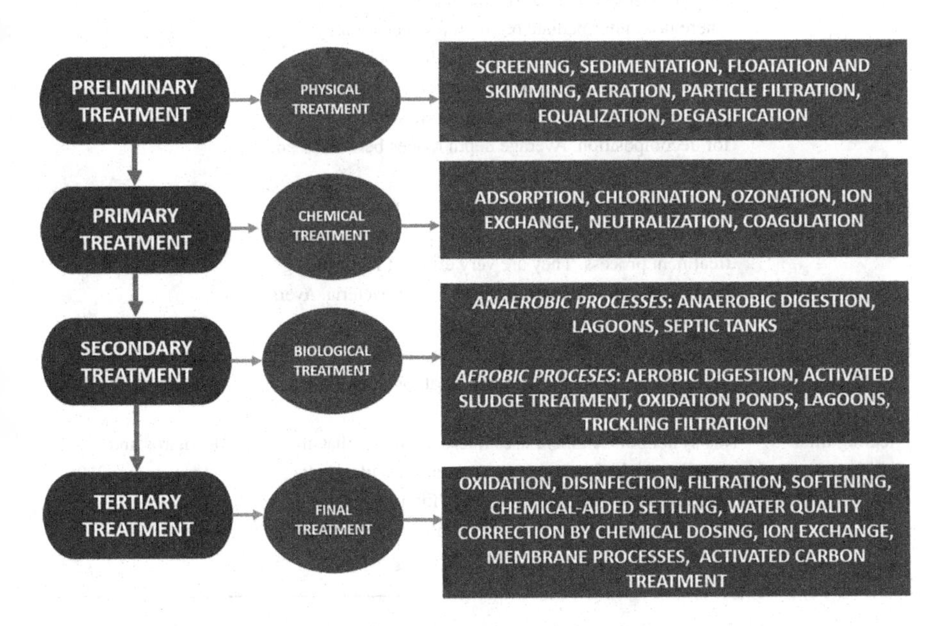

FIGURE 7.4 Categories of wastewater treatment methods.

TABLE 7.1
Engineered Structures and their Basic Mechanisms of Wastewater Remediation

Structures for Treatment	Mechanism	References
Aerobic ponds	These ponds utilize oxidative breakdown of the pollutants and therefore maintain an oxygen supply throughout their depth. Biological sources of oxygen like algae and bacteria may be used or mechanical devices for aeration can also be used. In the former case, the pond should be shallow compared to the latter. Average depth ranges from 30 to 45 cm.	Butler et al. (2017)
Anaerobic ponds	These structures carry out the treatment in the absence of oxygen throughout their entire depth. Anaerobic bacteria, present at the bottom of the pond, ferments the sludge producing CO_2 and CH_4. Therefore, the principle employed in this pond is sedimentation. The helminths settle down to the bottom on their own while the viruses and bacteria get attached to the settling solids and are thereby removed. They may also die due to lack of nutrition or be consumed by predators. Average depth ranges between 2 and 5 m. They are effective for treatment of industrial wastewater.	Butler et al. (2017)
Facultative ponds	These ponds have an upper aerobic layer composed of algae and aerobic bacteria while the lower layer (or the bottom) is anaerobic consisting of anaerobic bacteria. There is an intermediate region too which employs bacteria capable of surviving in both aerobic and anaerobic conditions. The treatment time is long because of the utilization of the photosynthetic services of algae for decomposition. Average depth ranges between 1 and 2 m.	Butler et al. (2017)
Maturation ponds	They are similar to facultative ponds but usually maintain anaerobic conditions and mainly depend on algae for the treatment process. They are very efficient in removing pathogens, nutrients and faecal coliform bacteria. Average depth ranges between 1 and 1.15 m.	Butler et al. (2017)
Activated sludge system	It mainly uses microorganisms to decompose the biodegradable compounds and/or pathogens present in wastewater.	Elimelech (2006); Doorn et al. (2006)
Trickling filter	This system of treatment uses microorganisms that are usually attached to large porous structures like sledge, rocks or plastic medium that act as filter media. They are very efficient in destroying the pathogens present in the wastewater as well as reducing the nitrogen levels.	Kornaros and Lyberatosa (2006)

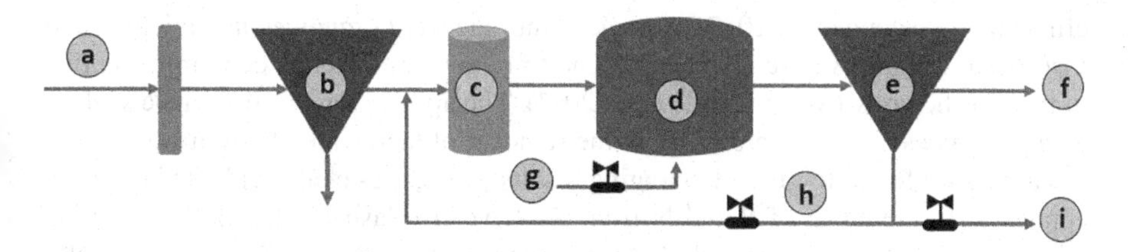

FIGURE 7.5 Schematic representation of the basic layout of activated sludge wastewater treatment process. [**a.** Influent, **b.** Primary settler, **c.** Anoxic reactor, **d.** Aerobic reactor, **e.** Secondary settler, **f.** Effluent, **g.** Air, **h.** Recycled sludge, **i.** Excess sludge]

7.5 BIOREMEDIATION OF WASTEWATER

Bioremediation has been defined in various ways, all of which suggest the use of living organisms such as bacteria, fungi and even plants with abilities to convert the toxic substances present in the wastewater to non-toxic forms or removing them altogether by degrading them (Van Hamme et al. 2003). The most common agents of bioremediation include plants, bacteria and fungi. Actually, the enzymes derived from these agents bring about the process of bioremediation. This can be achieved either by using the living organisms themselves or by extracting their enzymes and immobilizing them on silica and glass beads followed by their addition to wastewater to purify it (Ratnakar et al. 2016). The enzyme-mediated degradation products are usually non-toxic and/or can be removed easily by further processing (Karigar and Rao 2011).

An easy and affordable means of purifying wastewater under *in situ* or *ex situ* conditions is by phytoremediation. This eco-friendly technique employs the plants and the microorganisms in their rhizosphere to remove or immobilize the pollutants in wastewater (Delgadillo-López et al. 2011). The processes mainly include phytodegradation, phytoextraction and rhizofiltration (Ghosh and Singh 2005). Fast-growing native species with high biomass production and tolerance levels that can be easily harvested are usually preferred for phytoremediation (Mora-Ravelo et al. 2017). Some common examples include *Zea mays*, *Calendula officinalis*, *Brassica juncea* and *Medicago sativa*. The fundamental reason for using plants in bioremediation lies in their ability to absorb and store high amounts of heavy metals in their tissues (Arenas et al. 2011). However, plant species differ in their mechanisms to absorb, convert and store the toxic pollutants in their non-toxic forms (Delgadillo-López et al. 2011). Often, the sites of storage of these substances vary from species to species (Llugany et al. 2007).

Rhizofiltration technique basically filters the wastewater through the root mass of plants such as *Mentha deriva* and *Helianthus annuus* that are grown hydroponically (Cherian and Oliveira 2005). The roots of these plants either absorb or adsorb the contaminants dissolved in wastewater and store them in their tissues (Mora-Ravelo et al. 2017). However, reports show that it is not a very efficient technique. Comparatively, using macrophytes such as *Eichhornia crassipes*, *Salvinia minima* and *Myriophyllum aquaticum* has been shown to absorb heavy metals such as Pb, Cd and Cr more efficiently (Rezania et al. 2015). Aquatic plants such as *Potamogeton pectinatus* and *Callitriche stagnalis* have also been found to remove radioactive contaminants like uranium efficiently (Mora-Ravelo et al. 2017). The application of microalgae such as *Scenedesmus* sp. and *Chlorella* sp. in the removal of heavy metals from wastewater has also been found to give encouraging results (Canizares et al. 2013). This equally applies to the removal of P and N from wastewater (Nasir et al. 2015). Symbiotic relationships between algae and blue-green algae such as the *Azolla-Anabaena* relationship can also be exploited to remove metals like copper (Cu) and selenium (Se) from wastewater (Sánchez-Viveros et al. 2013). Table 7.2 gives a summary of some common plant-based wastewater remediation techniques.

7.6 MICROBIAL REMEDIATION OF WASTEWATER

Microorganisms have gained prominence in recent years as agents of wastewater bioremediation (Park et al. 2018). The basics of using microbes for bioremediation are derived from the fact that certain microbes can survive in some of the most difficult environments by virtue of their enzymes which they can use to decompose organic substances in their surroundings to derive nutrition and energy. Microbial remediation uses these extracellular enzymes to break down the organic matters present in wastewater (Kumar et al. 2011). Figure 7.6 describes the basic principle of enzymatic degradation of organic pollutants (by microbes) in wastewater that is exploited for microbial bioremediation. Despite its usefulness, the process is very slow and requires regular maintenance of a set of factors to make the process fruitful. Moreover, only a few bacteria have been recorded till date that can produce specific enzymes with capabilities of decomposing the pollutants (Bhandari et al. 2021). This has encouraged the production of genetically engineered microorganisms to augment the production of the necessary enzymes required to degrade the pollutants. The process has been further developed to utilize the energy and the biomass obtained during the decomposition of substances such as microplastics, polyhalogenated compounds, hydrocarbons and agrochemicals. The processes sometimes yield novel products that may have several important uses as well (Phale et al. 2019). Microbial remediation employs microorganisms such as algae, fungi, bacteria, rotifers and protozoa for degradation of the organic contaminants in wastewater. Aerobic bacteria such as *Rhodococcus* sp., *Pseudomonas* sp., *Sphingomonas* sp. and *Mycobacterium* sp. have been reported to produce enzymes that degrade hydrocarbons and pesticides. The anaerobic bacteria, on the other hand, produce enzymes that can degrade PCBs or dechlorinate trichloroethylene (TCE) (Sharma 2012).

Microbial enzymes that decompose the organic matters in wastewaters can broadly be classified into two groups – oxidoreductases and hydrolases (Ratnakar et al. 2016). The oxidoreductases can be used to degrade phenolic compounds, xenobiotics and azo dyes while the hydrolases are known to break down complex toxic substances into simpler forms (de Lourdes Moreno et al. 2013). Some of the most commonly used microbial enzymes in bioremediation include hydrolases, dehydrogenases, laccases, proteases, lipases, dehalogenases and cytochrome P450. Table 7.3 gives a summary of different microbial enzymes, their sources, the pollutants on which they act upon and their basic action mechanism.

TABLE 7.2
Different Procedures of Phytoremediation of Wastewater

Process	Mechanism	Examples of Plants Used	References
Phytoextraction	In this process the plants remove the pollutants (including heavy metals) from the wastewater and store it in their aerial parts that can be harvested.	*Lemna gibba*, *Ceratophyllum demersum*, *Potamogeton pusillus*, *Ipomoea aquatica*, *Eichhornia crassipes*, *Myriophyllum spicatum*	Farraji (2014), Khan (2018)
Phytomining	The metals extracted by the plants from the wastewater and accumulated in their aboveground parts when harvested, dried and converted into ash for bioextraction of metals on a commercial scale is known as phytomining.	*Typha latifolia* (used for boron)	Farraji (2014)
Phytostabilization	The plants serve to reduce the bioavailability and mobility of the pollutants in the wastewater to prevent them from leaching and polluting the groundwater. They do so mainly by adsorbing the contaminants (like heavy metals) and storing them in their roots or modifying them in some way.	*Brassica juncea*, *Helianthus annuus*, *Solanum lycopersicum*, *Hydrilla verticillata*, *Saponaria officinalis*, *Phragmites australis*	Farraji (2014), Khan (2018)
Phytodegradation and Rhizodegradation	The plants and/or the microbes associated with their rhizosphere can degrade the pollutants present in wastewater. They do so by various enzymes produced by the plant roots and/or the associated microbes.	*Hibiscus cannabinus*, *Plantago major*, *Salix babylonica*, *Sorghum bicolor*	Farraji (2014), Khan (2018)
Phytovolatilization	In this process the plants absorb the pollutants present in the wastewater and transform them into volatile compounds that are eventually lost by transpiration from their aerial parts.	*Brassica* spp.	Farraji (2014), Khan (2018)
Phycoremediation	This process uses macro and/or micro algae to remove pollutants from wastewater. It is applicable for both domestic and industrial wastewater.	*Desmococcus olivaceus*, *Chlorella vulgaris*	Farraji (2014)

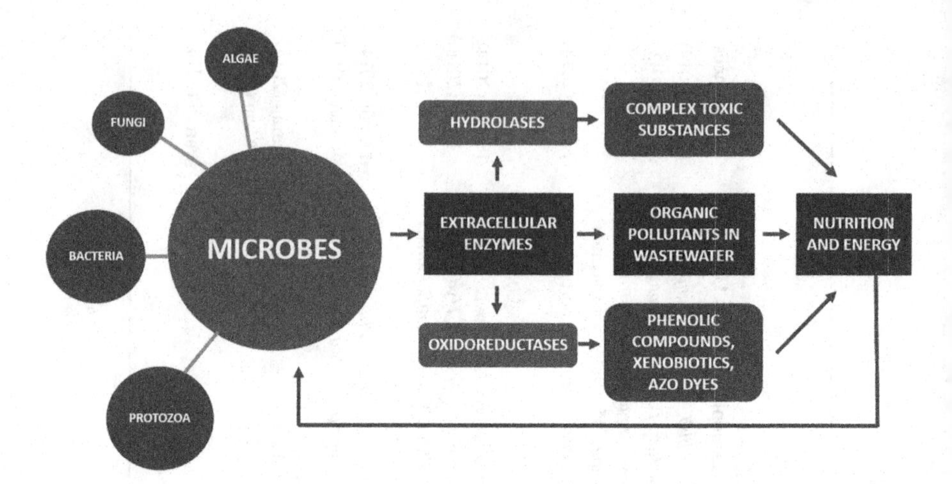

FIGURE 7.6 Basic pathway of enzymatic degradation of organic pollutants present in wastewater by microbes (for obtaining nutrition and energy) applied in microbial bioremediation of wastewater.

Hydrolases have diverse applications in the biomedical, chemical and feed additive industries, especially in the remediation of waste products from the food processing and plastic industries (Kumar and Sharma 2019). They offer additional advantages by being affordable, easily available, environment-friendly, devoid of cofactor selectivity and ability to withstand water-based solvents (Karigar and Rao 2011). Moreover, the hydrolases extracted from thermophilic bacteria can remain stable even at high temperatures and exhibit their remediating activities (Dang et al. 2018). Of all the hydrolases, lipase and protease have been found to be very effective in purifying wastewater. Lipase is not only involved in breaking down fatty substances but also finds application in detergent products to reduce the generation of phosphate-based substances that are potential pollutants (Saraswat et al. 2017). The activity of lipase has been used as an indicator of hydrocarbon degradation in the soil (Riffaldi et al. 2006). Similarly, some of the breakdown products of proteases also have economic value. For example, the keratinous wastes obtained from the livestock industry, after remediation, can be used as fertilizers and feed additives in the agricultural and livestock sectors, respectively (Nnolim et al. 2020). Protease enzymes used in the extraction of chitin from marine crustacean wastes also have several applications besides detoxification (Jellouli et al. 2011).

For removal of radioactive pollutants many bacteria use their oxidoreductase enzymes. The basics of these reactions lie in the conversion of the soluble oxidized form of the radioactive metal into an insoluble reduced form (Leung 2004). In some cases, fungi have been found to be more suitable than bacteria for detoxification because they can insert mycelia into their surroundings. As such, they can reach certain pollutants in the environments, especially soil, more efficiently than bacteria. This property of fungi can be utilized to degrade various halogenated phenolic compounds produced due to partial bleaching of pulp in the paper industry. The extracellular oxidoreductase enzymes produced by the fungi aid in the process (Rubilar et al. 2008).

TABLE 7.3
Some Enzymes Used for Microbial Remediation, Their Sources, Substrates and Mechanism of Action

Microorganism(s)	Enzyme	Reaction Mechanism	Pollutants Acted Upon	References
Bacillus cereus, *Flavobacterium* sp., *Nocardia* sp., *Pseudomonas* sp., *Achromobacter* sp.	Carbamate hydrolase, parathion hydrolase	Hydrolysis	Diazinon, Coumaphos, Carbofuran, Carbaryl or Parathion	Sutherland et al. (2002)
Bacillus licheniformis, *Brevibacillus* sp., *Bacillus cereus*, *Alicyclobacillus tengchogenesis*	Methyl parathion hydrolase (MPH), organophosphate hydrolase, organophosphate acid anhydrolase, amidohydrolase	Hydrolysis, Deamination, Dechlorination, Degradation of cyanuric acid	Organophosphate pesticides, chlorinated herbicides	Seeger et al. (2010), Littlechild (2015), Schenk et al. (2016)
Rhodococcus sp., *Nocardia* sp.	Nitrilases, nitrile hydratase, amidase	Hydrolysis	Cyanide-containing compounds	Rao et al. (2010)
Acinetobacter sp. M673, *Micrococcus* sp. YGJ1	Dialkyl phthalate ester hydrolases	Hydrolysis	Phthalates, phthalate esters (PEs), alkyl esters of phthalates	Jiao et al. (2013)
Fusarium solani f. pisi	Cutinase	Hydrolysis	Plastics, polycaprolactone	Singh et al. (2016)
Bacillus sp. BCBT21	Thermophilic chitinase, protease, xylanase, lipase	Hydrolysis	Biodegradable and oxo-biodegradable plastics	Dang et al. (2018)
Thermobifida fusca KW3	Polyester hydrolase	Hydrolysis	Synthetic polyesters	Wei et al. (2016)

(*Continued*)

TABLE 7.3 (*Continued*)
Some Enzymes Used for Microbial Remediation, Their Sources, Substrates and Mechanism of Action

Microorganism(s)	Enzyme	Reaction Mechanism	Pollutants Acted Upon	References
Pseudomonas sp. PKDE1, *Pseudomonas* sp. PKDE2, *Pseudomonas* sp. PKDM2	Mono-2-hexyl ethyl phthalate (MEHP) hydrolase	Reduction	di-(2-ethylhexyl phthalate) [DEHP]	Singh et al. (2017)
Streptomyces spp., *S. cyaneus*, *S. bikiniensis*, *S. ipomoeae*, *S. coelicolor*	Laccase	Oxidation	Lignin, phenolic compounds, aromatic amines, xenobiotic substances	Muthukumarasamy et al. (2015), Guan et al. (2018)
Sphingomonas sp.	Polyethylene glycol dehydrogenase	Catalyses the reversible conversion of alcohol to aldehyde or ketone	Xenobiotics, polyethylene glycol of various molecular weights	Kawai and Yamanaka (1989), Sugimoto et al. (2001)
Stenotrophomonas maltophilia	Polypropylene glycol dehydrogenase	Catalyses the reversible conversion of alcohol to aldehyde or ketone	Hydrophobic polymer medium-chain secondary alcohols, di- and tri-propylene glycols, polypropylene glycols	Tachibana et al. (2008)
Azoarcus evansii	3,4-dehydroadipyl-CoA semialdehyde dehydrogenase	Catalyses the oxidation of the aldehyde to a carboxylic acid	Several aromatic compounds	Gescher et al. (2006)

(*Continued*)

TABLE 7.3 (*Continued*)
Some Enzymes Used for Microbial Remediation, Their Sources, Substrates and Mechanism of Action

Microorganism(s)	Enzyme	Reaction Mechanism	Pollutants Acted Upon	References
Amycolatopsis sp., *Bacillus* sp., *Aspergillus* sp.	Microbial protease	Degrading α-ester bonds, poly (hydroxybutyrate) (PHB) depolymerase β-ester bonds, lipase c-ω bonds	Degradation of polymers	Kumar and Sharma (2019), Haider et al. (2019)
Pseudomonas sp., *Bacillus* sp.	Keratinase	Hydrolysis	Keratinous wastes from poultry	Mazotto et al. (2011)
Acinetobacter sp., *Mycobacterium* sp., *Rhodococcus* sp., *Pseudomonas* sp., *P. aeruginosa*	Lipase	Hydrolysis of the ester bond of triglycerides into fatty acids and glycerol	Oil residues, petroleum contaminants, fatty/greasy effluents, oil spills	Amara and Salem (2009), Verma et al. (2012), Casas-Godoy et al. (2012), Hassan et al. (2018)
Pseudomonas sp.	Amano lipase	Transesterification	Parabens	Wang et al. (2018b)
Bacillus sp. *GZT*, *Rhizobium* sp., *Pseudomonas umsongensis*	Dehalogenases like haloalkane dehydrogenase, halohydrin dehydrogenase	Dehalogenation reaction. Cleaves carbon-halogen bonds through hydrolytic, reductive and oxygenolytic cleavage reactions	Toxic halogenated compounds	Cairns et al. (1996), Allpress and Gowland (1998), Zu et al. (2012), Xue et al. (2018), Wang et al. (2018b), Wang et al. (2018)c
Pseudomonas sp., *Mycobacterium* sp., *Rhodococcus* sp., *Gordonia* sp.	Cytochrome P450	Mediates chemical transformations like aliphatic hydroxylations, dealkylations, dehalogenation and epoxidations	Xenobiotics	Anzenbacher and Anzenbacherová (2001), Chakraborty and Das (2016)

Laccases are oxidoreductases that are produced by a wide variety of organisms such as fungi, bacteria, plants and even insects. Both intracellular and extracellular laccases are produced by microorganisms which can not only oxidize phenolic compounds and their derivatives but also decarboxylate them (Rodríguez Couto and Toca Herrera 2006). The production of laccases has been reported to be accelerated by agricultural wastes such as banana peel, sawdust and rice bran (Muthukumarasamy et al. 2015). Interestingly, the enzyme is very stable and can withstand a wide range of fluctuations in temperature, pH and salinity (Guan et al. 2018). It is also tolerant to a range of organic solvents (Guan et al. 2018). It has important roles in preventing pollution of groundwater (Gianfreda et al. 1999). PAHs produced from industrial processes or incomplete combustion of fossil fuels attain their xenobiotic properties because of their poor solubility in water and slow rates of decomposition (Ihssen et al. 2015). Laccases are known to catalyse the oxidation of reduced aromatic and phenolic compounds present in the toxic wastes with simultaneous reduction of molecular oxygen to water (Mai et al. 2000). They can degrade pollutants such as various dyes and phenols from the textile industry (Sondhi et al. 2015). They are also very efficient in detoxification of complex distillery effluents and paper and pulp industry discharges rich in chlorolignin (Wang and Zhao 2017). These enzymes can depolymerize lignin and the microbes can use the products as nutrients or even repolymerize the breakdown products to humic substances (Kim et al. 2002). They have also been reported to aid in biobleaching of eucalyptus kraft pulps (Arias et al. 2003). Thus, they are very versatile and efficient detoxifying agents. Peroxidases are an equally important group of oxidoreductases that can aid in the oxidation of lignin (the main product of wood) and other phenolic compounds (Karigar and Rao 2011). The class II fungal peroxidases such as lignin peroxidase and manganese peroxidase produced by *Phanerochaete chrysosporium* are especially important to bring about degradation of lignin (Koua et al. 2009).

Oxygenases are the most extensively studied enzymes in relation to biodegradation (Karigar and Rao 2011). The reason probably lies in the wide range of substrates on which they can act and degrade. They do so by increasing the reactivity of the organic compounds or making them water-miscible. In some cases, they can cleave the aromatic rings of the organic molecules and insert oxygen atoms which decomposes them (Fetzner 2003). Monooxygenases are produced by both prokaryotes and eukaryotes that can catalyse a number of reactions including denitrification, hydroxylation, dehalogenation, ammonification, biotransformation, etc., that have been applied for biodegradation of toxic aromatic compounds (Arora et al. 2010). The mechanism of action of monooxygenases varies in the presence and absence of oxygen. They catalyse oxidative dehalogenation reactions when the supply of oxygen is sufficient. In the absence of enough oxygen, they catalyse reductive dehalogenation reactions. Whatever may be, the final products are labile and undergo chemical degradation (Jones et al. 2001). Dioxygenases find application in environmental bioremediation because of their multicomponent system that can be utilized to oxidize aromatic compounds (Dua et al. 2002). Catechol dioxygenases obtained abundantly from soil bacteria can be used to convert aromatic to aliphatic compounds and thereby their degradation (Que and Ho 1996).

Cellulases are a group of interesting enzymes consisting of a cocktail of endoglucanase, exoglucanase, β-glucosidase and some other auxiliary enzymes that catalyse the degradation of cellulose (Sun and Cheng 2002). These enzymes can be extracellular, cell bound or even cell envelope associated that are produced in low concentrations by bacteria and fungi (Adriano-Anaya et al. 2005). Besides bioremediation, they find application in the detergent industry, paper and pulp industry and also in the brewing of fruit juice (Karigar and Rao 2011). Some of the most common cellulases include the neutral and acidic cellulases by the fungi, *Humicola* sp. and *Trichoderma* sp. as well as the alkaline cellulases produced by *Bacillus* sp. bacteria (Leisola et al. 2006).

7.7 EFFICIENCY OF WASTEWATER REMEDIATION

Over the years, bioremediation techniques have gained popularity in degrading the organic wastes in water (Azubuike et al. 2016). However, not all processes are equally efficient. For example, earlier studies on wastewater remediation state that activated sludge is not a very efficient technique to remove total coliform bacteria. Lack of proper management has been highlighted as one of the most important reasons for this. As such, the process needs more careful handling (Aghalari et al. 2020). It has been suggested that treatments such as lime and chlorine coagulation, activated carbon filtration, oxidation pools and activated sludge can eliminate almost 50% to 90% of the viruses present in wastewater (Okoh et al. 2010). Stabilized pond systems have been found to be very efficient to remove the parasites and/or their eggs and cysts. However, they require large areas of land and emit odours because of the poor quality of effluents being drained into them and also because of the algae. This restricts their use in the vicinity of human habitations (Aghalari et al. 2020).

Some studies also give contradictory reports on efficiencies. For instance, according to Sharafi et al. (2016) parasite eggs and protozoan cysts could be removed by natural remediation of wastewater in wetland systems. However, some other reports claim that wetlands are very easy to maintain and are economical but they are not very efficient in removing the microbial contaminants (Aghalari et al. 2020). Application of the UV disinfection system has been found to be very efficient in removing the pathogenic organisms from the wastewater but have certain drawbacks like requirement of high-quality effluents which makes the process very expensive (Aghalari et al. 2020).

Since no single process of remediation is 100% efficient some precautions should be followed while using treated wastewater for irrigation of crops and vegetables. For example, in many cases it is advised that the application of treated wastewater should be suspended at least two weeks prior to harvest or, in case of fodder crops/grasslands the livestock animals should be allowed to graze only after a break of approximately two weeks of using treated wastewater for irrigation (Mora-Ravelo et al. 2017). These are precautionary steps to prevent the introduction of toxic substances into the food chain and should be rigorously maintained.

7.8 CONSTRAINTS OF WASTEWATER TREATMENT

Some important facts are to be considered while treating the wastewater. For example, the microorganisms can function properly only when certain conditions such as pH and dissolved oxygen are maintained because their enzymes function properly when the conditions are favourable (Ratnakar et al. 2016). Moreover, the presence of certain toxic substances has been found to slow down the process of microbial remediation or even halt the process altogether (Mora-Ravelo et al. 2017). For example, the activity of laccases can be inhibited by azide, cyanide, hydroxide and halides other than iodide (Xu 1996). These substances need some special treatments before pure water can be recovered. Additionally, nitrogen concentration affects laccase production in fungi. The laccase production is usually high at greater concentrations of nitrogen (Gianfreda et al. 1999). Another important constraint is that the process is very slow and only a few microbes produce the specific enzymes required for bioremediation (Bhandari et al. 2021). Though application of advanced technology (such as production of bioengineered microbes and/or their enzymes) has been suggested as remedies they are very expensive which restricts their application in resource-deprived countries.

7.9 FUTURE PERSPECTIVES

Wastewater management is essential not only for making it suitable to be used in the agricultural sector but also to protect the environment from its damaging effects. Though a lot has been done to take care of the organic, inorganic and biological components, literature raises doubt regarding the efficiency of the techniques. Sometimes clumping of more than one technique is required for proper wastewater remediation. The sludge generated in the process is also problematic as far as its reuse and/or disposal is concerned. The greatest disadvantages are the expenses involved and the slow rates of remediation that have prevented its mass acceptance. Future research should be focussed on development of cheaper and faster techniques to recycle wastewater and the sludge generated. Moreover, there should be a strict monitoring of the processes with stringent laws to ensure that the recommended standards of recycling wastewater for all purposes (including agriculture) are being followed. This can ensure a healthier environment for a healthier future.

7.10 CONCLUSION

The problem of water crisis will be getting worse in the coming years because of population explosion and global climate change. The pressure on the agriculture sector will increase which, in turn, will require an adequate supply of freshwater. Since the wastewater production will also increase to fulfil the demands of the growing population, recycling it in a proper way seems to offer two solutions at the same time – firstly, it can supplement the use of freshwater in agriculture (to some extent) and secondly, it can protect the environment from the toxic effects of untreated wastewater. Thus, this field of research needs more investment and encouragement.

7.11 DECLARATIONS

Funding: No funding was received for conducting this study.

Conflicts of interest/Competing interests: The author has no conflicts of interest to declare that are relevant to the content of this article.

REFERENCES

Abdel-Raouf, N., A. A. Al-Homaidan, and I. B. M. Ibraheem. 2012. "Microalgae and wastewater treatment." *Saudi Journal of Biological Sciences* 19: 257–275. doi:10.1016/j.sjbs.2012.04.005.

Adriano-Anaya, M., M. Salvador-Figueroa, J. A. Ocampo, and I. García-Romera. 2005. "Plant cell-wall degrading hydrolytic enzymes of *Gluconacetobacter diazotrophicus*." *Symbiosis* 40: 151–156.

Aghalari, Z., H.-U. Dahms, M. Sillanpää, J. E. Sosa-Hernandez, and R. Parra-Saldívar. 2020. "Effectiveness of wastewater treatment systems in removing microbial agents: a systematic review." *Globalization and Health* 16: 13. doi:10.1186/s12992-020-0546-y.

Ahmed, J., A. Thakur, and A. Goyal. 2021. "Industrial wastewater and its toxic effects." In *Biological Treatment of Industrial Wastewater*, edited by Maulin P. Shah, 1–14. The Royal Society of Chemistry. doi:10.1039/9781839165399-00001.

Alghobar, M. A., L. Ramachandra, and S. Suresha. 2014. "Effect of sewage water irrigation on soil properties and evaluation of accumulation of elements in grass crop in Mysore city, Karnataka, India." *American Journal of Environmental Protection* 3: 283–291. doi:10.11648/j.ajep.20140305.22.

Allpress, J. D., and P. C. Gowland. 1998. "Dehalogenases: environmental defence mechanism and model of enzyme evolution." *Biochemical Education* 26: 267–276.

Amara, A. A., and S. R. Salem. 2009. "Degradation of castor oil and lipase production by *Pseudomonas aeruginosa*." *American-Eurasian Journal of Agricultural and Environmental Science* 5: 556–563.

Anzenbacher, P., and E. Anzenbacherová. 2001. "Cytochromes P450 and metabolism of xenobiotics." *Cellular and Molecular Life Sciences* 58: 737–747. doi:10.1007/pl00000897.

Arenas, A. D., L.-M. Marcó, and G. Torres. 2011. "Evaluación de la planta Lemna minor como biorremediadora de aguas contaminadas con mercurio." *Avances en Ciencias e Ingeniería* 2: 1–11.

Arias, M. E., M. Arenas, J. Rodríguez, J. Soliveri, A. S. Ball, and M. Hernández. 2003. "Kraft pulp biobleaching and mediated oxidation of a nonphenolic substrate by laccase from *Streptomyces cyaneus* CECT 3335." *Applied and Environmental Microbiology* 69: 1953–1958. doi:10.1128/AEM.69.4.1953-1958.2003.

Arora, P. K., A. Srivastava, and V. P. Singh. 2010. "Application of monooxygenases in dehalogenation, desulphurization, denitrification and hydroxylation of aromatic compounds." *Journal of Bioremediation and Biodegradation* 1: 112. doi:10.4172/2155-6199.1000112.

Ayangbenro, A. S., and O. O. Babalola. 2017. "A new strategy for heavy metal polluted environments: a review of microbial biosorbents." *International Journal of Environmental Research and Public Health* 14: 94. doi:10.3390/ijerph14010094.

Azubuike, C. C., C. B. Chikere, and G. C. Okpokwasili. 2016. "Bioremediation techniques-classification based on site of application: principles, advantages, limitations and prospects." *World Journal of Microbiology & Biotechnology* 32: 180. doi:10.1007/s11274-016-2137-x.

Balakrishnan, H., and R. Velu. 2015. "Eco-friendly technologies for heavy metal remediation: pragmatic approaches." In *Environmental Sustainability* edited by P. Thangavel, and G. Sridevi, 205–215. New Delhi: Springer. doi:10.1007/978-81-322-2056-5_12.

Balkhair, K. S., and M. A. Ashraf. 2016. "Field accumulation risks of heavy metals in soil and vegetable crop irrigated with sewage water in western region of Saudi Arabia." *Saudi Journal of Biological Sciences* 23: S32–S44. doi:10.1016/j.sjbs.2015.09.023.

Bansal, A., O. Shinde, and S. Sarkar. 2018. "Industrial wastewater treatment using phycoremediation technologies and co-production of value-added products." *Journal of Bioremediation and Biodegradation* 9: 428. doi:10.4172/2155-6199.1000428.

Barakat, M. A. 2011. "New trends in removing heavy metals from industrial wastewater." *Arabian Journal of Chemistry* 4: 361–377. doi:10.1016/j.arabjc.2010.07.019.

Bazrafshan, E., L. Mohammadi, A. Ansari-Moghaddam, and A. H. Mahvi. 2015. "Heavy metals removal from aqueous environments by electrocoagulation process- a systematic review." *Journal of Environmental Health Science and Engineering* 13: 74. doi:10.1186/s40201-015-0233-8.

Becerra-Castro, C., P. S. Kidd, B. Rodríguez-Garrido, C. Monterroso, P. Santos-Ucha, and Á. Prieto-Fernández. 2013. "Phytoremediation of hexachlorocyclohexane (HCH)-contaminated soils using *Cytisus striatus* and bacterial inoculants in soils with distinct organic matter content." *Environmental Pollution* 178: 202–210. doi:10.1016/j.envpol.2013.03.027.

Bhandari, S., D. K. Poudel, R. Marahatha, S. Dawadi, K. Khadayat, S. Phuyal, S. Shrestha, S. Gaire, K. Basnet, U. Khadka, and N. Parajuli. 2021. "Microbial enzymes used in bioremediation." *Journal of Chemistry* 8849512: 1–17. doi:10.1155/2021/8849512.

Bhat, U. N., and A. B. Khan. 2011. "Heavy metals: an ambiguous category of inorganic contaminants, nutrients and toxins." *Research Journal of Environmental Sciences* 5: 682–690. doi:10.3923/rjes.2011.682.690.

Butler, E., Y.-T. Hung, M. S. Al Ahmad, R. Yu-Li Yeh, R. Lian-Huey Liu, and Y.-P. Fu. 2017. "Oxidation pond for municipal wastewater treatment." *Applied Water Science* 7: 31–51. doi:10.1007/s13201-015-0285-z.

Cairns, S. S., A. Cornish, and R. A. Cooper. 1996. "Cloning, sequencing and expression in *Escherichia coli* of two *Rhizobium* sp. genes encoding haloalkanoate dehalogenases of opposite stereospecificity." *European Journal of Biochemistry* 235: 744–749. doi:10.1111/j.1432-1033.1996.t01-1-00744.x.

Cañizares-Villanueva, R., Ma. del C. Montes-Horcasitas, and H. G. Perales-Vela. 2013. "Las microalgas en la biorremediación acuática: Una alternativa biotecnológica." In *Biorremediación de suelos y aguas*, edited by Ferrera-Cerrato Alarcón, 217–234. Mexico: Trillas.

Cao, H., H. Sun, and H. Yang. 2003. "A review: soil enzyme activity and its indication for soil quality." *Chinese Journal of Applied and Environmental Biology* 9: 105–109.

Casas-Godoy, L., S. D. Florence Bordes, G. Sandoval, and A. Marty. 2012. "Lipases: an overview." In *Lipases and Phospholipases. Methods in Molecular Biology* vol. 861, edited by G. Sandoval, 3–30. Humana Press. doi:10.1007/978-1-61779-600-5_1.

Chakraborty, J., and S. Das (2016). "Molecular perspectives and recent advances in microbial remediation of persistent organic pollutants." *Environmental Science and Pollution Research International* 23, no. 17: 16883–16903. doi:10.1007/s11356-016-6887-7.

Chen, P. H., K. C. Leung, and J. T. Wang. 2000. "Investigation of a ponding irrigation system to recycle agricultural wastewater." *Environment International* 26: 63–68. doi:10.1016/s0160-4120(00)00079-9.

Chen, G., J. Qin, D. Shi, Y. Zhang, and W. Ji. 2009. "Diversity of soil nematodes in areas polluted with heavy metals and polycyclic aromatic hydrocarbons (PAHs) in Lanzhou, China." *Environmental Management* 44: 163–172. doi:10.1007/s00267-008-9268-2.

Cherian, S., and M. Margarida Oliveira. 2005. "Transgenic plants in phytoremediation: recent advances and new possibilities." *Environmental Science and Technology* 39: 9377–90. doi:10.1021/es051134l.

Chibuike, G. U., and S. C. Obiora. 2014. "Heavy metal polluted soils: effect on plants and bioremediation methods." *Applied and Environmental Soil Science* 752708: 1–12. doi:10.1155/2014/752708.

Chowdhury, S., N. N. Bala, and P. Dhauria. 2012. Bioremediation-a natural way for cleaner environment. *International Journal of Pharmaceutical, Chemical and Biological Sciences* 2: 600–611.

Dang, T. C. H., D. T. Nguyen, H. Thai, T. Chinh Nguyen, T. T. Hien Tran, V. Hung Le, V. H. Nguyen, X. Bach Tran, T. P. Thao Pham, and T. Giang Nguyen. 2018. "Plastic degradation by thermophilic *Bacillus* sp. BCBT21 isolated from composting agricultural residual in Vietnam." *Advances in Natural Sciences: Nanoscience and Nanotechnology* 9: 015014. doi:10.1088/2043-6254/aaabaf.

de Lourdes Moreno, M., D. Pérez, M. T. García, and E. Mellado. 2013. "Halophilic bacteria as a source of novel hydrolytic enzymes." *Life* 3: 38–51. doi:10.3390/life3010038.

Delgadillo-López, A. E., C. A. González-Ramírez, F. Prieto-García, J. R. Villagómez-Ibarra, and O. Acevedo-Sandoval. 2011. "Fitorremediación: una alternativa para eliminar la contaminación." *Tropical and Subtropical Agroecosystems* 14: 597–612.

Dheri, G. S., M. S. Brar, and S. S. Malhi. 2007. "Heavy-metal concentration of sewage-contaminated water and its impact on underground water, soil, and crop plants in alluvial soils of Northwestern India." *Communications in Soil Science and Plant Analysis* 38: 1353–1370. doi:10.1080/00103620701328743.

Dixit, R. W., D. Malaviya, K. Pandiyan, U. B. Singh, A. Sahu, R. Shukla, B. P. Singh, J. P. Rai, P. K. Sharma, H. Lade, and D. Paul. 2015. "Bioremediation of heavy metals from soil and aquatic environment: an overview of principles and criteria of fundamental processes." *Sustainability* 7, no. 2: 2189–2212. doi:10.3390/su7022189.

Doorn, M. R. J., S. Towprayoon, S. Maria, M. Vieira, W. Irving, C. Palmer, R. Pipatti, and C. Wang. 2006. "Wastewater treatment and discharge." In *2006 IPCC Guidelines for National Greenhouse Gas Inventories vol. 5. (Waste)*, 1–6. WMO, UNEP.

Dsikowitzky, L., and J. Schwarzbauer. 2013. "Organic contaminants from industrial wastewaters: identification, toxicity and fate in the environment." In *Pollutant diseases, remediation and recycling*, edited by Eric Lichtfouse, Jan Schwarzbauer, and Didier Robert, 45–101. Switzerland: Springer International Publishing. doi:10.1007/978-3-319-0238 7-8_2.

Du, B., J. Gong, and J. L. Li. 2010. "Study on organic pollution of soil and water environment by sewage irrigation in Taiyuan City." *Yangtze River* 41: 58–61.

Dua, M., A. Singh, N. Sethunathan, and A. Johri. 2002. "Biotechnology and bioremediation: successes and limitations." *Applied Microbiology and Biotechnology* 59: 143–152. doi:10.1007/s00253-002-1024-6.

Elekwachi, C. O., J. Andresen, and T. C. Hodgman. 2014. "Global use of bioremediation technologies for decontamination of ecosystems." *Journal of Bioremediation and Biodegradation* 5: 225. doi:10.4172/2155-6199.1000225.

Elimelech, M. 2006. "The global challenge for adequate and safe water." *Journal of Water Supply: Research and Technology-Aqua* 55: 3–10. doi:10.2166/aqua.2005.064.

El-Nahhal, Y., Y. Awad, and J. Safi. 2013. "Bioremediation of acetochlor in soil and water systems by cyanobacterial mat." *International Journal of Geosciences* 4, no. 5: 880–890. doi:10.4236/ijg.2013.45082.

Fang, Y. D. 2011. "Research on the current situation, prevention and treatment of farmland wastewater irrigation in China." *Journal of Agricultural Resources and Environment* 5: 1–6.

Farraji, H. 2014. "Wastewater treatment by phytoremediation methods." In *Wastewater Engineering: Types, Characteristics and Treatment Technologies*, edited by H. Aziz, 205–218. Malaysia: IJSRPUB.

Ferrera-Cerrato, R., N. G. Rojas-Avelizapa, H. M. Poggi-Varaldo, A. Alarcón, and R. O. Cañizares-Villanueva. 2006. "Procesos de biorremediación de suelo y agua contaminados por hidrocarburos del petróleo y otros compuestos orgánicos." *Revista Latinoamericana de Microbiología* 48: 179–187.

Fetzner, S., and F. Lingens. 1994. "Bacterial dehalogenases: biochemistry, genetics, and biotechnological applications." *Microbiological Reviews,* 58, no. 4: 641–685. doi:10.1128/mr.58.4.641-685.1994.

Fetzner, S. 2003. "Oxygenases without requirement for cofactors or metal ions." *Applied Microbiology and Biotechnology* 60: 243–257. doi:10.1007/s00253-002-1123-4.

Flores-Magdaleno, H., O. R. Mancilla-Villa, E. Mejía-Saenz, Ma. del, C. Olmedo-Bolaños, and A. L. Bautista-Olivas. 2011. "Heavy metals in agricultural soils and irrigation wastewater of Mixquiahuala, Hidalgo, Mexico." *African Journal of Agricultural Research* 6, no. 24: 5505–5511.

Friedel, J. K., T. Langer, C. Siebe, and K. Stahr. 2000. "Effects of long-term waste water irrigation on soil organic matter, soil microbial biomass and its activities in Central Mexico." *Biology and Fertility of Soils* 31: 414–421. doi:10.1007/s003749900188.

Ge, H. L., L. Chen, J. L. Zhang, and W. S. Huang. 2009. "Effect of long-term sewage irrigation on rhizosphere soil microbial populations of wheat." *Water Saving Irrigation* 5: 14–15.

Genuis, S. J., and K. L. Kelln. 2015. "Toxicant exposure and bioaccumulation: a common and potentially reversible cause of cognitive dysfunction and dementia." *Behavioural Neurology* 620143. doi:10.1155/2015/620143.

Gescher, J., W. Ismail, E. Ölgeschläger, W. Eisenreich, J. Wörth, and G. Fuchs. 2006. "Aerobic benzoyl-coenzyme A (CoA) catabolic pathway in *Azoarcus evansii*: conversion of ring cleavage product by 3, 4-dehydroadipyl-CoA semialdehyde dehydrogenase." *Journal of Bacteriology* 188, no. 8: 2919–27. doi:10.1128/JB.188.8.2919-2927.2006.

Ghosh, M., and S. P. Singh. 2005. "A review on phytoremediation of heavy metals and utilization of its byproducts." *Applied Ecology and Environmental Research* 3, no. 1: 1–18.

Gianfreda, L., F. Xu, and J.-M. Bollag. 1999. "Laccases: a useful group of oxidoreductive enzymes." *Bioremediation Journal* 3, no. 1: 1–26. doi:10.1080/10889869991219163.

Grabow, G. O. K. 1997. "Hepatitis viruses in water: update on risk and control." *Water SA* 23: 379–386.

Guan, Z.-B., Q. Luo, H.-R. Wang, Y. Chen, and X.-R. Liao. 2018. "Bacterial laccases: promising biological green tools for industrial applications." *Cellular and molecular life sciences* 75, no. 19: 3569–3592. doi:10.1007/s00018-018-2883-z.

Haider, T. P., C. Völker, J. Kramm, K. Landfester, and F. R. Wurm. 2019. "Plastics of the future? The impact of biodegradable polymers on the environment and on society." *Angewandte Chemie International Edition* 58, no. 1: 50–62. doi:10.1002/anie.201805766.

Hassan, Sahar W. M., H. H. Abd El Latif, and S. M. Ali. 2018. "Production of cold-active lipase by free and immobilized marine *Bacillus cereus* HSS: application in wastewater treatment." *Frontiers in Microbiology* 9: 2377. doi:10.3389/fmicb.2018.02377.

Hernández, J. P., M. R. Befani, N. G. Boschetti, C. E. Quintero, E. L. Díaz, M. Lado, and A. Paz-González. 2015. "Impact of soil resistance to penetration in the irrigation interval of supplementary irrigation systems at the Humid Pampa, Argentina." *Journal of Cardiac Failure* 12: 87–88.

Hussain, M., A. R. Butt, F. Uzma, R. Ahmed, S. Irshad, A. Rehman, and B. Yousaf. 2019. "A comprehensive review of climate change impacts, adaptation, and mitigation on environmental and natural calamities in Pakistan." *Environmental Monitoring and Assessment* 192, no. 1: 48. doi:10.1007/s10661-019-7956-4.

Ihssen, J., R. Reiss, R. Luchsinger, L. Thöny-Meyer, and M. Richter. 2015. "Biochemical properties and yields of diverse bacterial laccase-like multicopper oxidases expressed in *Escherichia coli*." *Scientific Reports* 5, no. 1: 10465. doi:10.1038/srep10465.

IPCC (Intergovermental Panel on Climate Change). "Climate change 2013: the physical science basis." In *Contribution of Working Group I to the Fifth Assessment Report of the Intergovernmental Panel on Climate Change*, edited by T. F. Stocker, D. Qin, G.-K. Plattner, M. Tignor, S. K. Allen, J. Boschung, A. Nauels, Y. Xia, V. Bex and P. M. Midgley, 1535. Cambridge: Cambridge University Press.

Jan, A. T., M. Azam, K. Siddiqui, A. Ali, I. Choi and Q. Mohd, R. Haq. 2015. "Heavy metals and human health: mechanistic insight into toxicity and counter defense system of antioxidants." *International Journal of Molecular Sciences* 16, no. 12: 29592–29630. doi:10.3390/ijms161226183.

Jaromin-Gleń, K., T. Kłapeć, G. Łagód, J. Karamon, J. Malicki, A. Skowrońska, and A. Bieganowski. 2017. "Division of methods for counting helminths' eggs and the problem of efficiency of these methods." *Annals of Agricultural and Environmental Medicine* 24, no. 1: 1–7. doi:10.5604/12321966.1233891.

Jellouli, K., O. Ghorbel-Bellaaj, H. B. Ayed, L. Manni, R. Agrebi, and M. Nasri. 2011. "Alkaline-protease from *Bacillus licheniformis* MP1: purification, characterization and potential application as a detergent additive and for shrimp waste deproteinization." *Process Biochemistry* 46, no. 6: 1248–1256.

Jeong, H., C. Seong, T. Jang, and S. Park. 2016. "Classification of wastewater reuse for agriculture: a case study in South Korea." *Irrigation and Drainage* 65: 76–85. doi:10.1002/ird.2053.

Jiao, Y., X. Chen, X. Wang, X. Liao, L. Xiao, A. Miao, J. Wu, and L. Yang. 2013. "Identification and characterization of a cold-active phthalate esters hydrolase by screening a metagenomic library derived from biofilms of a wastewater treatment plant." *PloS One* 8, no. 10: e75977. doi:10.1371/journal.pone.0075977.

Jones, J. P., E. J. O'Hare, and L.-L. Wong. 2001. "Oxidation of polychlorinated benzenes by genetically engineered CYP101 (cytochrome P450(cam))." *European Journal of Biochemistry* 268: 1460–1467. doi:10.1046/j.1432-1327.2001.02018.x.

Karigar, C. S. and S. S. Rao. 2011. "Role of microbial enzymes in the bioremediation of pollutants: a review." *Enzyme Research* 805187. doi:10.4061/2011/805187.

Kaur, P., S. Singh, V. Kumar, N. Singh, and J. Singh. 2018. "Effect of rhizobacteria on arsenic uptake by macrophyte *Eichhornia crassipes* (Mart.) Solms." *International Journal of Phytoremediation* 20, no. 2: 114–120. doi:10.1080/15226514.2017.1337071.

Kavitha, R. V., V. K. Murthy, R. Makam, and K. A. Asith. 2012. "Physico-chemical analysis of effluents from pharmaceutical industry and its efficiency study." *International Journal of Engineering Research and Applications* 2, no. 2: 103–110.

Kawai, F., and H. Yamanaka. 1989. "Inducible or constitutive polyethylene glycol dehydrogenase involved in the aerobic metabolism of polyethylene glycol." *Journal of Fermentation and Bioengineering* 67: 300–302.

Khan, N., and A. Bano. 2016. "Role of plant growth promoting rhizobacteria and Ag-nano particle in the bioremediation of heavy metals and maize growth under municipal wastewater irrigation." *International Journal of Phytoremediation* 18, no. 3: 211–221. doi:10.1080/15226514.2015.1064352.

Khan, N. 2018. "Natural ecological remediation and reuse of sewage water in agriculture and its effects on plant health." In *Sewage*, edited by I. X. Zhu. Intechopen. doi:10.5772/intechopen.75455.

Kim, J.-S., J.-W. Park, S.-E. Lee, and J.-E. Kim. 2002. "Formation of bound residues of 8-hydroxybentazon by oxidoreductive catalysts in soil." *Journal of Agricultural and Food Chemistry* 50, no. 12: 3507–3511. doi:10.1021/jf011504z.

Kornaros, M., and G. Lyberatos. 2006. "Biological treatment of wastewaters from a dye manufacturing company using a trickling filter." *Journal of Hazardous Materials* 136, no. 1: 95–102. doi:10.1016/j.jhazmat.2005.11.018.

Koua, D., L. Cerutti, L. Falquet, C. J. A. Sigrist, G. Theiler, N. Hulo, and C. Dunand. 2009. "PeroxiBase: a database with new tools for peroxidase family classification." *Nucleic Acids Research* 37: D261–D266. doi:10.1093/nar/gkn680.

Kumar, A., and S. Sharma. 2019. *Microbes and Enzymes in Soil Health and Bioremediation*. Singapore: Springer.

Kumar, L., G. Awasthi, and B. Singh. 2011. "Extremophiles: a novel source of industrially important enzymes." *Biotechnology* 10, no. 2: 121–135. doi:10.3923/biotech.2011.121.135.

Kumar, V., N. Upadhyay, V. Kumar, S. Kaur, J. Singh, S. Singh, and S. Datta. 2014. "Environmental exposure and health risks of the insecticide monocrotophos-a review." *Journal of Biodiversity and Environmental Sciences* 5, no. 1: 111–120.

Lan, M. J., M. S. Li, G. J. Zhao, and L. Rui. 2000. "Effects of eutrophic sewage irrigation on soil-holding capacity." *Journal of Shihezi University* 28: 497–500.

Lederberg, J. 1997. "Infectious disease as an evolutionary paradigm." *Emerging Infectious Diseases* 3: 417–423.

Leisola, M., J. Jokela, O. Pastinen, and O. Turunen. 2006. *Industrial Use of Enzymes-Essay, Laboratory of Bioprocess Engineering*. Espoo: Helsinki University of Technology.

Leung, M. 2004. "Bioremediation: techniques for cleaning up a mess." *BioTeach Journal* 2: 18–22.

Li, L. Q. 2001. "Study on properties of physics and chemistry of cinnamon soil with wastewater irrigation." *Journal of Shanxi Agricultural University* 21: 73–75.

Li, Q., J. Tang, T. Wang, D. Wu, C. A. Busso, R. Jiao, and X. Ren. 2018. "Impacts of sewage irrigation on soil properties of farmland in China: a review." *ϕYTON* 87: 40–50.

Littlechild, J. A. 2015. "Archaeal enzymes and applications in industrial biocatalysts." *Archaea* 147671: 1–10. doi:10.1155/2015/147671.

Liu, S. Q. 1996. "Relationship between soil Pb and Cd pollution and enzyme activities in wastewater irrigated area of Baoding city." *Acta Pedologica Sinica* 33, no. 2: 175–182.

Liu, L., and G. H. Lu. 2002. "Nitrogen wastewater irrigation study and its contamination risk analysis." *Advances in Water Science* 13: 313–318.

Liu, L., M. Bilal, X. Duan, and H. M. N. Iqbal. 2019. "Mitigation of environmental pollution by genetically engineered bacteria - Current challenges and future perspectives." *The Science of the Total Environment* 667: 444–454. doi:10.1016/j.scitotenv.2019.02.390.

Llugany, M., R. Tolrà, C. Poschnrieder, and J. Barceló. 2007. "Hiperacumulación de metales: ¿una ventaja para la planta y para el hombre?" *Ecosistema* 16: 4–9.

Lottermoser, B. G. 2012. "Effect of long-term irrigation with sewage effluent on the metal content of soils, Berlin, Germany." *Environmental Geochemistry and Health* 34, no. 1: 67–76. doi:10.1007/s10653-011-9391-5.

Lucho-Constantino, C. A., M. Álvarez-Suárez, R. I. Beltrán-Hernández, F. Prieto-García, and H. M. Poggi-Varaldo. 2005. "A multivariate analysis of the accumulation and fractionation of major and trace elements in agricultural soils in Hidalgo State, Mexico irrigated with raw wastewater." *Environment International* 31, no. 3: 313–323. doi:10.1016/j.envint.2004.08.002.

Ma, Q., and G. X. Zhao. 2010. "Effects of different land use types on soil nutrients in intensive agricultural region." *Journal of Natural Resources* 25: 1834–1844.

Mai, C., W. Schormann, O. Milstein, and A. Hüttermann. 2000. "Enhanced stability of laccase in the presence of phenolic compounds." *Applied Microbiology and Biotechnology* 54, no. 4: 510–514. doi:10.1007/s002530000452.

Mani, D., and C. Kumar. 2014. "Biotechnological advances in bioremediation of heavy metals contaminated ecosystems: an overview with special reference to phytoremediation." *International Journal of Environmental Science and Technology* 11: 843–872. doi:10.1007/s13762-013-0299-8.

Mara, D., and R. G. A. Feachem. 2003. "Unitary environmental classification of water-and excreta-related communicable diseases." In *Handbook of Water and Wastewater Microbiology*, edited by D. Mara, and N. Horan, 185–192. Academic Press. doi:10.1016/B978-012470100-7/50012-1.

Masciandaro, G., R. Iannelli, M. Chiarugi, and E. Peruzzi. 2015. "Reed bed systems for sludge treatment: case studies in Italy." *Water Science and Technology* 72: 1043–1050. doi:10.2166/wst.2015.309.

Masto, R. E., P. K. Chhonkar, D. Singh, and A. K. Patra. 2009. "Changes in soil quality indicators under long-term sewage irrigation in a sub-tropical environment." *Environmental Geology* 56:1237–1243.

Mazotto, A. M., A. C. N. de Melo, A. Macrae, A. S. Rosado, R. Peixoto, S. M. L. Cedrola, S. Couri, R. B. Zingali, A. L. V. Villa, L. Rabinovitch, J. Q. Chaves, and A. B. Vermelho. 2011. "Biodegradation of feather waste by extracellular keratinases and gelatinases from *Bacillus* spp." *World Journal of Microbiology & Biotechnology* 27: 1355–1365. doi:10.1007/s11274-010-0586-1.

Megharaj, M., B. Ramakrishnan, K. Venkateswarlu, N. Sethunathan, and R. Naidu. 2011. "Bioremediation approaches for organic pollutants: a critical perspective." *Environment International* 37: 1362–1375. doi:10.1016/j.envint.2011.06.003.

Meli, S., M. Porto, A. Belligno, S. A. Bufo, A. Mazzatura, and A. Scopa. 2002. "Influence of irrigation with lagooned urban wastewater on chemical and microbiological soil parameters in a citrus orchard under Mediterranean condition." *The Science of the Total Environment* 285: 69–77. doi:10.1016/s0048-9697(01)00896-8.

Mora-Ravelo, S. G., A. Alarcón, M. Rocandio-Rodríguez, and V. Vanoye-Eligio. 2017. "Bioremediation of wastewater for reutilization in agricultural systems: a review." *Applied Ecology and Environmental Research* 15, no. 1: 33–50. doi:10.15666/aeer/1501_033050.

Mtshali, J. S., A. T. Tiruneh, and A. O. Fadiran. 2014. "Characterization of sewage sludge generated from wastewater treatment plants in Swaziland in relation to agricultural uses." *Resources and Environment* 4, no. 4: 190–199.

Muthukumarasamy, N. P., B. Jackson, A. J. Raj, and M. Sevanan. 2015. "Production of extracellular laccase from *Bacillus subtilis* MTCC 2414 using agroresidues as a potential substrate." *Biochemistry Research International* 765190. doi:10.1155/2015/765190.

Naidoo, S., and A. O. Olaniran. 2014. "Treated wastewater effluent as a source of microbial pollution of surface water resources." *International Journal of Environmental Research and Public Health* 11, no. 1: 249–70. doi:10.3390/ijerph110100249.

Nasir, N. M., N. S. A. Bakar, F. Lananan, S. H. A. Hamid, S. S. Lam, and A. Jusoh. 2015. "Treatment of African catfish, *Clarias gariepinus* wastewater utilizing phytoremediation of microalgae, *Chlorella* sp. with *Aspergillus niger* bio-harvesting." *Bioresource Technology* 190: 492–498. doi:10.1016/j.biortech.2015.03.023.

Nicholls, R. J., N. Marinova, J. A. Lowe, S. Brown, P. Vellinga, D. de Gusmão, J. Hinkel, and R. S. J. Tol. 2011. "Sea-level rise and its possible impacts given a 'beyond 4°C world' in the twenty-first century." *Philosophical Transactions. Series A, Mathematical, Physical, and Engineering Sciences* 369: 161–181. doi:10.1098/rsta.2010.0291.

Nnolim, N. E., A. I. Okoh, and U. U. Nwodo. 2020. "*Bacillus* sp. FPF-1 produced keratinase with high potential for chicken feather degradation." *Molecules* 25, no. 7: 1505. doi:10.3390/molecules25071505.

Noyola, R. A. 1999. "Una experiencia en el Desarrollo de tecnología biológica para el tratamiento de aguas residuales." *Interciencia* 24, no. 3: 169–172.

Okeyo, A. N., N. Nontongana, T. O. Fadare and A. I. Okoh. 2018. "*Vibrio* species in wastewater final effluents and receiving watershed in South Africa: implications for public health." *International Journal of Environmental Research and Public Health* 15, no. 6: 1266. doi:10.3390/ijerph15061266.

Okoh, Anthony I., Thulani Sibanda and Siyabulela S. Gusha. 2010. "Inadequately treated wastewater as a source of human enteric viruses in the environment." *International Journal of Environmental Research and Public Health* 7, no. 6: 2620–2637. doi:10.3390/ijerph7062620.

Olmos, M. A., and M. T. Alarcón Herrera. 2014. "Impacto ambiental generado por el uso de osmosis inversa en la remoción de arsénico para la obtención de agua potable." *Revista Ambiental* 5, no. 1: 1–5.

Osuolale, O., and A. Okoh. 2017. "Human enteric bacteria and viruses in five wastewater treatment plants in the Eastern Cape, South Africa." *Journal of Infection and Public Health* 10, no. 5: 541–547. doi:10.1016/j.jiph.2016.11.012.

Park, J.-H., Y.-J. Kim, B.-K., and K.-H. Seo. 2018. "Spread of multidrug-resistant *Escherichia coli* harboring integron via swine farm waste water treatment plant." *Ecotoxicology and Environmental Safety* 149: 36–42. doi:10.1016/j.ecoenv.2017.10.071.

Phale, P. S., A. Sharma, and K. Gautam. 2019. "Microbial degradation of xenobiotics like aromatic pollutants from the terrestrial environments." In *Pharmaceuticals and Personal Care Products: Waste Management and Treatment Technology*, edited by M. N. V. Prasad, M. Vithanage, and A. Kapley, 259–278. Butterworth-Heinemann: Elsevier.

Piao, S., P. Ciais, Y. Huang, Z. Shen, S. Peng, J. Li, L. Zhou, H. Liu, Y. Ma, Y. Ding, P. Friedlingstein, C. Liu, K. Tan, Y. Yu, T. Zhang, and J. Fang. 2010. "The impacts of climate change on water resources and agriculture in China." *Nature* 467, no. 7311: 43–51. doi:10.1038/nature09364.

Pilon-Smits, E.. 2005. "Phytoremediation." *Annual Review of Plant Biology* 56: 15–39. doi:10.1146/annurev.arplant.56.032604.144214.

Plevich, O. J., S. A. R. Delgado, C. Saroff1, J. C. Tarico, R. J. Crespi, and O. M. Barotto. 2012. "El cultivo de alfalfa utilizando agua de perforación, agua residual urbana y precipitaciones." *Revista Brasileira de Engenharia Agrícola e Ambiental* 16, no. 12: 1353–1358.

Poerio, T., E. Piacentini, and R. Mazzei. 2019. "Membrane Processes for Microplastic Removal." *Molecules* 24, no. 22: 4148. doi:10.3390/molecules24224148.

Que Jr., L., and Raymond Y. N. Ho. 1996. "Dioxygen activation by enzymes with mononuclear non-heme iron active sites." *Chemical Reviews* 96, no. 7: 2607–2624. doi:10.1021/cr960039f.

Ramírez Zamora, R. M., A. Chávez Mejia, R. Domínguez Mora, and A. Durán Moreno. 2004. "Performance of basaltic dust issued from an asphaltic plant as a flocculant additive for wastewater treatment." *Water science and technology: a journal of the International Association on Water Pollution Research* 49, no. 1: 147–54.

Rao, M. A., R. Scotti, and L. Gianfreda. 2010. "Role of enzymes in the remediation of polluted environments." *Journal of Soil Science and Plant Nutrition* 10, no. 3: 333–353. doi:10.4067/S0718-95162010000100008.

Raschid-Sally, L., and P. Jayakody. 2008. *Drivers and characteristics of wastewater agriculture in developing countries: results from a global assessment.* IWMI Research Reports H041686: International Water Management Institute.

Ratnakar, A., S. Shankar, and Shikha. 2016. An overview of biodegradation of organic pollutants. *International Journal of Scientific and Innovative Research* 4, no. 1: 73–91.

Rattan, R. K., S. P. Datta, P. K. Chhonkar, K. Suribabu, and A. K. Singh. 2005. "Long-term impact of irrigation with sewage effluents on heavy metal content in soils, crops and groundwater-a case study." *Agriculture Ecosystems and Environment* 109: 310–322. doi:10.1016/j.agee.2005.02.025.

Rezania, S., M. Ponraj, A. Talaiekhozani, S. E. Mohamad, M. F. Md Din, S. Mat Taib, F. Sabbagh, and F. Md Sairan. 2015. "Perspectives of phytoremediation using water hyacinth for removal of heavy metals, organic and inorganic pollutants in wastewater." *Journal of Environmental Management* 163: 125–133. doi:10.1016/j.jenvman.2015.08.018.

Riffaldi, R., R. Levi-Minzi, R. Cardelli, S. Palumbo, and A. Saviozzi. 2006. "Soil biological activities in monitoring the bioremediation of diesel oil-contaminated soil." *Water, Air, and Soil Pollution* 170: 3–15. doi:10.1007/s11270-006-6328-1.

Rodríguez Couto, S., and J. L. T. Herrera. 2006. "Industrial and biotechnological applications of laccases: a review." *Biotechnology Advances* 24, no. 5: 500–513. doi:10.1016/j.biotechadv.2006.04.003.

Rooney, C. P., F.-J. Zhao, and S. P. McGrath. 2006. "Soil factors controlling the expression of copper toxicity to plants in a wide range of European soils." *Environmental Toxicology and Chemistry* 25: 726–732. doi:10.1897/04-602r.1.

Rubilar, Olga, M. C. Diez, and L. Gianfreda. 2008. "Transformation of chlorinated phenolic compounds by white rot fungi." *Critical Reviews in Environmental Science and Technology* 38, no. 4: 227–68. doi:10.1080/10643380701413351.

Safary, S. A. A., and S. Hajrasoliha. 1995. "Effects of North Isfahan sewage effluent on the soils of Borkhar region and composition of alfalfa." *Paper Presented at the 5th Soil Science Congress*. Karaj, Iran: Agricultural vocational school.

Sánchez-Viveros, G., R. Ferrera-Cerrato, and A. Alarcón. 2013. "Potencial del simbiosistema Azolla-Anabaena Azollae en la destoxificación de aguas contaminadas." In *Biorremediación de suelos y aguas*, edited by A. Alarcón and R. Ferrera-Cerrato, 235–256. México: Trillas.

Sarabia-Meléndez, I. F., R. Cisneros-Almazán, J. Aceves-De Alba, H. M. Durán-García, and J. Castro-Larragoitia. 2011. "Calidad del agua de riego en suelos agrícolas y cultivos del Valle de San Luis Potosí, México." *Revista internacional de contaminación ambiental* 27, no. 2: 103–13.

Saraswat, R., V. Verma, S. Sistla, and I. Bhushan. 2017. "Evaluation of alkali and thermotolerant lipase from an indigenous isolated *Bacillus* strain for detergent formulation." *Electronic Journal of Biotechnology* 30: 33–38. doi:10.1016/j.ejbt.2017.08.007.

Satcher, D. 1995. Emerging infections: getting ahead of the curve. *Emerging Infectious Diseases* 1, no. 1: 1–6. doi:10.3201/eid0101.950101.

Schenk, G., I. Mateen, T.-K. Ng, M. M. Pedroso, N. Mitić, M. Jafelicci, R. F.C. Marques, L. R. Gahan, and D. L. Ollis. 2016. "Organophosphate-degrading metallohydrolases: structure and function of potent catalysts for applications in bioremediation." *Coordination Chemistry Reviews* 317: 122–31. doi:10.1016/j.ccr.2016.03.006.

Seeger, M., M. Hernández, V. Méndez, B. Ponce, M. Córdova, and M. González. 2010. "Bacterial degradation and bioremediation of chlorinated herbicides and biphenyls." *Journal of Soil Science and Plant Nutrition* 10, no. 3: 320–32. doi:10.4067/S0718-95162010000100007.

Sharafi, K., M. Moradi, A. Azari, H. Sharafi, and M. Pirsaheb. 2016. "Comparative evaluation of parasitic removal in municipal wastewater using constructed wetland and extended aeration-activated sludge system in full scale: Kermanshah province, Iran." *International Journal of Health and Life Sciences* 2, no. 2: e74091.

Sharma, S. 2012. "Bioremediation Features, Strategies, and Applications." *Asian Journal of Pharmacy and Life Science* 2: 202–13.

Siebe, C., and E. Cifuentes. 1995. "Environmental impact of wastewater irrigation in central Mexico: an overview." *International Journal of Environmental Health Research* 5, no. 2: 161–173. doi:10.1080/09603129509356845.

Singh, R., N. Gautam, A. Mishra, and R. Gupta. 2011. "Heavy metals and living systems: an overview." *Indian Journal of Pharmacology* 43: 246–53. doi:10.4103/0253-7613.81505.

Singh, R., M. Kumar, A. Mittal, and P. K. Mehta. 2016. "Microbial enzymes: industrial progress in 21st century." *3 Biotech* 6, no. 2: 174. doi:10.1007/s13205-016-0485-8.

Singh, N., V. Dalal, J. K. Mahto, and P. Kumar. 2017. "Biodegradation of phthalic acid esters (PAEs) and in silico structural characterization of mono-2-ethylhexyl phthalate (MEHP) hydrolase on the basis of close structural homolog." *Journal of Hazardous Materials* 338: 11–22. doi:10.1016/j.jhazmat.2017.04.055.

Singh, S., V. Kumar, A. Chauhan, S. Datta, A. B. Wani, N. Singh, and J. Singh. 2018. "Toxicity, degradation and analysis of the herbicide atrazine." *Environmental Chemistry Letters,* 16, no. 1: 211–237. doi:10.1007/s10311-017-0665-8.

Singh S., S. Singh, V. Kumar, S. Kumar, D. S. Dhanjal, R. Romero, S. Datta, P. Bhadrecha, and J. Singh. 2020. "Microbial Remediation for Wastewater Treatment." (pp. 57 - 72). In *Microbial Technology for Health and Environment*, edited by P. K. Arora, vol. 22. Singapore: Springer Nature.

Sondhi, S., P. Sharma, N. George, P. S. Chauhan, N. Puri, and N. Gupta. 2015. "An extracellular thermo-alkali-stable laccase from *Bacillus tequilensis* SN4, with a potential to biobleach softwood pulp." *3 Biotech* 5: 175–185. doi:10.1007/s13205-014-0207-z.

Sugimoto, M., M. Tanabe, M. Hataya, S. Enokibara, J. A. Duine, and F. Kawai. 2001. "The first step in polyethylene glycol degradation by sphingomonads proceeds via a flavoprotein alcohol dehydrogenase containing flavin adenine dinucleotide." *Journal of Bacteriology* 183: 6694–6698. doi:10.1128/JB.183.22.6694-6698.2001.

Sun, Y. and J. Cheng. 2002. "Hydrolysis of lignocellulosic materials for ethanol production: a review." *Bioresource Technology* 83, no. 1: 1–11. doi:10.1016/S0960-8524(01)00212-7.

Sutherland, T. D., R. J. Russell, and M. Selleck. 2002. "Using enzymes to clean up pesticide residues." *Pesticide Outlook* 13: 149–151.

Tachibana, S., N. Naka, F. Kawai, and M. Yasuda. 2008. "Purification and characterization of cytoplasmic NAD-dependent polypropylene glycol dehydrogenase from *Stenotrophomonas maltophilia*." *FEMS microbiology letters* 288, no. 2: 266–272. doi:10.1111/j.1574-6968.2008.01363.x.

Tchobanoglous, G., and L. F. Burton. 2005. *Ingeniería de aguas residuales: tratamiento, vertido y reutilización*. New York: Metcalf y Eddy (Inc.), Editorial McGraw Hill.

van Hamme, J. D., A. Singh, and O. P. Ward. 2003. "Recent advances in petroleum microbiology." *Microbiology and Molecular Biology Reviews: MMBR* 67, no. 4: 503–549. doi:10.1128/MMBR.67.4.503-549.2003.

van Leeuwen, J. (Hans), A. Sridhar, A. Kamel Harrata, M. Esplugas, S. Onuki, L. Cai, and J. A. Koziel. 2009. "Improving the biodegradation of organic pollutants with ozonation during biological wastewater treatment." *Ozone: Science and Engineering* 31, no. 2: 63–70. doi:10.1080/01919510802668380.

Vassilev, A., J.-P. Schwitzguebél, T. Thewys, D. van der Lelie, and J. Vangronsveld. 2004. "The use of plants for remediation of metal-contaminated soils." *The Scientific World Journal* 4: 9–34. doi:10.1100/tsw.2004.2.

Verma, S., J. Saxena, R. Prasanna, V. Sharma, and L. Nain. 2012. "Medium optimization for a novel crude-oil degrading lipase from *Pseudomonas aeruginosa* SL-72 using statistical approaches for bioremediation of crude-oil." *Biocatalysis and Agricultural Biotechnology* 1, no. 4: 321–329. doi:10.1016/j.bcab.2012.07.002.

Wan, L., M. R. Zhang, S. Lu, and K. Hu. 2015. "Study progress on effect of polluted water irrigation on soil and problem analysis." *Ecology and Environmental Sciences* 24: 906–10.

Wang, G. L., and W. J. Lin. 2003. "Contamination of soil from sewage irrigation and its remediation." *Journal of Agro-environmental Science* 22: 163–166.

Wang, T.-N., and M. Zhao. 2017. "A simple strategy for extracellular production of CotA laccase in *Escherichia coli* and decolorization of simulated textile effluent by recombinant laccase." *Applied Microbiology and Biotechnology* 101, no. 2: 685–696. doi:10.1007/s00253-016-7897-6.

Wang, Y., M. Qiao, Y. Liu, and Y. Zhu. 2012. "Health risk assessment of heavy metals in soils and vegetables from wastewater irrigated area, Beijing-Tianjin city cluster, China." *Journal of Environmental Sciences (China)* 24, no. 4: 690–698. doi:10.1016/s1001-0742(11)60833-4.

Wang, C., Q. Zhang, L. Jiang, and Z. Hou. 2018a. "The organic pollutant characteristics of lurgi coal gasification wastewater before and after ozonation." *Journal of Chemistry* 1461673. doi:10.1155/2018/1461673.

Wang, D., A. Li, H. Han, T. Liu, and Q. Yang. 2018b. "A potent chitinase from *Bacillus subtilis* for the efficient bioconversion of chitin-containing wastes." *International Journal of Biological Macromolecules* 116: 863–868. doi:10.1016/j.ijbiomac.2018.05.122.

Wang, Y., Y. Feng, X. Cao, Y. Liu, and S. Xue. 2018c. "Insights into the molecular mechanism of dehalogenation catalyzed by D-2-haloacid dehalogenase from crystal structures." *Scientific Reports* 8: 1454. doi:10.1038/s41598-017-19050-x.

Wei, R., T. Oeser, J. Schmidt, R. Meier, M. Barth, J. Then, and W. Zimmermann. 2016. "Engineered bacterial polyester hydrolases efficiently degrade polyethylene terephthalate due to relieved product inhibition." *Biotechnology and Bioengineering* 113, no. 8: 1658–1665. doi:10.1002/bit.25941.

Westerhoff, P., S. Lee, Y. Yang, G. W. Gordon, K. Hristovski, R. U. Halden, and P. Herckes. 2015. "Characterization, recovery opportunities, and valuation of metals in municipal sludges from US wastewater treatment plants nationwide." *Environmental Science and Technology* 49, no. 16: 9479–9488. doi:10.1021/es505329q.

Wuana, R. A., and F.E. Okieimen. 2011. "Heavy metals in contaminated soils: a review of sources, chemistry, risks and best available strategies for remediation." *ISRN Ecology* 402647. doi:10.5402/2011/402647.

Xu, Feng. 1996. "Catalysis of novel enzymatic iodide oxidation by fungal laccase." *Applied Biochemistry and Biotechnology* 59: 221–230. doi:10.1007/BF02783566.

Xu, G. 2007. "Effect of sewage irrigation with molasses alcohol wastewater on soil quality." PhD diss., Guangxi University, Nanning.

Xue, F., X. Ya, Q. Tong, Y. Xiu, and H. Huang. 2018. "Heterologous overexpression of *Pseudomonas umsongensis* halohydrin dehalogenase in *Escherichia coli* and its application in epoxide asymmetric ring opening reactions." *Process Biochemistry* 75: 139–145.

Xue, Z. J. 2012. *"Assessment of soil quality and pollution risk in main sewage-irrigated area of Hebei province."* Agricultural University of Hebei Province, Baoding.

Zhang, J., H. W. Zhang, Z. C. Su, X. Y. Li, and C. G. Zhang. 2007. "The effect of long-term organic wastewater irrigation on the soil nitrogen-fixing bacteria population." *Journal of Agro-Environment Science* 26: 662–666.

Zhang, Y. L., J. L. Dai, and R. Q. Wangab. 2008. "Effects of long-term sewage irrigation on agricultural soil microbial structural and functional characterizations in Shandong, China." *European Journal of Soil Biology* 44, no. 1: 84–91. doi:10.1016/j.ejsobi.2007.10.003.

Zhao, L., X. K. Diao, D. C. Yuan, and W. Tang. 2011. "Enhanced classification based on probabilistic extreme learning machine in wastewater treatment process." *Procedia Engineering* 15: 5563–5567. doi:10.1016/j.proeng.2011.08.1032.

Zheng, C., L. Zhao, X. Zhou, Z. Fu and A. Li. 2013. "Treatment technologies for organic wastewater." In *Water Treatment*, edited by W. Elshorbagy, and R. K. Chowdhury, pp. 249–286. IntechOpen. doi:10.5772/52665.

Zu, L., G. Li, T. An, and P.-K. Wong. 2012. "Biodegradation kinetics and mechanism of 2, 4, 6-tribromophenol by *Bacillus* sp. GZT: a phenomenon of xenobiotic methylation during debromination." *Bioresource Technology* 110: 153–159. doi:10.1016/j.biortech.2012.01.131.

8 Application of Nano-biocatalysts and Nanozymes in Wastewater Treatment

Nikhi Verma and Arindam Mitra

8.1 INTRODUCTION

A limited resource on this planet earth is water, which is consistently recycled by the water cycle. Different varieties of pollutants contaminate water and cause water pollution and inorganic, organic materials, industrial effluents, microbes, and many other harmful substances. Besides, water plays a pivotal role in the global economy by being a crucial and valuable resource for life (Butt 2020). Water pollution not only affects human health and causes adverse effects on the environment but also adds to the social and economic burden. Various non-commercial and commercial technologies are increasingly used to address the issue of water pollution (He et al. 2012).

Nanotechnology is a promising, advanced, and cutting-edge technology proven for treating wastewater. Nanotechnology is one of the exclusive and approachable technologies for wastewater treatment. According to Butt (2020), nanomaterials have numerous applications in various areas of science and are contemplated as essential materials for reclamation of wastewater and polluted sites. Nanomaterials are thought to be noteworthy because of their high reactivity, ratio of large surface area (SA) to volume (V), their nanosize, and a strong affinity. In addition, for the effective removal of contaminants and remediation of wastewater, nanomembranes and nanofibers are the best advanced types of nanomaterials. Lately, many advances have occurred in nanomaterials like nanosorbents, nanomembranes, nanomotors, and nanophotocatalysts. Moreover, a few polymers are imprinted and effective in treating wastewater (Yaqoob et al. 2020). The role of nanomaterials in wastewater treatment has been shown in Figure 8.1.

The utilization of engineered nanomaterials for the remediation of water is quite restricted at the moment for examining their capability to behave as disinfectants, filters, reactive agents, and effective adsorbents. However, they exhibit assurance for in-depth water treatment and remediation of the environment (Adeleye et al. 2016). Furthermore, Prachi and coworkers (He et al. 2012) mentioned that nanoparticles have high interacting and reacting capabilities and have a great rate of absorbance due to their small size. In addition, nanoparticles are also proven to be economical as they can conserve energy because of their nanosize, eventually bringing cost savings.

DOI: 10.1201/9781003517238-8

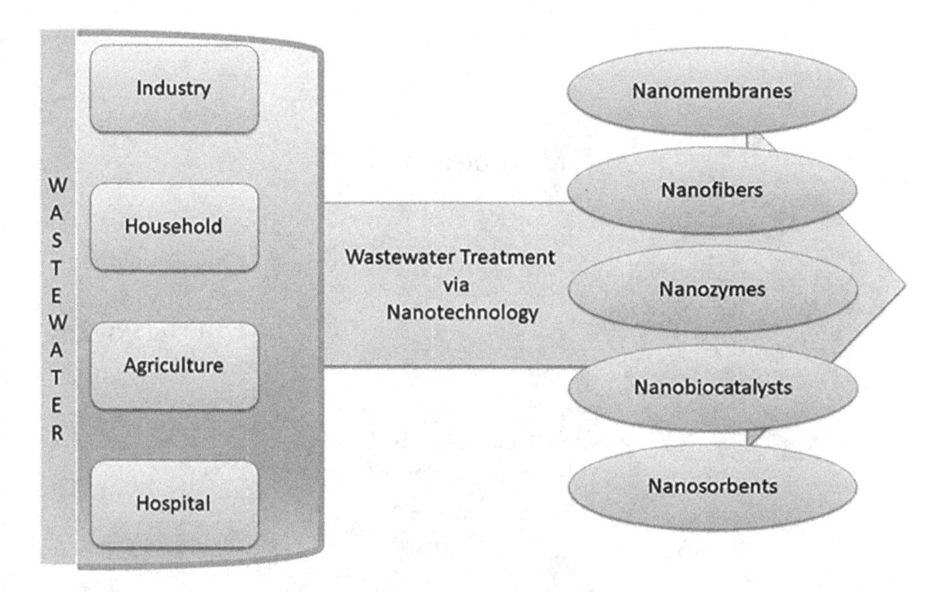

FIGURE 8.1 Role of nanomaterials in wastewater treatment.

Nanoparticles have massive applications in wastewater treatment from in-depth to any other sites that are usually skipped by the conventional technologies.

Apart from the utilization of nanoparticles, nanomembranes, nanostructures, etc., there are nanomaterials available that exhibit enzyme-like activity known as nanozymes and are very promising as they help in the treatment of wastewater and are also eco-friendly and effective, whereas nanobiocatalyst is one of the emerging innovations resulting from advanced nanotechnology and biotechnology. Nano-biocatalysts help remove artificial (synthetic) dyes from wastewater efficaciously. Figure 8.2 represents various nanotechnology-based approaches for wastewater treatment.

8.2 ROLE OF ENZYMES IN WASTEWATER TREATMENT

Clean and safe water is essential for all the living beings present on this globe for their survival. Many parts on this earth do not have safe water, whereas some regions do not even have water. As contaminated water leads to the threat of diseases, there has been the norm of wastewater treatment in several places. Recently, enzymes have gained prominence in wastewater treatment due to their excellent potency. For the remediation of wastewater, biological systems are being employed well. Initially, the realization for using the enzyme for wastewater treatment was narrow. But lately, many research and studies have been done in this sector to treat wastewater. Today, a pivotal role is played by enzymes for wastewater treatment.

Microorganisms have provided and are still continuously serving as one of the leading and beneficial sources for several enzymes (Adrio 2008). Active enzymes are produced by a distinct variety of bacteria and fungus. Enzymes are essential as they

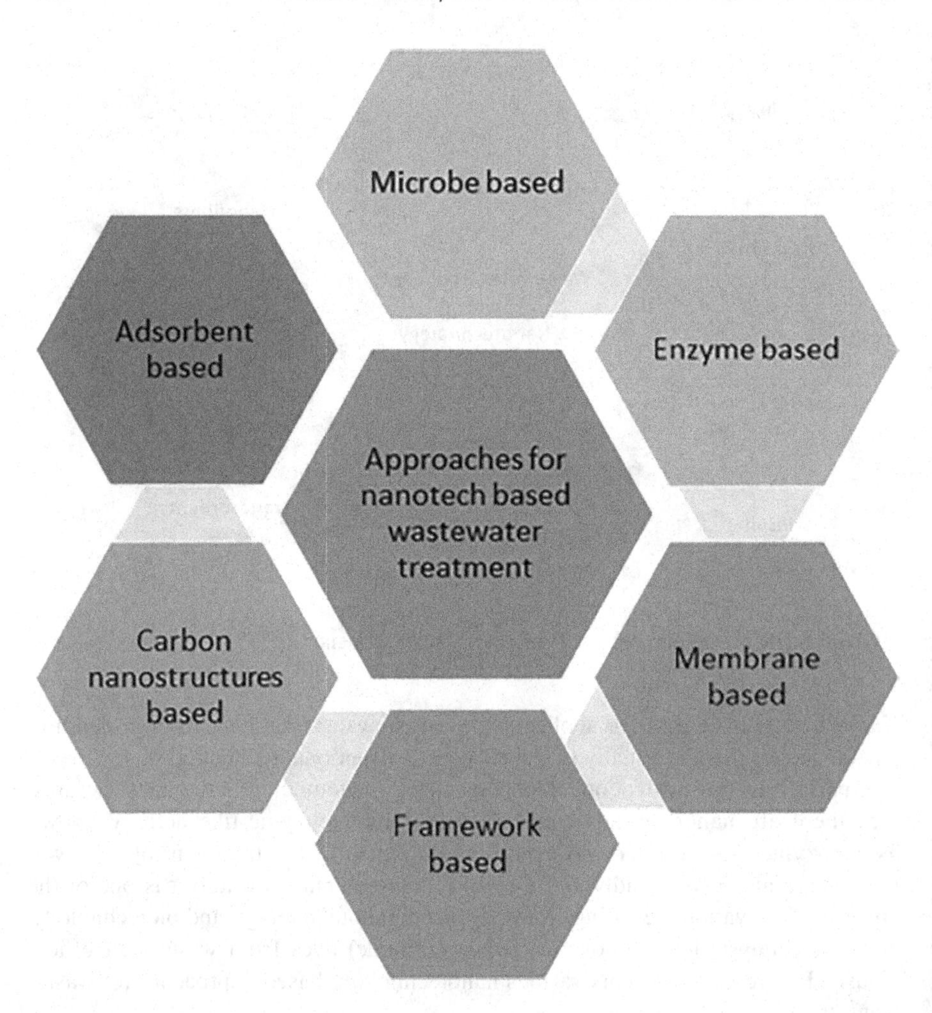

FIGURE 8.2 Approaches for nanotechnology-based wastewater treatment.

can function under mild pH, temperature, and environmental conditions. Enzymes do not require any substrate functional group shielding. Moreover, their half-life is long, and they work on synthetic substrates as well (Johnson 2013). Besides, the selection of the enzymes can be made according to the modification, i.e., chemical or genetic, to improve their main properties like specificity of substrate, stability, and particular activity (Pandey et al. 2017). Biological treatment seems to be a promising methodology for wastewater treatment to achieve profits from Certified Emission Reduction (CER) credits. This is popularly called carbon credits because of methane gas generation due to anaerobic digestion, which can further be used as renewable energy (Tchobanoglous 2003). Furthermore, this method also has several benefits like low cost and being environment friendly (Sponza 2005). The treatment of waste is classified into the above-mentioned two typical biological methods as it includes chemical processes that work based on biological catalysts.

8.2.1 Types of Enzymes

8.2.1.1 Oxidoreductases

Several higher plants, fungi, and bacteria detoxify the toxic organic substances through oxidative coupling mediated by oxidoreductases (Gianfreda 2010).

8.2.1.2 Oxygenases

Oxygenases predominantly are intracellular enzymes. They mainly take part in metabolism and biosynthesis. Besides its utilization in pharmaceutical companies for pharmaceutical fabrication and specialty chemical (effect chemicals or specialties) synthesis, oxygenase enzymes are also allied with hydrocarbon biodegradation. In general, this process is known as biotransformation (Unuofin, Okoh, and Nwodo 2019). The use of oxygenase enzymes for biotransformation is due to three primary purposes:

i. Its chemical analogs do not exist
ii. Molecular oxygen may be utilized as an inexpensive, eco-friendly oxidant in contrast to harsh oxidants (chemical)
iii. They are engaged in natural product modification

Dehalogenation reactions for halogenated alkanes and alkynes (methanes, ethanes, and ethylenes) are mediated by oxygenases only affiliated with multifunctional enzymes (Huber et al. 2003).

8.2.1.3 Peroxidases

The enzyme peroxidases are often mentioned as heme-carrying proteins. They exist in all living forms, i.e., from decomposers to consumers. They mainly act in the process of lignification (Unuofin, Okoh, and Nwodo 2019). The impedance of mediator peroxidases helps catalyze the process and oxidize the phenolic and lignin compounds at the cost of hydrogen peroxide (H_2O_2) (Pandey et al. 2017). Another application of enzyme peroxidases is its utilization in the treatment of contaminated water sources through chemical compounds such as dyes, aromatic compounds, organic peroxides (H_2O_2), and phenols. Peroxidases work by transferring electrons to such compounds which ultimately leads to the disintegration of these harmful compounds into smaller non-dangerous compounds.

8.2.1.4 Polyphenol Oxidases

They are present ubiquitously in nature. Their secretion is predominantly associated with the fungi pathogenicity, mechanical distortion, and physiological reaction toward stress in plants (Unuofin, Okoh, and Nwodo 2019). This category of enzymes helps treat wastewater by eliminating mainly the phenolic contaminants. They break down the impurities consistently until they form irresolvable molecular pigments that can be extracted and discarded effortlessly. Because of the phenolic compounds differentiation they oxidize, polyphenol oxidase enzymes are further classified into two main classes.

8.2.1.4.1 Tyrosinases

Tyrosinases play a fundamental role in melanin synthesis from l-tyrosine in humans and animals. They are type-3 cuprodinucleate metalloproteins. They form insoluble agglomerates by converting phenolic contaminants.

8.2.1.4.2 Laccases

The enzyme laccases comprise a group of multicopper oxidases produced by a few insects, bacteria, plants, and fungi that catalyze the process of oxidation for a broad variety of phenolic and aromatic substrates resulting in reduction of molecular oxygen to water (Arora 2010).

8.2.1.5 Lipases

Enzymes like lipases may catalyze several reactions like hydrolysis, aminolysis, esterification, interesterification, and alcoholysis (Prasad 2011). In addition, lipase enzyme has a diagnostic application in bioremediation. Moreover lipases have many prospective applications in industries of food, cosmetic, chemical, paper making, and detergent fabrication (Husain et al. 2013).

8.2.1.6 Proteases

The enzyme proteases belong to the class of enzymes which hydrolyze peptide bonds. They hydrolyze in aqueous environment, whereas they synthesize in non-aqueous environment. Furthermore, protease has extensive array of uses in several industries like leather, food, pharmaceutical, and detergent (Geevarghese 2010).

8.2.1.7 Cellulases

The enzyme cellulases mainly work on cellulose and its degradation results in monosaccharides (reducing sugars) which further can be fermented by microbes (bacteria or yeasts) (Sun and Cheng 2002). Cellulases help eliminate cellulose microfibrils produced while washing and utilizing cloths that are cotton-based. Besides this, cellulases in paper and pulp industries aid in ink removal during paper recycling.

8.3 NANOTECHNOLOGY AND TYPES OF NANOMATERIALS AS CARRIER OF BIOCATALYSTS

Nanotechnology: A discipline of nanoscience. Its phenomenon is based on nanometer-scale extent. The manmade nanoscopic structures are known as nanomaterials (Chaturvedi, Dave, and Shah 2012). It is one of the best alternatives to treat wastewater compared to conventional methods as it is a promising one and can eliminate emerging pollutants effectively (Adeleye et al. 2016). Nanotechnology could be the foundation and may support the progress of subsequent generation's water supply approach. To be exact the constituents of structure whose single dimension is so tiny that is even less than 100 nm are called nanoparticles (Amin, Alazba, and Manzoor 2014). A few researchers mentioned that several nanomaterials have been developed like nanowires, particles, colloids, nanotubes, quantum dots, and films (Edelstein and Cammaratra 1998, Lubick 2008). In the application for wastewater treatment, several different types of nanomaterials have been developed that are environment friendly and effective and inexpensive. They have distinctive functionalities for prospective sterilization of water from different sources like surface, ground, drinking, or industries (Brumfiel 2003, Gupta et al. 2015, Theron 2008). Nanotechnology can cut down the expense rate as mentioned by various sectors to alleviate these contaminants by fabricating environment-friendly nanomaterials (Mandeep and Shukla 2020).

Several ultra-modern oxidation methodologies, valorization, and electrochemical processes have been employed to reduce effluent toxicity coming from wastewater and make its utilization sustainable (Gupta 2020). However, these methodologies are not pocket-friendly to every industry. Therefore, green nanomaterials' synthesis from different microbes and other organisms' extract have paved the way for environmentally friendly reconstruction of pollutants. Green nanoparticles such as iron are being used for remediation because of their redox capability while reacting with magnetic susceptibility, innocuous nature, and water (Bolade, Williams, and Benson 2020). Nanomaterials that are membrane-associated are also the practical procedure for eliminating effluent. Nanomaterials enhance membrane permeability, fouling resistance, temperature, and mechanical strength to degrade pollutants. Nanocatalysts play a crucial role in the degradation reaction's improvement (Corsi 2018). Different nanomaterials for wastewater treatment are discussed in Tables 8.1 and 8.2.

TABLE 8.1
Nanoadsorbent and Nanocatalyst Approach for Wastewater Treatment

Approaches	Properties	Applications	Advantages	References
		Nanoadsorbents		
Oxide-based nanoparticles (Fe_2O_3, MnO, ZnO)	High Brunauer-Emmett-Teller (BET) surface area, less solubility	Water disinfection, removal of heavy metals	Minimum environmental impacts, no secondary pollutants	Anjum et al. (2016)
Carbon-based nanotubes	High surface area for adsorption, nanoporous nature	Removal of heavy metals, organic and inorganic pollutants	Extremely cheap, portable	Mines et al. (2015)
Graphene-based nanoadsorbents	High surface area, mechanical strength, light weight, flexibility, and chemically stable	Removal of heavy metals, antifouling and antibacterial properties	Low cost, reusable	Alnoor et al. (2020)
		Nanocatalysts		
Photocatalyst	High surface area, electronic, catalytic, and antimicrobial properties	Degrade persistent organic pollutants and volatile compounds	Target recalcitrant compounds, consumes less time	Wigginton et al. (2012)
Fentocatalyst				Hering et al. (2013)

TABLE 8.2
Approaches of Nanomembranes, Bioenzymes, and Zeolites for Wastewater Treatment

Approaches	Properties	Applications	Advantages	References
		Nanomembranes		
Carbon nanotube membranes	Low mass density, extremely high strength and tensile modulus, high flexibility	Desalination, removal of heavy metals and pollutants	Highly economical, efficient, and simple design	Anjum et al. (2016)
Electrospun nanofiber membranes	Higher porosity and surface-to-volume ratio	Remove particles from biotreated wastewater	Less energy consumption, less expensive, and lighter process	
Hybrid nanomembranes	Adsorptions, photocatalysis	Removal of oil, heavy metals, antimicrobial properties	No addition of chemicals	
		Bioenzymes		
Nanozymes	Biocatalytic activity, other enzyme-like properties	Removal of pollutants	Low cost, eco-friendly, high stability, robust catalytic performance, simple synthesis	Singh (2019)
Chitosan	High hydrophilicity, flexible nature, presence of surface functional groups	Antimicrobial properties, removal of dyes, heavy metals, phenols	Biodegradable, nontoxic	Bhatnagar and Sillanpaa (2009)
Laccase	Monomeric, dimeric, and tetrameric glycoprotein, thermal stability	Biodegradation of aromatic contaminants, oxidize both phenolic and nonphenolic lignin-related compounds	Abundant in nature	Unofin (2019), Shraddha et al. (2011)
Zeolites	Highly porous and crystalline	Removal of organic and inorganic ions, heavy metals, dyes, humic substances, and radioactive elements	Economical	Mines et al. (2015)

8.3.1 Types of Nanomaterials

8.3.1.1 Nanoadsorbents

Nanoparticles have been considered for their capability to work as adsorbents recently. Nanoadsorbents own two fundamental properties: extrinsic functionalization and innate surface. In addition, their chemical, physical, and materialistic effects are also associated with their external surface, superficial size, and peculiar conformation (Mirkin 1996). The factors responsible for influencing the process of adsorption in the aqueous conditions are large surface area, chemical and adsorption activity, high energy of surface binding, exterior position of atoms, and shortage of inner diffusion resistance (Khajeh, Laurent, and Dastafkan 2013). Nanoparticles are extensively utilized as adsorbents to eliminate harmful pollutants from wastewater. In addition, nano adsorbents can eliminate organic as well as inorganic contaminants. They are further classified chiefly as metal, metal oxide, or carbon-based nanoparticles.

Many types of nanoparticles are carbon-based. These include activated carbon, graphene, carbon nanotubes (CNTs), and somewhat fullerene (Kumari 2019). Adsorbents like CNTs act on hazardous chemicals from wastewater of fabricating and pharmaceutical industries, whereas nanomagnets altered from activated carbon are being employed to eliminate fluoride ions. Sorption and separation by magnets were used for the heavy metal elimination (Nogueira 2019).

Moreover, a nanocollector is graphene based to remove nickel (Ni) ions. Thus, the process involved is known as ion flotation. This nanocollector is cost-effective, sturdy, and effective. Nanocollectors are highly effective as they are capable of bringing about 100% Ni ions removal from wastewater (Hoseinian et al. 2020). Furthermore, nano adsorbents based on metal and metal oxide play a fundamental role in eliminating contaminants from wastewater. Researchers have unveiled that laying magnetic nanoparticles along with other substances showed an addition in adsorption capability (Mandeep and Shukla 2020).

8.3.1.2 Nanocatalysts

Nanoparticles may be utilized as mediators or catalysts and stimulate reactants in new forms like water. There are three significant catalysis classes: enzyme-based catalysis, homogeneous, and heterogeneous. Enzyme-based is extremely efficient. At the same time, homogeneous and heterogeneous have their advantages and disadvantages. Due to this, there is a critical demand for a stimulant framework that must be productive as homogeneous catalysts and adequately recoverable as heterogeneous catalysts. Nanocatalyst is the one which has exhibited the framework of both, that is, homogeneous as well as heterogeneous (Somwanshi, Somvanshi, and Kharat 2020).

The framework of nanocatalysis allows for fast, selective chemical transformation with high yield integrated with ease of separation of catalyst and recovery. For the treatment of wastewater, nanocatalysts, mainly inorganic substances like metal oxides and semiconductors, are receiving significant attention. Several different types of nanocatalysts are being used for the treatment of wastewater. These are as follows.

8.3.1.2.1 Photocatalysts

Nanoparticle photocatalytic reaction relies on the relationship of light energy and nanoparticles. They are of considerable interest because of their large and high-level photocatalytic actions for different contaminants (Akhavan 2009). Generally, these photocatalysts are composed of semiconductor metal that can break down several constant organic pollutants present in wastewater, such as detergents, dyes, vaporous organic complex, and pesticides (Lin et al. 2014). Moreover semiconductor nanocatalysts are extremely powerful against both types of organic compounds for degradation, that is, halogenated and non-halogenated.

For photocatalysis, titanium dioxide (TiO_2) has been used extensively because its rate of reactivity is high beneath ultraviolet (UV) light. Several researches have revealed the photocatalytic action of numerous catalysts that are synthesized, the potential of which relies on different characteristics like pH, energy of band gap, dose, size of particles, and concentration of pollutant. Likewise, zinc oxide (ZnO) and cadmium sulfide (CdS) nanoparticles have also captured thorough attention as photocatalysts for remediation of industrial dyes found in wastewater. Further Yu et al. (2003) added that photocatalysis is a likely methodology for treatment and disinfection of different types of wastewater. Moreover, they further added that it can inactivate infectious microorganisms like bacteria.

8.3.1.2.2 Electrocatalysts

One of the most trending topics for the treatment of wastewater leading to generation of electricity directly is by electrocatalysis method in microbial fuel cell (MFC). In MFC, electrocatalyst plays a significant part in functioning of a fuel cell (Cheng-Chuan, Lin and Chen 2015). The utilization of nanomaterials like electrocatalysts can enhance fuel cell working by attaining broad surface area and consistent dispersal of catalyst within the reaction forum (Liu et al. 2005). Immense study and exploration had been organized on the development of nano-electrocatalyst supported by carbon for their implementation in fuel cells.

Though platinum (Pt) may be utilized as an electrocatalyst, it has several drawbacks that restrain its usage. For example, as Pt is an expensive metal, its availability is also limited. In addition, it is costly, which reduces the interest of using it as a catalyst. Nevertheless, the nanoparticles of palladium (Pd) can replace platinum (Pt) as it is found abundantly in the earth, approximately 50% more. For example, in ethanol's fuel cell, the nanocatalyst of Pd can cut the price of anodes because of its ampleness.

8.3.1.2.3 Fenton Catalysts

Oxidation of organic contaminants by using Fenton's reaction for the treatment of wastewater has been applied extensively. But its main limitation is the loss of catalyst matter continuously along with discharge of liquid waste. Another drawback is its necessity of acidic environment (pH 3) for optimum working. To overcome these difficulties, Fenton's reagent based on nanomaterial has been utilized. The nanoferrites with measured crystal clear size, dispersal, and chemical framework can be achieved by either auto-combustion or sol gel method (Kurian 2014).

Iron oxide (FeO) nanoparticles that can be separated by magnet may also be employed as Fenton catalysts to eliminate various kinds of pollutants. Furthermore, on comparing with methods like filtration and decantation, the separation through magnets is very fast and effective (Ambashta and Sillanpää 2010). Nanocatalysts have shown solidarity with regard to unlimited oxidation of contaminant and organic intermediary products, whereas in other research the nano nickel-zinc-ferrite catalyst was studied for the breakdown of 4-chlorophenol.

8.3.1.3 Nanomembranes

Among the recently developed techniques for the treatment of wastewater, filtration through membrane produced by nanomaterials is the best effectual approach. Nanomembranes are a distinctive variety of membrane made with distinct nanofibers used to remove undesirable nanoparticles existing in the liquid phase. The membrane carries a porous base along with complex sheet. The concept of nanotechnology surpasses the ultra-modern execution of membranes used for water treatment and permits new performance like high reactivity, catalytic function, and weak resistance (Pendergast and Hoek 2011). The principal purposes responsible for acquiring this methodology are its advantages in respect of effectual decontamination, standard treated water, and the need for less plant area (Jun Hee Jang 2015). Besides, it is of low cost, with a simple and systematic design compared to the rest of the treatment methods.

The separation technique by using nanomembrane in wastewater treatment is for the successful elimination of heavy metals, dyes, and other pollutants. In addition to separation of particles from wastewater, nanomaterials in membrane play an essential part in the chemical degradation of organic pollutants. These kinds of membranes' constitution are from linear nanomaterials like nanoribbons, nanotubes, and nanofibers (Liu et al. 2014). Moreover for the particular filtration and elimination of nanoparticle, a membrane produced with carbonaceous nanofibers (CNFs) exhibited excellent selective filtration beneath high pressure (Liang 2010). Besides CNFs, CNTs have antimicrobial properties that reduce the formation of biofilm and fouling and minimize the possibility of mechanical failures (Ocampo-Sosa et al. 2015). The particles of silver metal that are said to be antimicrobial material are also doped along with polymer to form a polymeric membrane to avoid bacterial attachment and hinder biofilm formation on membrane surface. It is set out for virus inactivation and can impede membrane biofouling (Ronen et al. 2015).

The main aim of utilizing membranes is to segregate the hazardous particles from different water supplies. The filters of nanomembranes were also employed to estimate the water safety level (Jawaid 2018). The benefits of using nanomembranes for filtration as compared to conventional methods are that throughout the complete procedure of filtration, another ion is needed by the calcium (Ca) and magnesium (Mg) to compensate, for which predominantly sodium (Na^+) ions were used as an exchanger. While there are no requirements of any type of ion exchangers in case of nanomembranes, the use of nanomembrane also has disadvantages, such as after using the membrane a few times, the membrane starts fouling which is one of the crucial concerns as it makes this procedure costly and inefficacious. Another drawback is stability of membrane. As the stability of the membrane starts decreasing its effectiveness and thus the rate of outcome decreases as compared to earlier.

8.4 IMMOBILIZATION OF LACCASE ON NANOPARTICLES

The enzyme laccases have been found in the exudates of Japanese lacquer tree since 1833. The tree is named *Toxicodendron vernicifluum* formerly known as *Rhus vernicifera*. Laccases also occur naturally in different microbes (fungi, bacteria, and insects) and plants. In bacteria, they could be indulged in pigmentation, morphogenesis, sporulation, bioassimilation, and many other processes. Besides, laccases help in catalyzation of different substrates, especially phenolic compounds. This enzyme's characteristics indicate its potential to begin and manage the biodegradation of several aromatic pollutants found in wastewater.

Laccases showed its applications in different types of industries such as pulp and paper, food, and textile. Lately, it is also utilized in the composition of biofuel cells and biosensors and as clinical diagnostic device and agent of bioremediation for the removal of pesticides, herbicides, and some explosives from soil. In insects it has been reported as the chief enzyme for cuticular hardening. There are two types of this enzyme found predominantly: laccase-1 and laccase-2. In nanotechnology, laccases have been used competently because of their capability to catalyze the reaction responsible for electron transfer without any supplementary co-factor.

Biomolecule immobilization techniques such as layer by layer, self-build monolayer method, and micro-patterning may be employed to preserve laccase enzymatic activity. Immobilization is an integral approach to boost both its stability and activity of enzyme. Immobilization of enzymes in combination with nanostructures improves its properties, which permits the enzyme to develop resistance and stability. The immobilization of enzymes on nanoparticles is usually robust and stable compared to their free complements because of their predetermined configuration. The distinctive internal characteristics of nanoparticles contribute to their applicability in a wide range of fields (Lee and Au-Duong 2018). Schematic representation of laccase immobilization on nanoparticles is shown in Figure 8.3.

8.4.1 Laccase Immobilization on Magnetite Nanoparticles

A new immobilization technique was developed based upon dopamine (DA) self-polymerization for laccase immobilization on nanoparticles. The immobilization of laccase was done on magnetic nanoparticles (Fe_3O_4). For the optimization of conditions required in immobilization involving reaction pH, enzyme concentration, and DA, a middle combined surface technique was employed. Over the past few years, the laccase enzyme has been immobilized on several distinct supports like activated carbon, titanium oxide (TiO_2), porous glass, microsphere, membranes, etc.

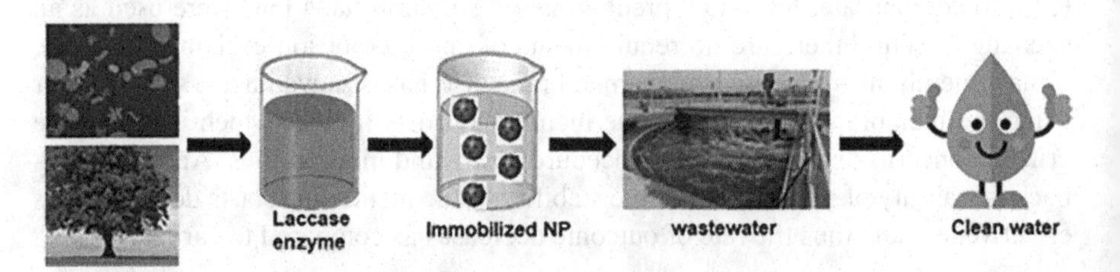

FIGURE 8.3 Immobilization of laccase on nanoparticles.

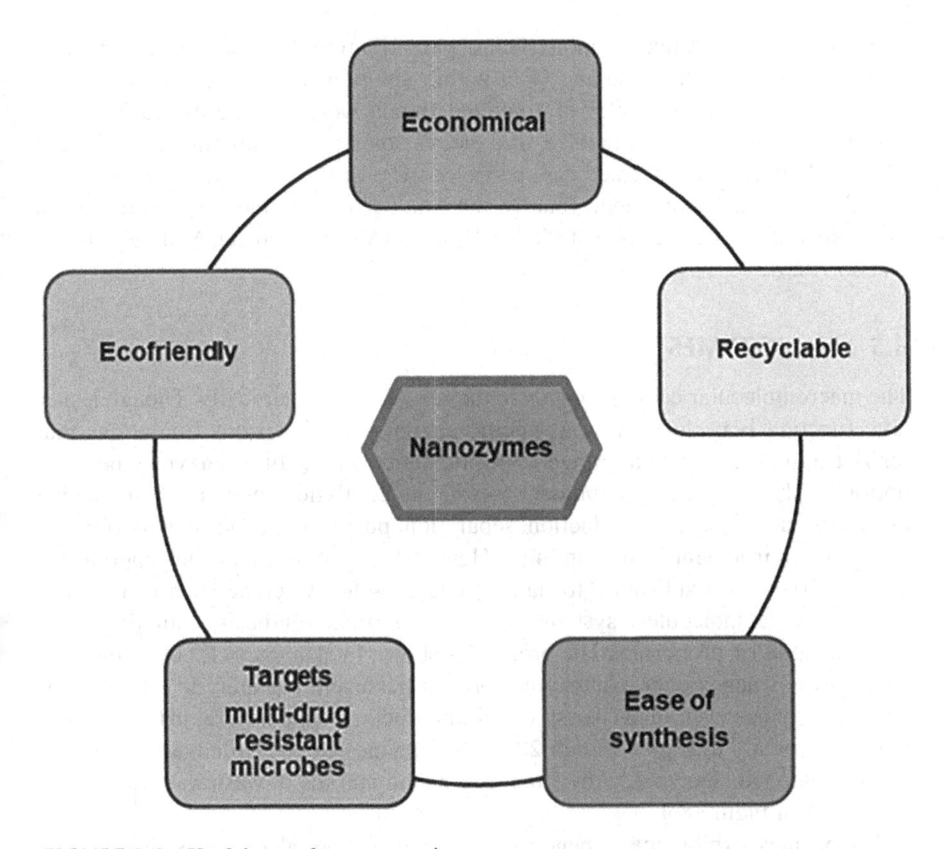

FIGURE 8.4 Usefulness of nanozymes in wastewater treatment.

Among them the most propitious support considered were magnetic nanoparticles due to properties like large surface area, precise design, and more. In addition, the enzymes that are immobilized upon Fe_3O_4 magnetic nanoparticles manage their own distinctive function and can get recycled and separated easily.

The Fe_3O_4 magnetite nanoparticles were developed utilizing ordinary method as reported by J.G and coworkers (2007). In short, Fe(acec)3 and diethylene glycol are mixed and vigorously stirred simultaneously beneath nitrogen stream at 230°C for 5 hours. Further, when the mixture gets cooled below room temperature (RT) then ethyl acetate ($C_4H_8O_2$) was added. This step was performed to precipitate the Fe_3O_4 nanoparticles formed. Finally, washing of the precipitate of above-formed nanoparticles was done five times using distilled water (dH_2O) and finally stored within distilled water only as homogeneous colloidal suspension (Zhang et al. 2017).

8.4.2 Laccase Immobilization on Titania Nanoparticles

The enzyme laccase immobilized the nanoparticles of titania. Thus the nanoparticles formed were identified by different methods like SEM and FTIR. The resultant nanoparticles were known as immobilized laccase onto titania nanoparticle (ILTN). ILTN were utilized for the breakdown of a few anionic dyes such as Direct Red 31 (DR31), Direct Green 6 (DG6), and Acid Blue 92 (AB92).

First of all, the titania nanoparticles were silanized by submerging them in 3-amino propyl triethoxy silane (4%) within solution of acetone for 24 hours at 45°C. Then the nanoparticles of silanized titania were thoroughly washed with dH_2O and then further immersed in aqueous solution of glutaraldehyde (5%) at RT for 2 hours. Again, formed nanoparticles were washed with dH_2O followed by heat drying at 60°C for 1 hour. Finally, the nanoparticles formed (5 g) were kept in given quantity of free laccase at RT for 48 hours (Mohajershojaei, Mahmoodi, and Khosravi 2015).

8.5 NANOZYMES

The macromolecular catalysts catalyze the reactions biochemically. Though a specific function is performed by a specific enzyme, it has various limitations that restrict its usage in a wide range of applications. The natural enzymes need an optimum physiological environment to execute catalytic function. Their notable drawbacks are expensive production, separation, purification, and short stability in stringent environmental surroundings. Hence in recent times various approaches have emerged as a substitute to natural enzymes to overcome such difficulties. These include molecules, systems, and nanoparticle synthesis, imitating their internal catalytic properties. The properties shown by nanoparticles of an enzyme are known as nanozymes. Nanozymes proffer persistent biocatalytic activity over natural enzymes even under harsh conditions such as temperature, pH, and resistance to protease digestion (Singh 2019). Nanozymes can be called the "next-generation manmade enzyme." Advantages of nanomaterials in wastewater treatment are shown in Figure 8.4.

Nanozymes exhibit many benefits compared to natural enzymes; hence, this leads to prompt diversification of artificial biocatalyst development. Further benefits involve an easy way of synthesis, higher stability, vital catalytic function, inexpensive and sleek surface moderation of nanomaterials. Therefore, nanozymes are enormously being studied to promote a broad range of applications, such as applications in immunoassays, diagnosis of diseases, therapies, biosensing, elimination of contaminants, oxidative stress shielding, growth of cells or tissues. Considerably the nanozymes have been further classified into two broad groups. These are as follows.

8.5.1 Nanozymes: Antioxidant

In the living environment, antioxidants are needed to safeguard the cells or tissues from injury inflicted by the plethora of free radicals formed at the time of biochemical reactions occurring usually in the body. There is a well-developed endogenous antioxidant system within the human body which is mainly managed through free radical foraging enzymes like superoxide dismutase (SOD), glutathione reductase, glutathione peroxidase, catalase, peroxiredoxins, and so on. Among antioxidants which are mainly inorganic, cerium oxide nanoparticles (CeNPs) are described to reveal cleansing of superoxide radicals and hydrogen peroxide (H_2O_2) degeneration under different conditions, namely *in vivo*, *in vitro*, and a few animal models (Cassandra Korsvik 2007; Heckert 2008; Karakoti et al. 2009).

8.5.1.1 SOD Imitative Nanoparticles

The enzyme SOD is one of the specific antioxidant enzymes. There are fewer types of nanoparticles developed for the execution of activity by superoxide anion scavenging. One of a kind is CeNPs. Dhall and Self (2018) reported that when the ratio of surface is high with Ce^{+3}/Ce^{+4}, the atoms of Ce from CeNPs show SOD imitative activity, while when the ratio of Ce^{+3}/Ce^{+4} is less, it results in catalase imitative activity. The enzyme SOD plays a crucial shielding part in superoxide anion scavenging, but its expensive rate of synthesis and short stability lead to the chance to develop an effectual substitute.

8.5.1.2 Catalase Imitative Nanoparticles

The surplus of cellular H_2O_2 is decomposed by the biological enzyme catalase that catalyzes the reaction and results in water (H_2O) and molecular oxygen (O). Usually, the superoxide radicals' disproportionation by enzyme SOD results in hydrogen peroxide (H_2O_2) generation due to the crucial role of H_2O_2 in the generation of hydroxyl radicals and in biological signaling. Besides, it is a stable species and has low reactivity within the cytoplasm. It is studied and understood that H_2O_2 undergoes Fenton reaction in the existence of transition ions of metal and thus produces hydroxyl radicals that are harmful against biological molecules. Hence the cytoplasmic H_2O_2 present in excess must be converted into H_2O and O by utilizing catalase enzyme. Nevertheless, when the functional catalase enzyme is absent, too much of H_2O_2 may cause a variety of diseases like vitiligo, diabetes, and acatalasemia. That is why a substitute to biological catalase is essential and researchers have developed various kinds of nanoparticles showing activity similar to catalase. The emerged nanoparticles involve gold, iron oxide, cerium oxide, and cobalt oxide nanoparticles.

8.5.2 Nanozymes: Pro-oxidant

Pro-oxidant nanozymes introduce nanozyme activities which bring about oxidative stress by generating free radicals in cells of mammals or inhibiting the antioxidant network. The ordinary drugs known to produce free radicals are the painkiller paracetamol and anticancerous methotrexate and, hence, are believed to be pro-oxidants. Likewise, transition metals namely copper (Cu), iron (Fe), and so on are also revealed to go under Haber-Weiss and Fenton reactions and thus eventually generate free radicals excessively.

8.5.2.1 Peroxidase Imitative Nanoparticles

Peroxidases occurring naturally comprise of a huge family. They principally use H_2O_2 for the oxidation of peroxidase substrates. The enzyme peroxidase is of significant importance due to their action as the free radicals behave as detoxifying agents. For instance, glutathione peroxidase is one such example which provides a defense against myeloperoxidase, an invading pathogen (Strzepa 2017). In addition, the enzyme Horseradish peroxidase (HRP) is one of the popular enzymes known for its usage in clinical and bioanalytical chemistry. Nanozymes exhibiting peroxidase imitative activity can be catalyzed in two steps:

$$3H_2O_2 + \text{Peroxidase imitative nanozymes} \rightarrow 6HO^{\bullet}$$

$$6HO^{\bullet} + 2\text{Peroxidase substrate (TMB/OPD/ABTS)}_{\text{Red}\cdot} \rightarrow 2\text{(TMB/OPD/ABTS)}_{\text{Ox.}} + 6H_2O$$

8.5.2.2 Oxidase Imitative Nanoparticles

The enzyme oxidase catalyzes the reaction of oxidation. This includes the substrate oxidation by molecular oxygen (O) that is further converted into either H_2O or H_2O_2. The enzyme oxidase does not need H_2O_2 as required by enzyme peroxidase. In fact, in return they produce superoxide radicals or H_2O_2. Because of the production of superoxide radicals and H_2O_2 *in situ*, nanozymes and enzyme oxidase mimic the oxidase activity. This activity may further efficiently oxidize the achromic substrates into corresponding chromic products that make them perfect for detecting chemical or biological molecules. Moreover, various nanomaterials have been revealed to show enzyme oxidase-like activity in recent times.

8.6 CHITOSAN NANOPARTICLE FOR ENCAPSULATION OF ENZYMES

Chitosan, a derivative of chitin, is a renewable biopolymer. It is produced by partial *N*-deacetylation of chitin. It is used for encapsulation of enzymes for application in the wastewater treatment. The biodegradable, biocompatible, cost-effectiveness, and nontoxic properties of chitosan make it a widely used choice for these treatments. Chitosan is the supreme choice for enzyme immobilization due to its properties like improved resistance to chemical degradation and reducing disturbance of metal ions to an enzyme. Specifically, the amino group of chitosan makes it a better fit for encapsulation of enzymes. Chitosan nanoparticles have been used to encapsulate laccase, an enzyme with biodegradation of phenolic compounds. Chitosan nanoparticles-encapsulated peroxides are thermally stable and suitable for use in wastewater treatment (Alarcon-Payan, Koyani, and Vazquez-Duhalt 2017). Such encapsulated stable laccase have been demonstrated to have enhanced decolorization of Congo Red dye. Besides chitosan-based nanoparticles, cellulose-based nanoparticles are also effective in removal of heavy metals and dyes because of superior adsorbent properties (Olivera et al. 2016). Chitosan nanofibers are also extensively used in the wastewater treatment for removal of heavy metals and colored dyes (Nthunya et al. 2017). The immobilized laccase conjugated to silica-coated magnetic nanoparticles was also effective for phenol biotransformation with recovery of nano-biocatalyst (Moreira et al. 2017). Nano-biocatalysts are assessed based on their bioprocessing abilities and performances such as improved catalytic activity, stability, and reutilization. The structure and properties of nanoparticles influence the performance of enzymes.

8.7 AMALGAMATION OF MICROBES WITH NANOTECHNOLOGY

Combining microbes with nanotechnology makes the wastewater treatment process sustainable, cost-effective, and eco-friendly. Compared to the chemical synthesis of nanoparticles, the green synthesis of nanoparticles has some advantages. *Aspergillus* spp. have been used to create iron oxide nanoparticles. More than 90% of heavy metals from wastewater could be chemically adsorbed and removed by these biofriendly nanoparticles with regeneration potential. Another study has identified the use of exopolysaccharides from *Chlorella vulgaris* to develop iron oxide nanoparticles, which could remove a significant amount of phosphate and ammonium from wastewater. Copper nanoparticles synthesized from *Escherichia coli* were shown to degrade azo dyes and effluent from textile industry. *Pseudoalteromonas* spp.-mediated synthesis of iron sulfur nanoparticles could degrade Naphthol Green B dye. Microorganisms could also provide enzymes that can catalyze the remediation of pollutants from wastewater and even generate valuable products. The biohydrogen production can be increased by combining the mixed culture of bacteria and different combinations of nanoparticles.

8.8 CONCLUSIONS AND FUTURE PERSPECTIVES

Nanomaterials have been an excellent choice for wastewater treatment due to their increased surface area, catalytic properties, reuse potential, and stability. Various types of nanozymes are also used in wastewater treatments. Furthermore, combining enzymes or microbes with nanoparticles has paved the way for the sustainable and eco-friendly treatment of wastewater pollutants. Nanotechnology is also cost-effective and can be used globally to improve water quality.

REFERENCES

Adeleye, A. S., J. R. Conway, K. Garner, Y. Huang, Y. Su, and A. A. Keller. 2016. "Engineered nanomaterials for water treatment and remediation: costs, benefits, and applicability." *Chemical Engineering Journal* 286:640–662. doi: 10.1016/j.cej.2015.10.105.

Adrio, A. L. D., and L. Jose. 2008. "Contributions of microorganisms to industrial biology." *Molecular Biotechnology* 38: 41–55.

Akhavan, O. 2009. "Lasting antibacterial activities of Ag-TiO_2/Ag/a-TiO_2 nanocomposite thin film photocatalysts under solar light irradiation." *Journal of Colloid and Interface Science* 336 (1):117–124. doi: 10.1016/j.jcis.2009.03.018.

Alarcon-Payan, D. A., R. D. Koyani, and R. Vazquez-Duhalt. 2017. "Chitosan-based biocatalytic nanoparticles for pollutant removal from wastewater." *Enzyme and Microbial Technology* 100:71–78. doi: 10.1016/j.enzmictec.2017.02.008.

Ambashta, R. D., and M. Sillanpää. 2010. "Water purification using magnetic assistance: a review." *Journal of Hazardous Materials* 180 (1):38–49. doi: 10.1016/j.jhazmat.2010.04.105.

Amin, M. T., A. A. Alazba, and U. Manzoor. 2014. "A review of removal of pollutants from water/wastewater using different types of nanomaterials." *Advances in Materials Science and Engineering* 2014:1–24. doi: 10.1155/2014/825910.

Arora, P. K.. 2010. "Application of monooxygenases in dehalogenation, desulphurization, denitrification and hydroxylation of aromatic compounds." *Journal of Bioremediation & Biodegradation* 01 (03). doi: 10.4172/2155-6199.1000112.

Bolade, O. P., A. B. Williams, and N. U. Benson. 2020. "Green synthesis of iron-based nanomaterials for environmental remediation: a review." *Environmental Nanotechnology, Monitoring & Management* 13:100279. doi: 10.1016/j.enmm.2019.100279.

Brumfiel, G.. 2003. "Nanotechnology: a little knowledge." *Nature* 424 (6946):246–248. doi: 10.1038/424246a.

Butt, B.Z. 2020. "Nanotechnology and waste water treatment." In: S. Javad, ed. *Nanoagronomy*, pp. 153–177, Springer, Cham.

Cammaratra, R. C, and A. S Edelstein (ed.). 1998. "Nanomaterials: synthesis, properties and applications." ed. Boca Raton: CRC press. https://www.taylorfrancis.com/books/mono/10.1201/9781482268591/nanomaterials-edelstein-cammaratra.

Cassandra, K., S. Patil, S. Seal, W. T. Self. 2007. "Superoxide dismutase mimetic properties exhibited by vacancy engineered ceria nanoparticles." *Chemical Communication* (10). doi: 10.1039/B615134E.

Chaturvedi, S., P. N. Dave, and N. K. Shah. 2012. "Applications of nano-catalyst in new era." *Journal of Saudi Chemical Society* 16 (3):307–325. doi: 10.1016/j.jscs.2011.01.015.

Chen-Chuang, C., C.-L. Lin, and L.-C. Chen. 2015. Functionalized carbon nanomaterial supported palladium nano-catalysts for electrocatalytic glucose oxidation reaction. doi: 10.1016/j.electacta.2014.11.116.

Corsi, I., M. Winther-Nielsen, R. Sethi, C. Punta, C. Della Torre, G. Libralato, G. Lofrano, L. Sabatini, M. Aiello, L. Fiordi, F. Cinuzzi, A. Caneschi, D. Pellegrini, and I. Buttino. 2018. "Ecofriendly nanotechnologies and nanomaterials for environmental applications: key issue and consensus recommendations for sustainable and ecosafe nanoremediation." *Ecotoxicology and Environmental Safety* 154:237–244. doi: 10.1016/j.ecoenv.2018.02.037.

Dhall, A., and W. Self. 2018. "Cerium oxide nanoparticles: a brief review of their synthesis methods and biomedical applications." *Antioxidants* 7 (8):97. doi: 10.3390/antiox7080097.

Geevarghese, A.K. and P.I. Beena. 2010. A solvent tolerant thermostable protease from a psychotrophic isolate obtained from pasteurized milk. In: *Developmental Microbiology and Molecular Biology*, pp. 113–119.

Gianfreda, L., F. Xu, and J.-M. Bollag. 2010. "Laccases: a useful group of oxidoreductive enzymes." *Bioremediation*: 1–26. doi: 10.1080/10889869991219163.

Gupta, G. K., and P. Shukla 2020. "Insights into the resources generation from pulp and paper industry wastes: challenges, perspectives and innovations." *Bioresource Technology* 297:122496. doi: 10.1016/j.biortech.2019.122496.

Gupta, V. K., I. Tyagi, H. Sadegh, R. S. Ghoshekand, A. S. H. Makhlouf, and B. Maazinejad. 2015. "Nanoparticles as adsorbent; a positive approach for removal of noxious metal ions: a review." *Science, Technology and Development* 34 (3):195–214. https://docsdrive.com/pdfs/std/std/2015/195-214.pdf

He, Z., Q. Wang, Y. Hu, J. Liang, Y. Jiang, R. Ma, Z. Tang, and Z. Huang. 2012. "Use of the quorum sensing inhibitor furanone C-30 to interfere with biofilm formation by Streptococcus mutans and its luxS mutant strain." *International Journal of Antimicrobial Agents* 40 (1):30–35. doi: 10.1016/j.ijantimicag.2012.03.016.

Heckert, E. G., A. S. Karakoti, S. Seal, and W. T. Self. 2008. The role of cerium redox state in the SOD mimetic activity of nanoceria. In: *Biomaterials*, pp. 2705–2709.

Hoseinian, F.S., B. Rezai, E. Kowsari, A. Chinnappan, and S. Ramakrishna. 2020. Synthesis and characterization of a novel nanocollector for the removal of nickel ions from synthetic wastewater using ion flotation. *Separation and Purification Technology* 240: 116639. doi: 10.1016/j.seppur.2020.116639.

Huber, B., L. Eberl, W. Feucht, and J. Polster. 2003. "Influence of polyphenols on bacterial biofilm formation and quorum-sensing." *Zeitschrift für Naturforschung* 58 (11–12):879–84.

Husain, F. M., I. Ahmad, M. Asif, and Q. Tahseen. 2013. "Influence of clove oil on certain quorum-sensing-regulated functions and biofilm of Pseudomonas aeruginosa and Aeromonas hydrophila." *Journal of Biosciences* 38 (5):835–44.

Jang, J. H., J. Lee, S. Y. Jung, D. C. Choi, Y. J. Won, K. H. Ahn, P. K. Park, and C. H. Lee. 2015. "Correlation between particle deposition and the size ratio of particles to patterns in nano- and micro-patterned membrane filtration systems." *Separation and Purification Technology* 156:608–616. doi: 10.1016/j.seppur.2015.10.056.

Jawaid, M., and Khan, M. M. 2018. "Polymer-based nanocomposites for energy and environmental applications." In: M. Jawaid, and M. M. Khan, ed. Woodhead Publishing. https://1lib.in/book/3497895/83fa8c.

Johnson, E. A. 2013. "Biotechnology of non-Saccharomyces yeasts-the ascomycetes." *Applied Microbiology and Biotechnology* 97:503–517. doi: 10.1007/s00253-012-4497-y.

Karakoti, A. S., S. Singh, A. Kumar, M. Malinska, S. V. Kuchibhatla, K. Wozniak, W. T. Self, and S. Seal. 2009. "PEGylated nanoceria as radical scavenger with tunable redox chemistry." *Journal of the American Chemical Society* 131 (40):14144–5. doi: 10.1021/ja9051087.

Khajeh, M., S. Laurent, and K. Dastafkan. 2013. "Nanoadsorbents: classification, preparation, and applications (with emphasis on aqueous media)." *Chemical Reviews* 113 (10):7728–7768. doi: 10.1021/cr400086v.

Kumari, P., M. Alam, and W. A. Siddiqi. 2019. "Usage of nanoparticles as adsorbents for waste water treatment: an emerging trend." *Sustainable Materials and Technologies* 22 (1): e00128.

Kurian, M., D. S. Nair, and A. M. Rahnamol. 2014. "Influence of the synthesis conditions on the catalytic efficiency of NiFe2O4 and ZnFe2O4 nanoparticles towards the wet peroxide oxidation of 4-chlorophenol." *Reaction Kinetics, Mechanisms and Catalysis*. doi: 10.1007/s11144-013-0667-x.

Lee, C.-K., and A.-N. Au-Duong. 2018. "Enzyme immobilization on nanoparticles: recent applications." In H.N. Chang (ed.): *Emerging Areas in Bioengineering*, 67–80.

Liang, H.-W., L. Wang, P.-Y. Chen, H.-T. Lin, L.-F. Chen, D. He, S.-H. Yu. 2010. Carbonaceous nanofiber membranes for selective filtration and separation of nanoparticles. Advanced Materials. 22: 4691–4695.

Lin, S.-T., M. Thirumavalavan, T.-Y. Jiang, and J.-F. Lee. 2014. "Synthesis of ZnO/Zn nano photocatalyst using modified polysaccharides for photodegradation of dyes." *Carbohydrate Polymers* 105:1–9. doi: 10.1016/j.carbpol.2014.01.017.

Liu, T., B. Li, Y. Hao, and Z. Yao. 2014. "MoO3-nanowire membrane and Bi2Mo3O12/MoO3 nano-heterostructural photocatalyst for wastewater treatment." *Chemical Engineering Journal* 244:382–390. doi: 10.1016/j.cej.2014.01.070.

Liu, Y., X. Chen, J. Li, and C. Burda. 2005. "Photocatalytic degradation of azo dyes by nitrogen-doped TiO2 nanocatalysts." *Chemosphere* 61 (1):11–18. doi: 10.1016/j.chemosphere.2005.03.069.

Lubick, N., and K. Betts. 2008. "Silver rocks have cloudy lining. Court bans widely used flame retardant." *Environmental Science & Technology* 42 (11):3910.

Mandeep, and P. Shukla. 2020. "Microbial nanotechnology for bioremediation of industrial wastewater." *Frontiers in Microbiology* 11:590631. doi: 10.3389/fmicb.2020.590631.

Mohajershojaei, K., N. M. Mahmoodi, and A. Khosravi. 2015. "Immobilization of laccase enzyme onto titania nanoparticle and decolorization of dyes from single and binary systems." *Biotechnology and Bioprocess Engineering* 20 (1):109–116. doi: 10.1007/s12257-014-0196-0.

Moreira, M. T., Y. Moldes-Diz, S. Feijoo, G. Eibes, J. M. Lema, and G. Feijoo. 2017. "Formulation of laccase nanobiocatalysts based on ionic and covalent interactions for the enhanced oxidation of phenolic compounds." *Applied Sciences* 7 (8):851.

Mirkin, C. A., R. L. Letsinger, R. C. Mucic, J. J. Storhoff. 1996. "A DNA-based method for rationally assembling nanoparticles into macroscopic materials." *Nature* 382 (6592):607–9. doi: 10.1038/382607a0.

Nogueira, H. P., S. H. Toma, A. T. Silveira, A. A. Carvalho, A. M. Fioroto, and K. Araki 2019. "Efficient Cr (VI) removal from wastewater by activated carbon superparamagnetic composites." *Microchemical Journal* 149:104025. doi: 10.1016/j.microc.2019.104025.

Nthunya, L. N., M. L. Masheane, S. P. Malinga, E. N. Nxumalo, and S. D. Mhlanga. 2017. "Environmentally benign chitosan-based nanofibres for potential use in water treatment." *Cogent Chemistry* 3 (1):1357865. doi: 10.1080/23312009.2017.1357865.

Ocampo-Sosa, A. A., M. Fernandez-Martinez, G. Cabot, C. Pena, F. Tubau, A. Oliver, and L. Martinez-Martinez. 2015. "Draft genome sequence of the quorum-sensing and biofilm-producing Pseudomonas Aeruginosa Strain Pae221, belonging to the epidemic high-risk clone sequence type 274." *Genome Announc* 3 (1). doi: 10.1128/genomeA.01343-14.

Olivera, S., H. B. Muralidhara, K. Venkatesh, V. K. Guna, K. Gopalakrishna, and K. Y. Kumar. 2016. "Potential applications of cellulose and chitosan nanoparticles/composites in wastewater treatment: a review." *Carbohydr Polym* 153:600–618. doi: 10.1016/j.carbpol.2016.08.017.

Pandey, K., B. Singh, A. K. Pandey, I. J. Badruddin, S. Pandey, V. K. Mishra, and P. A. Jain. 2017. "Application of microbial enzymes in industrial waste water treatment." *International Journal of Current Microbiology and Applied Sciences* 6 (8):1243–1254. doi: 10.20546/ijcmas.2017.608.151.

Parsons JG, Peralta-Videa JR & Gardea-Torresdey JL 2007 Chapter 21 Use of plants in biotechnology: Synthesis of metal nanoparticles by inactivated plant tissues, plant extracts, and living plants. Developments in Environmental Science, Vol. 5 (Sarkar D, Datta R & Hannigan R, eds.), pp. 463–485. Elsevier.Pendergast, M. M., and E. M. V. Hoek. 2011. "A review of water treatment membrane nanotechnologies." *Energy & Environmental Science* 4 (6):1946–1971. doi: 10.1039/C0EE00541J.

Prasad, M. P., and K. Manjunath. 2011. "Comparative study on biodegradation of lipid-rich wastewater using lipase producing bacterial species." *Indian Journal of Biotechnology* 10:121–124.

Ronen, A., W. Duan, I. Wheeldon, S. Walker, and D. Jassby. 2015. "Microbial attachment inhibition through low-voltage electrochemical reactions on electrically conducting membranes." *Environmental Science & Technology* 49 (21):12741–12750. doi: 10.1021/acs.est.5b01281.

Singh, S. 2019. "Nanomaterials exhibiting enzyme-like properties (nanozymes): current advances and future perspectives." *Frontiers in Chemistry* 7:46. doi: 10.3389/fchem.2019.00046.

Somwanshi, S. B., S. B. Somvanshi, and P. B. Kharat. 2020. "Nanocatalyst: a brief review on synthesis to applications." *Journal of Physics: Conference Series* 1644 (1):012046. doi: 10.1088/1742-6596/1644/1/012046.

Sponza, D. T., and A. Uluköy. 2005. "Treatment of 2,4-dichlorophenol (DCP) in a sequential anaerobic (upflow anaerobic sludge blanket) aerobic (completely stirred tank) reactor system." *Process Biochemistry* 40 (11):3419–3428. doi: 10.1016/j.procbio.2005.01.020.

Strzepa, A, K. A. Pritchard, and B. N. Dittel. 2017. "Myeloperoxidase: a new player in autoimmunity." *Cellular Immunology* 317:1–8. doi: 10.1016/j.cellimm.2017.05.002.

Sun, Y., and J. Cheng. 2002. "Hydrolysis of lignocellulosic materials for ethanol production: a review." *Bioresource Technology* 83 (1):1–11. doi: 10.1016/S0960-8524(01)00212-7.

Tchobanoglous, G., F. L. Burton, and H. D. Stensel. 2003. *Wastewater Engineering Treatment and Reuse*. McGraw Hill Companies. https://www.powells.com/book/wastewater-engineering-treatment-reuse-4th-edition-9780070418783.

Theron, J., J. A. Walker, and T. E. Cloete. 2008. "Nanotechnology and water treatment: applications and emerging opportunities." *Critical Reviews in Microbiology* 34 (1):43–69. doi: 10.1080/10408410701710442.

Unuofin, J. O., A. I. Okoh, and U. U. Nwodo. 2019. "Aptitude of oxidative enzymes for treatment of wastewater pollutants: a laccase perspective." *Molecules* 24 (11). doi: 10.3390/molecules24112064.

Yaqoob, A. A., T. Parveen, K. Umar, and M. N. Mohamad Ibrahim. 2020. "Role of nanomaterials in the treatment of wastewater: a review." *Water* 12 (2):495. doi: 10.3390/w12020495.

Yu, J. C., W. Ho, J. Lin, H. Yip, and P. Wong. 2003. "Photocatalytic activity, antibacterial effect, and photoinduced hydrophilicity of TiO2 films coated on a stainless steel substrate." *Environmental Science & Technology* 37 (10):2296–2301. doi: 10.1021/es0259483.

Zhang, D., M. Deng, H. Cao, S. Zhang, and H. Zhao. 2017. "Laccase immobilized on magnetic nanoparticles by dopamine polymerization for 4-chlorophenol removal." *Green Energy & Environment* 2 (4):393–400. doi: 10.1016/j.gee.2017.04.001.

9 Application of Potential Extremophilic Bacteria for Treatment of Xenobiotic Compounds Present in Effluents

Priya Banerjee, Komal Sharma, Ashok Kumar Nadda, and Aniruddha Mukhopadhayay

9.1 INTRODUCTION

Chemical compounds synthesized naturally or anthropogenically are known as xenobiotics (Gupta et al. 2022). The increase in standards of living and higher consumer demand have amplified the discharge of such xenobiotics into the environment. The technological and industrial expansion of different sectors (achieved to meet the requirements of modern society) have resulted in an increased complexity and toxicity of wastes produced as a result of the same (Gavrilescu et al. 2015; Giovanella et al. 2020). All pollutants discharged by such industries pose a concern to both public and environmental health. Examples of xenobiotics discharged with industrial effluents are heavy metals, nuclear wastes, pesticides, polyaromatic hydrocarbons (PAHs), etc. (Azubuike, Chikere, and Okpokwasili 2016). Rising concentrations of these compounds in different streams of effluents discharged in the environment has led to increased pollution of adjacent ecosystems receiving the same (Giovanella et al. 2020). These xenobiotics enter aquatic systems through different routes of direct and indirect discharge like abiotic and biotic movement, atmospheric deposition, food chain transfer and land runoff (Gupta et al. 2022). Pathways of xenobiotic movement in the environment are shown in Figure 9.1.

Xenobiotic compounds appear ubiquitously in almost all streams of wastewater at trace levels as a result of inefficient removal of the same via wastewater treatment plants (Gupta et al. 2022). Physicochemical techniques that are conventionally applied for remediation of polluted environments have proven to be less effective owing to high costs of operation and production of toxic by-products (Chen et al. 2015). Techniques like adsorption, advanced oxidation and membrane filtration have reportedly been efficient for removal of such xenobiotics from wastewaters

 DOI: 10.1201/9781003517238-9

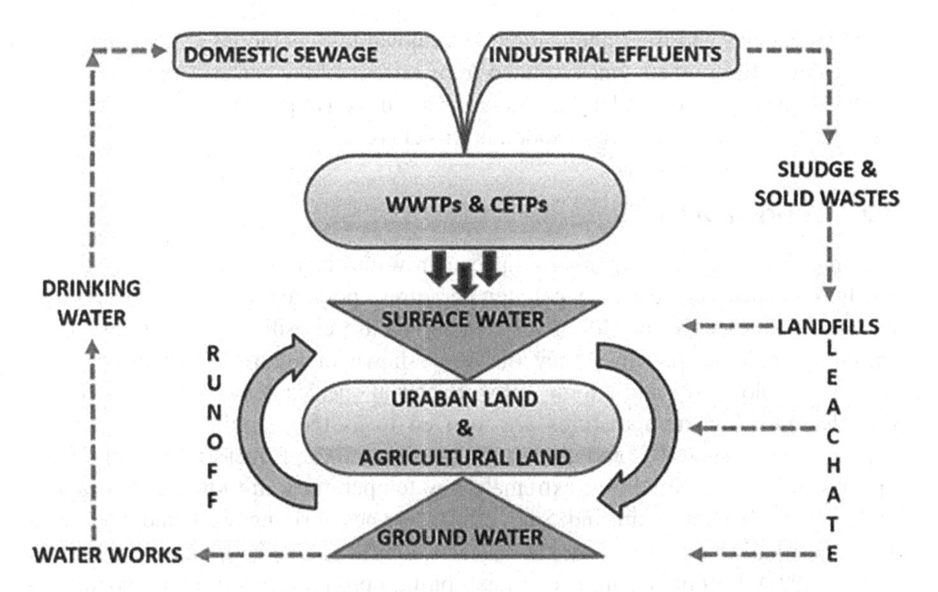

FIGURE 9.1 Movement of xenobiotic compounds in the environment. (*Source*: Reproduced with permission from Gupta et al. 2022).

(Gupta et al. 2022). Nevertheless, these processes have limited large-scale applications due to their complex nature, expensive maintenance, high energy requirement and production of toxic sludge (Gupta et al. 2022). Thus, recent studies have focused on more cost- and energy-efficient processes for removal of such xenobiotics from wastewaters.

In comparison to other processes, techniques of bioremediation have been widely preferred for effluent treatment, due to its cost-effective and eco-friendly nature (Tapadar et al. 2021). Bioremediation-based processes are also capable of efficient conversion of recalcitrant xenobiotics present in effluents to non-toxic compounds (Gillespie and Philp 2013; Deshmukh, Khardenavis, and Purohit 2016). Various algal, bacteriological, fungal and plant-based bioremediation systems are presently being investigated for xenobiotic removal and nutrient recovery from different types of effluents (Ashok et al. 2019; Gupta et al. 2016, 2018). However, in comparison to other microorganisms, bacterial species demonstrate a general tolerance towards a wide range of habitats as well as faster growth rates. Therefore, bacteria-mediated wastewater treatment is considered to be more stable in comparison to other microorganisms (Margot et al. 2013).

However, industrial effluents often yield non-conducive environments for growth and activity of bacterial species due to high salinity or extreme temperature or pH of the same. Waste degrading enzymes applied for effluent treatment are also unable to operate under such extreme conditions (Tapadar et al. 2021). Hence, organisms able to survive and grow under extreme environmental conditions (extremophiles) are presently being investigated as a whole and as a source of enzymes for efficient treatment of industrial effluents. The present study discusses different extremophilic bacterial species and their application for removal of xenobiotics from effluents. It also reviews different enzymes isolated from extremophilic bacteria and their application for xenobiotic degradation. A detailed discussion of process pathways, benefits and limitations has also been included in this chapter.

9.2 EXTREMOPHILES

Extremophiles are microorganisms that can grow and survive in a variety of extreme conditions, including extreme pH, temperatures, pollutant concentrations, salinity, etc. (Giovanella et al. 2020). Extremophiles are classified on the basis of the extreme conditions in which they thrive as shown in Figure 9.2 (Aragaw 2020; Bonch-Osmolovskaya and Atomi 2015). Bacterial species demonstrating optimum growth under high temperatures are referred to as thermophiles (50–55°C) and hyperthermophiles (60°C and above) (Haque et al. 2022; Huber and Stetter, 2021). Species growing optimally at extremely low temperatures are known as psychrophiles (<15°C) (Rathinam and Sani 2018). Species thriving well under extreme acidic or alkaline conditions are known as acidophiles and alkaliphiles, respectively (Yadav 2021). Barophiles are species exhibiting optimum growth under conditions of high pressure (Arora and Panosyan 2019). Organisms growing best under conditions of extreme salinity are known as halophiles (Daoud and Ali 2020). Most acidophilic, halophilic and hyperthermophilic species belong to the Archaea group (González and Terrón 2021). Groups of cyanobacteria are also highly adapted to

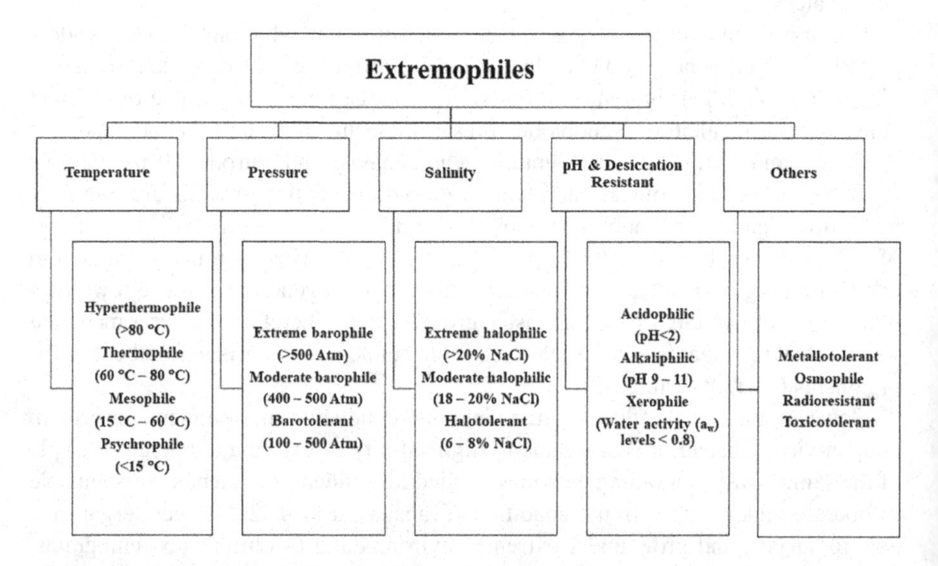

FIGURE 9.2 Classification of extremophiles. (Reproduced with permission from Rathinam and Sani (2018).)

TABLE 9.1
Properties of Extremophilic Microorganisms Preferred for as a Platform Strain for Biotechnological Processes Carried Out on Industrial Scale

Growth Conditions	Specifications	Type of Extremophile
Low or high temperature (T)	T < 25°C or T > 45°C	Psychrophiles or thermophiles
Low or high pH	pH < 5 or pH > 8.5	Acidophiles or alkaliphiles
High osmotic pressure (P)	NaCl > 30 g/L	Halophiles
Utilization of unusual substrate(s)	Long-chain fatty acids (C > 14), cellulose, chitin, rubbers, methane, methanol, H_2, CO_2	Xerophiles or methanotrophs or cellulose or gaseous substrate utilizers
In the presence of toxic compounds	Short-chain length alcohols (ethanol, propanol, butanol), short-chain length fatty acids (formate, propionate, butanoate), heavy metals (Cu, Zn) or toxic compounds (unsaturated fatty acids or aldehydes), among others	Xerophiles or various resistant microorganisms
Above combinations	High T + high pH High T + low pH High pH + high osmotic P High pH + unusual substrate(s), among others	Thermophiles: high T + low pH Halophiles: high pH + high P

Source: Reproduced with permission from Chen and Jiang (2018).

adverse environmental conditions and have been found to inhabit Antarctic ice as well as continental hot springs (Duleba 2019). Properties of extremophiles preferred for industrial biotechnological applications are shown in Table 9.1.

9.3 BIOREMEDIATION OF XENOBIOTICS USING EXTREMOPHILIC BACTERIA

Xenobiotic compounds frequently appear in industrial wastewater as they are widely used in several industries. Those effluents bear dissolved and suspended organics and inorganics, heavy metals and colourants (Aragaw and Angerasa 2020). Different kinds of wastes and contaminants are produced from industrial activities, mining activities, oil extraction or accidental oil spills. All these activities release several pollutants in the environments, such as triazines, azo dyes, cyclic biphenyls, nitroaromatic compounds, PAHs, aliphatic and aromatic halogenated compounds, organic sulphonic acid, and many more having xenobiotic structural features (Godheja et al. 2016). Industrial waste chemicals have a diverse spectrum of physicochemical properties and can be found in a number of physical environments

(Meckenstock et al. 2015). These effluents, even after treatment, when discharged into the water bodies may exert adverse effects on the receiving water environment (Pehlivanoglu-Mantas et al. 2008).

Several studies have investigated bioremediation for treatment of contaminated locations. Enzymes obtained from extremophilic bacteria reportedly play a significant role as metabolic catalysts in such processes (Shukla and Singh 2020). Extremophiles possess different molecular mechanisms for surviving in such contaminated conditions. Acidophiles possess potassium antiporter (responsible for releasing protons in response to extracellular medium), ATP synthase and chaperones. On the other hand, alkaliphiles exhibit antiporter-mediated electrochemical gradients (of Na^+ and H^+) for proton accumulation, Na^+-mediated solute uptake and cytochrome c-552 (which enhances terminal oxidation via electron and H^+ accumulation). Thermophiles possess upregulated glycolysis proteins and lipids (having iso-branched fatty acids and long-chain dicarboxylic fatty acids). Halophiles have high salt tolerance and possess chloride transporters. They are also capable of potassium uptake within cells via bacteriorhodopsin and ATP synthase. Halophiles also exhibit low-salt strategy like *de novo* synthesis. Isolation of extremophiles and their subsequent application in bioremediation have been shown in Figure 9.3. Pollutants and extremophiles used for remediation of the same have been enlisted in Table 9.2.

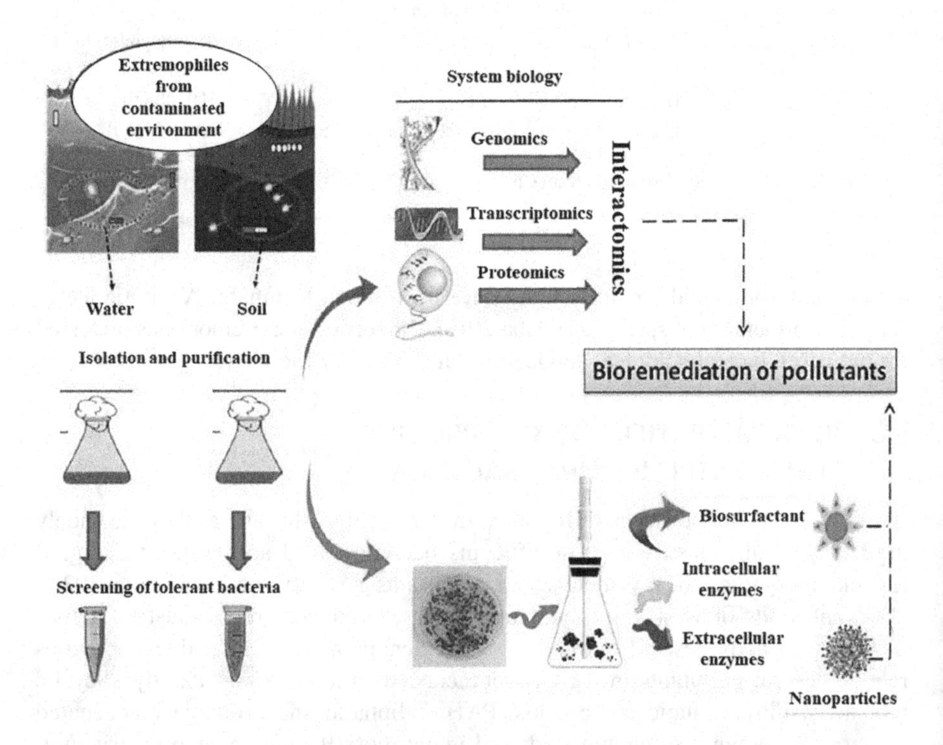

FIGURE 9.3 Isolation of extremophiles and their application in bioremediation. (Reproduced with permission from Shukla and Singh (2020) © 2020 Bentham Science Publishers.)

TABLE 9.2
Utilization of Extremophiles for Biodegradation of Pollutants.

Bacteria	Extremophile Type	Target Pollutant
Streptomyces albiaxialis	Halophile	Crude oil (30%)
Haloferax sp. strain D1227		Benzoic acid, cinnamic acid and 3-phenylpropionic acid
Dietzia maris		Paraffin and other petroleum derivatives
Bacterial consortia: *Marinobacter* ssp., *Erwinia ananas* and *Bacillus* spp.		Petroleum hydrocarbons
Haloarcula sp. D1		4-hydroxybenzoic acid
Cellulomonas spp., *Bacillus marisflavi, D. maris* and *Halomonas eurihalina*		Diesel
Halomonas campisalis		Phenol and catechol
Halomonas organivorans		Benzoic, p-hydroxybenzoic, cinnamic, salicylic, phenylacetic, phenyl-propionic, p-coumaric, ferulic and p-aminosalicylic acids
Marinobacter, Chromohalobacter, Salinicoccus and *Halobacillus*		Hydrocarbon and organic pollutants
Halomonas elongate		Ferulic acid
Acinetobacter strain		Fuel oil
Fusarium sp.		Crude oil
Shewanella putrefaciens MR-1	Psychrophile	Uranium
Natronocella acetinitrilica	Acidophile	Acetonitrile and propionitrile
Ochrobactrum sp.	Alkaliphile	Crude oil

Source: Reproduced with permission from Shukla and Singh 2020 © 2020 Bentham Science Publishers.

9.4 EXTREMOZYMES APPLIED FOR BIOREMEDIATION

Extremozymes are enzymes obtained from extremophiles that accelerate catalysis at specific rates (Rathinam and Sani 2018). These extremozymes have yielded extremely promising results in the fields of bioenergy, biotechnology, detergents, diagnostics, food processing, tanning, therapeutics, as well as other industrial applications. Recent studies have focused on the application of such extremophiles for bioremediation as they are able to tolerate extreme environmental conditions. Moreover, these extremophiles secrete extremozymes that are capable of functioning under adverse environmental conditions without denaturing. Extremozymes exhibit stability under extreme conditions of pH, temperature, ionic strength and exposure to organic solvents. Hence, these extremozymes are considered appropriate for treatment of environments polluted with recalcitrant contaminants. Owing to their extremophilic origin, extremozymes are resistant to detergents and solvents, sustain long storage periods, remain active over broad temperature ranges and are robust in nature (Kaushik et al. 2021). Extremophiles also possess specific protein adaptations (known as chaperone systems) that can function in the environment without succumbing to denaturation. These proteins are also functional under conditions that render mesophilic proteins or enzymes nonfunctional. Therefore, these enzymes are considered extremely convenient for immobilization and subsequent application in filtration devices used for bioremediation of effluents.

Use of extremophilic organisms in industrial processes has rapidly grown in the two previous decades. Over 3000 enzymes isolated from extremophiles have been identified. Extremozymes like alkane hydroxylase, dioxygenase, haloalkane dehalogenase, laccase, oxidoreductase, etc., have been reportedly applied for removal and treatment of effluents and wastes (Kaushik et al. 2021). Extremozymes are highly stable and cultures of extremophiles producing them are rarely contaminated. These extremozymes are broadly categorized into different types of polymer degrading enzymes like amylases, cellulases, proteases, pullulanases and xylanases (Dumorné et al. 2017). These enzymes have also been used widely as catalysts in chemical, food, paper and pulp, pharmaceutical and waste treatment industries (Shukla and Singh 2020). Extremophiles also produce important products like compatible solutes, cyclodextrins and polyunsaturated fatty acids.

Extremozymes, especially those obtained from marine organisms, have been successfully used in several bioremediation applications. Few such reports compiled by Shukla and Singh (2020) are stated as follows. Oxidoreductase obtained from *Bacillus safensis* (CFA-06) was responsible for degradation of aromatic compounds. Laccase secreted by *Thermus thermophilus* HB27 reportedly degraded PAHs. Phosphotriesterase-like lactonase obtained from *Geobacillus stearothermophilus* acted upon organophosphate. Dioxygenase derived from *Nocardioides* sp. strain KP7 was found to degrade polyaromatic compounds. Alkane hydroxylases/cytochrome P450 obtained from organisms like *Alcanivorax borkumensis* SK2T, *Bacillus licheniformis* ATCC 14580T, *Halomonas ventosae* Al12T and *Idiomarina baltica* DSM 15154T were found to target alkanes. Haloalkane dehalogenases produced by *Pseudomonas stutzeri* DEH130 and *Alcanivorax dieselolei* strain B-5 were reportedly responsible for the degradation of halogenated pollutants. Few such extremozymes reported by Gunjal et al. (2021) are elaborately discussed as follows.

9.4.1 Pectinases

Pectin polysaccharides are present in abundance in fruits. Hence, effluents discharged by fruit juice manufacturing units are rich in pectin. This enzyme facilitates the release of methane and in turn reduces the efficiency of treatment of the effluents bearing the same. Nevertheless, pectinolytic enzyme-based activated sludge treatment has reportedly been effective for the treatment of such effluents (Gundala and Chinthala, 2017). Studies have also previously reported the effective treatment of such pectin-loaded effluent using alkalophilic *Bacillus* spp.

9.4.2 Cellulases

Endoglucanase, exoglucanase and β-glucosidase are three groups of cellulases used for wastewater treatment. Endoglucanase targets low crystallinity of the cellulose fibre and forms free chain terminals. Exoglucanase or cellobiohydrolase acts upon the free chain terminals and removes cellobiose units from the same. β-Glucosidase then hydrolyses these cellobiose moieties to form glucose units. Cellulase Puradax HA obtained from alkalophilic *Bacillus* spp. have been effectively applied for removal of colour preservatives, stains and dyes like indigo removal from textile effluents.

9.4.3 Laccases

Laccase is a multi-copper-bearing extremozyme belonging to the class oxidoreductase. Laccases occurring in white rot fungi and moulds are high and low redox potential laccases, respectively. Laccases utilize oxygen as co-substrate. Oxygen is reduced to water by laccases when the latter oxidizes a substrate. Laccases reportedly decarboxylate compounds like aryl diamines, lignins, phenol and phenolic compounds, polyamines, etc., by targeting their methoxyl groups. These enzymes are also capable of depolymerizing lignin. These enzymes are extremely stable in extracellular fluids and are, therefore, widely applied in bioremediation.

Previous studies have reported PAH degradation using laccases. PAHs are mutagenic and carcinogenic compounds introduced into the environment through emissions, vehicular exhausts, incomplete combustion of organic matter, etc. This extremozyme oxidizes the aromatic rings present in PAHs resulting in their complete degeneration to yield water and carbon dioxide.

Non-phenolic and xenobiotic compounds are also oxidized by laccases. A previous study had reported the laccase-based biodegradation of a mixture containing compounds like 2-chlorophenol, 2,4-dichlorophenol, 2,4,6-trichlorophenol and pentachlorophenol. Laccases used in this study were derived from *Trametes pubescens*. In other studies laccases have also been reportedly used for the treatment of azo dye-bearing effluents. Azo dyes are primarily used in textile industries. Almost 7×10^5 tons of azo dye are produced every year of which approximately 1.5 tons are discharged along with industrial effluents. Around 10%–15% of the dyes used are lost due to inefficient dyeing processes. These dyes are carcinogenic to humans and exert toxic impacts on aquatic life. Laccases derived from alkali-halotolerant *Pseudomonas resinovorans* have been reportedly used for degradation of Bezaktiv Blue-S matrix 150 and Tubantin Brown GGL. Laccases were also used for degradation of remazol dyes (black-5, blue-19 and orange-1692).

9.4.4 Microbial Dioxygenases

Dioxygenase group of enzymes facilitate addition of oxygen to their substrate. These enzymes have been widely applied for bioremediation due to their ability to oxidize aromatic compounds into aliphatic products. Fe (II) and Fe (III) are utilized by extradiol and intradiol cleaving enzymes, respectively. Dioxygenase obtained from

Nocardioides sp. KP7 isolated from Kuwait beach reportedly facilitated phenanthrene degradation and detoxification of oil spills.

9.4.5 Monooxygenases

Monooxygenases demonstrate stereo selectivity on different substrates and are therefore widely used for bioremediation purposes. These enzymes are primarily involved in the degradation of hydrocarbons. Monooxygenases reportedly degrade aromatic and aliphatic compounds by acting substrates as reducing agents.

9.4.6 Peroxidases

Peroxidases are widely used in bioremediation processes for treatment of solids and effluents contaminated with oil and other hydrocarbons. Several phenolic compounds are reportedly oxidized by these enzymes. Peroxidase-based degradation of hydrocarbons is known to occur via mechanisms like Cα-Cβ cleavage, hydroxylation of benzylic methylene groups and oxidation of benzyl alcohols and phenolic groups. Of different types of peroxidases, lignin and manganese peroxidases are significant ones. Reactions are catalysed by lignin peroxidases in the presence of H_2O_2. Manganese peroxidases facilitate the oxidation of Mn^{2+} to Mn^{3+}. Both lignin and manganese peroxidases reportedly degrade polychlorinated biphenyls and dyes like amaranth, crystal violet, malachite green, orange (G and I) and remazol brilliant blue R.

9.4.7 Lipases

Lipases are widely applied for treatment of oil contamination. Lipases obtained from yeast strains and *Moraxella* sp. (isolated from water samples collected from Antarctica) were found to degrade oil. Lipases are reportedly used for treatment of saline effluents and oilfield wastes. Lipases are also used for treatment of effluents bearing high concentrations of oil and proteins like those discharged from dairies and tanneries, respectively. Polyester wastes and biofilms formed in water cooling units are also degraded by these enzymes.

9.4.8 Esterases

Esterases hydrolyse ester bearing compounds such as carbamates, organophosphates, pyrethroids, etc. Esterase extremozymes obtained from thermophilic microorganisms have been widely used for treatment of xenobiotics and other toxic chemicals.

9.4.9 Nitrilase

Nitrilases are commercially applied for surface modification, waste treatment and synthesis of carboxylic acids. Nitrilases act on compounds like cyanoglycosides, cyanolipids, phenylacetonitrile, ricinine, etc., known as nitrile compounds owing to the presence of cyano group (-CN). Nitrilase facilitates the synthesis of carboxylic acid and ammonia (by-product) or amide products from nitrile compounds by acting upon the carbon-nitrogen triple bonds. Nitrilases have been reported in various microorganisms and plants. These enzymes are widely used for their environmentally benign nature and efficiency of producing amides and carboxylic acid.

In terms of activity, thermophilic nitrilase is considered more useful in comparison to mesophilic nitrilase. Synthesis of thermophilic nitrilase at elevated temperatures offers advantages like enhanced rate of transfer, solubility of substrate, reduction of viscosity and lowered chances of contamination. Thermophilic bacteria such as *Acidovorax facilis* 72W12, *Bacillus pallidus* Dac 521 and *Geobacillus pallidus* RAPc8 (NRRL: B-59396)97 have been reported to produce nitrilase. Nitrilase isolated from hyperthermophilic anaerobic archaeon *Pyrococcus* sp. M24D13 has been used for bioremediation of cyanide (Dennett and Blamey, 2016).

9.4.10 Tyrosinases

Effluents bearing phenolic compounds are discharged by industries associated with conversion of coal, production of plastic and resin, petroleum refining, etc. Toxic effects of phenol hinder biological processes of effluent treatment, such as the activated sludge process. Enzymes like horseradish peroxidase have been applied since the 1980s for oxidation of phenolic compounds present in effluents. Enzyme-based treatment of phenol bearing effluents has been considered beneficial as it also yields valuable by-products at a lower cost of purification. β-Tyrosinase obtained from thermophile *Symbiobacterium* sp. SMH-1 yielded promising results regarding bioconversion of phenol present in effluents discharged by resin manufacturing industry to L-tyrosine (a value-added product).

9.4.11 Catalase

Strong oxidants like H_2O_2 are widely used in textile and semiconductor industries. Effluents contaminated with H_2O_2 need to be treated prior to discharge in order to prevent toxicity to flora and fauna exposed to the same. Effluents are subjected to chemical treatments prior to activated sludge treatment for removal of H_2O_2. Greater quantities of catalase or catalase producers have been reportedly effective for treatment of effluents discharged by textile industries.

9.5 POTENTIAL OF EXTREMOPHILES FOR BIOREMEDIATION

The fact that extremophilic microorganisms can grow under a wide range of extreme conditions makes them good candidates for bioremediation. The biological processes have many advantages from the aspect of environmental, economic and practical applications to remediate polluted sites. The immobilization, mobilization and transformation of metals/metalloids, as well as the adsorption and biodegradation of organic contaminants, are the main remediation processes that can be mediated by the action of several microorganisms, especially extremophiles surviving in harsh environments containing high concentrations of pollutants (Donati et al. 2019).

Extremophiles demonstrate a variety of adaptations by keeping their cellular proteins stable and functioning, allowing them to deal with such extreme environments. Extremozymes, which perform the same enzymatic tasks as their non-extreme counterparts, are one example of this adaptation. Under extreme conditions, extremozymes can catalyse chemical reactions inside the cell (Tapadar et al. 2021). Extremophiles adapt their membrane structure to cope with the harsh physiological circumstances. The lipid makeup of the cytoplasmic membrane of thermophiles changes in response to variations in external temperature (Koga 2012).

Extremophiles from Archaea, bacteria, and eukaryotes all adopt similar tactics to survive and flourish in extreme environments, including secreting a variety of indigenous metabolic products (extremolytes), extremozymes, and primary and secondary products. These released compounds enable proteins to change their conformations and movements in response to any stressful situation (Babu, Chandel, and Singh 2015).

In addition to other natural microorganisms, ammonia-degrading extremophiles are now one of the leading possibilities for wastewater treatment. Other impurities that can be removed include sulphur, manganese, iron and runoff pollutants (hydrocarbons, fertilizers). Because industrial effluents contain high salt concentrations as well as other organic compounds and heavy metals, polyextremophilic bacteria with increased metal resistance, complex colours and high salt concentrations can be used to treat industrial and other comparable pollutants. Polyextremophilic microorganisms can be found in industrial effluent and waste sites, and they can be identified and separated. Effluents contain diverse mixtures of colours, metals and other organic chemicals, as well as high salt levels. Some industrial effluents are extremely acidic or basic (Kaushik et al. 2021).

9.6 EFFLUENT TREATMENT USING EXTREMOPHILIC BACTERIA

Extremophilic microbes have evolved to be highly tolerant of poisonous materials and also convert unstable harmful contaminants into sufficiently stable useful resources for their cellular metabolism. Extremophiles play a significant role in the bioremediation of harsh environments since they are adapted to these habitats (Khemili-Talbi et al. 2015). Hydrocarbons can be converted or mineralized through the biodegradation process that happens in certain severe habitats. Extremophiles such as black yeasts, can adapt to different levels of environmental stress and tolerate a wide range of pH, temperature and xenobiotic compounds. Hence, they have been widely applied for treatment of industrial wastes and novel environmental recovery technologies (Arulazhagan et al. 2017; Blasi et al. 2016). According to Azubuike, Chikere, and Okpokwasili (2016), the application of a competent microbial consortium (bioaugmentation) in conjunction with a biostimulation method (which entails the enhancement of working conditions), can significantly improve hydrocarbon degradation.

9.6.1 Using Extremozymes

Cytochrome P450s, laccases, hydrolases, dehalogenases, dehydrogenases, proteases and lipases are extremophilic enzymes that have showed promising results in the degradation of polymers, aromatic hydrocarbons, halogenated chemicals, dyes, detergents, agrochemical compounds and others. Various mechanisms, such as oxidation, reduction, elimination and ring-opening, favour this type of bioremediation. Significant improvement of pollution degradation is frequently achieved using genetically modified microorganisms. These organisms produce a large variety of recombinant enzymes which are subsequently used in novel ecofriendly effluent treatment technologies (Bhandari et al. 2021).

9.6.2 Using Thermophiles/Acidophiles Alkaliphilic Bacteria/Halophiles/ Bacteria Surviving Under Other Extreme Conditions

Acidophilic microorganisms that thrive in low pH environments have been employed as host strains for heavy metal detoxification via biomining processes such as bioleaching and bio-oxidation in severely acidic environments. The creation of bioremediation procedures using *Acidothiobacillus* strains, which are the most prevalent acidophilic and chemolithotrophic microorganisms, has been reported in various publications. *Acidothiobacillus ferrooxidans*, for example, has been used in industrial-scale bioleaching (Jafari et al. 2019). A microbial consortium can be used to provide more efficient decontamination of harmful heavy metals, with one of its main advantages being the ability to synergize diverse enzyme systems and metabolic pathways of specific bacteria. On polluted port sediment, bioaugmentation of heavy metals was recently done using an acid mine drainage (AMD)-isolated acidophilic microbe consortium. An acidophilic microbial consortium composed of *Acidothiobacillus thiooxidans*, *At. ferrooxidans*,

TABLE 9.3
Extremozymes That Are Used in Bioremediation

S. No.	Extremozymes	Microbes	Function	References
1	Oxidoreductase	*Bacillus safensis* (CFA-06)	Degrades aromatic compounds	Da Fonseca et al. (2015)
2	Laccase	*Thermus thermophilus* HB27	Degradation of polyaromatic hydrocarbons	Das (2014)
3	Haloalkane dehalogenases	*Pseudomonas stutzeri* DEH130, *Alcanivorax dieselolei* B-5	Halogenated pollutants degradation	Donato et al. (2018)
4	Dioxygenase	*Nocardioides* sp. KP7	Polyaromatic compounds degradation	Shukla and Singh (2020)
5	Alkane hydroxylases/ Cytochrome P450	*Alcanivorax borkumensis* SK2T *Bacillus licheniformis* ATCC 14580T *Halomonas ventosae* Al12T *Idiomarina baltica* DSM 15154T	Alkane degradation	Shukla and Singh (2020)
6	Phosphotriesterase-like lactonase	*Geobacillus stearothermophilus*	Degradation of organophosphate	Das (2014)

Acidiphilium cryptum and *Leptospirillum ferrooxidans* proved successful in extracting more than 90% of Cu^{2+}, Cd^{2+}, Hg^{2+} and Zn^{2+} (Navarro, von Bernath, and Jerez 2013).

In high-salt environments, halophilic bacteria offer significant benefits in the remediation of harmful contaminants. Bioremediation with marine bacteria, for example, is a promising approach for removing hazardous heavy metals from saltwater since these bacteria can survive at high salt concentrations. There have been a few examples of hazardous heavy metals being removed using a variety of marine microorganisms. *Vibrio harveyi*, for example, demonstrated a strong adsorption capacity and an excellent ability to accumulate cadmium cations inside the cell (up to 23.3 mg Cd^{2+}/g of dry cells) (Abd-Elnaby, Abou-Elela, and El-Sersy 2011).

9.7 USING FRESH WATER AND MARINE BACTERIA

The microbial populations present in industrial wastewater (rich in ammonia, phenol and with high salinity) treatments are closely related to *Methanobrevibacter smithii*, the predominant methanogen present in human intestines (Gómez-Silván et al. 2010). The majority of species in the *Haloferacease* and *Halobacteriaceae* families can grow anaerobically in a wide range of salinity (Bonete et al. 2015). Consequently, these microorganisms might be applied for bioremediation in saline and hypersaline wastewater treatments because of their high tolerance to salt, metals and organic pollutants (Nájera-Fernández et al. 2012). Bacterial laccases are the enzymes that can catalyse the oxidation of phenolic and non-phenolic aromatic compounds and have unique properties such as high stability at 40°C, for pHs ranging from 5.5 to 9.0, high activity in the presence of chloride and high decolourization capability towards azo dyes. Hence, such extremophilic microorganisms producing extremozymes find applications in bioremediation of textile dyes bearing effluents (Fang et al. 2012).

9.8 UNDERLYING PATHWAYS/MECHANISMS

Extremophiles have developed many physiological adaptation mechanisms to cope with these extreme conditions, including membrane fluidity regulation, the action of molecular chaperones and the creation of antifreeze compounds (Collins and Margesin 2019). They can, for example, change the lipid composition of the membrane, increasing the number of polyunsaturated fatty acids and polar/non-polar carotenoids while lowering the size of the lipid head groups (Maayer et al. 2014). Cold-shock resistance is also regulated by temperature-induced enzymes such as cold-shock proteins (Csps) and heat-shock proteins (Hsps), which regulate signalling cascades that preserve damaged proteins and cofactors (Białkowska et al. 2020). Furthermore, antifreeze proteins and polysaccharides found in biofilm (such as trehalose, mannitol and exopolysaccharides) can operate as cryoprotectants (Yoshimune et al. 2005). Schematic representation of pathways involved in PAH and dye degradation is shown in Figure 9.4 (A and B respectively).

Special mechanisms of resistance have developed in a few bacterial species (shown in Figure 9.5) to adapt to different types of contaminants. These mechanisms

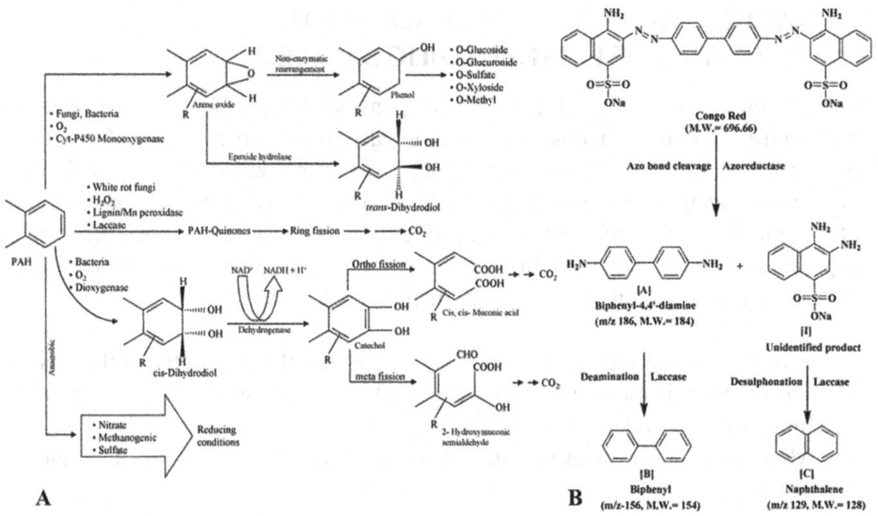

FIGURE 9.4 Schematic representation of pathways involved in PAH (a) and dye (b) degradation. (Reproduced with permission from Giovanella et al. (2020) © 2019 Elsevier B.V.)

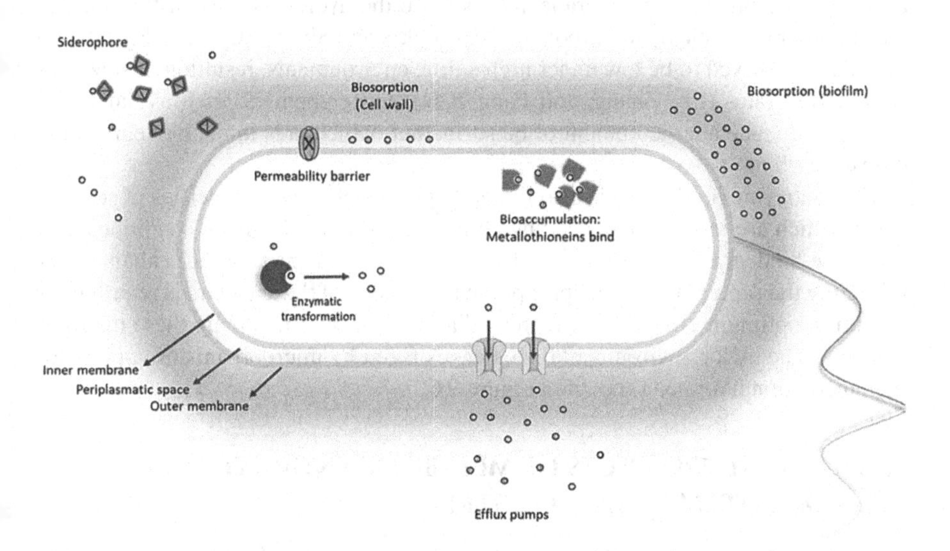

FIGURE 9.5 Metallic pollutant resistance mechanisms developed by bacteria. (Reproduced with permission from Giovanella et al. (2020) © 2019 Elsevier B.V.)

have rendered the species promising for application in bioremediation processes. Hence, these species can be explored for designing clean technologies for treatment of metallic and organic pollution and promoting the mitigation of environmental impacts (Giovanella et al. 2017).

9.9 FACTORS AFFECTING BIOREMEDIATION POTENTIAL OF EXTREMOPHILIC BACTERIA

Different biotic and abiotic factors are responsible for promoting the growth of extremophilic microorganisms. These factors also affect other biological processes taking place in a microbial community. The process of bioremediation occurs in multiphasic heterogeneous surroundings which influence the rates of reactions involved in the same. Dearth of information regarding the same may reduce the efficiency of the process when implemented (Srivastava et al. 2014). There are various factors that affect the bioremediation process which include the following:

a. Nature, state, concentration, solubility and bioavailability of the pollutants
b. Different environmental factors such as pH, temperature, soil type, nutrient and moisture, and redox potential
c. Microbe biodiversity at site, genetic competence, growth, metabolism and physiology

Biodegradation involving extremophiles is also influenced by the hydrophobic nature of the hydrocarbon, bioavailability and the surrounding environment. Microbes easily adapt to alterations in environmental parameters. Few microorganisms have the ability to alter the outer membrane and increase the hydrophobicity of the cell surface in order to enable hydrocarbon uptake (Shukla and Singh 2020). Biodegradation of PAH is observed to be low under high-saline environments, resulting in slow biodegradation rates (Lu, Zhang, and Fang 2011). Extremophiles, such as halophilic bacteria, produce a large amount of biosurfactant, which aids in the biodegradation of hydrocarbons.

Extremozyme production is also linked to physicochemical properties of soil and water, which are in turn influenced by climatic conditions. Climatic conditions may affect the biodegradation efficiency of these microorganisms by either enhancing or inhibiting the same. Therefore, appropriate knowledge of these factors is required for obtaining optimum efficiency of bioremediation processes involving these microorganisms. The major environmental challenges faced by microorganisms applied for bioremediation have been shown in Figure 9.6.

9.10 FUTURE PROSPECTS OF METABOLIC ENGINEERING OF EXTREMOPHILIC BACTERIA

Despite the fact that extremozymes were discovered several decades ago, researchers are still focusing their efforts on genetically engineering existing enzymes (for increasing their activity) and screening novel enzymes from a variety of sources, for the characteristics needed for industrial and biotechnological applications (Zhu et al. 2020). In recent years, the term "omics," which includes meta-genomics, meta-proteomics, meta-transcriptomics and metabolomics, has become increasingly essential in the development of novel enzymes with improved activity for biorefinery (Annamalai, Rajeswari, and Balasubramanian 2016). Furthermore, bioinformatics and algorithms are critical in the design of in situ mutagenesis and gene

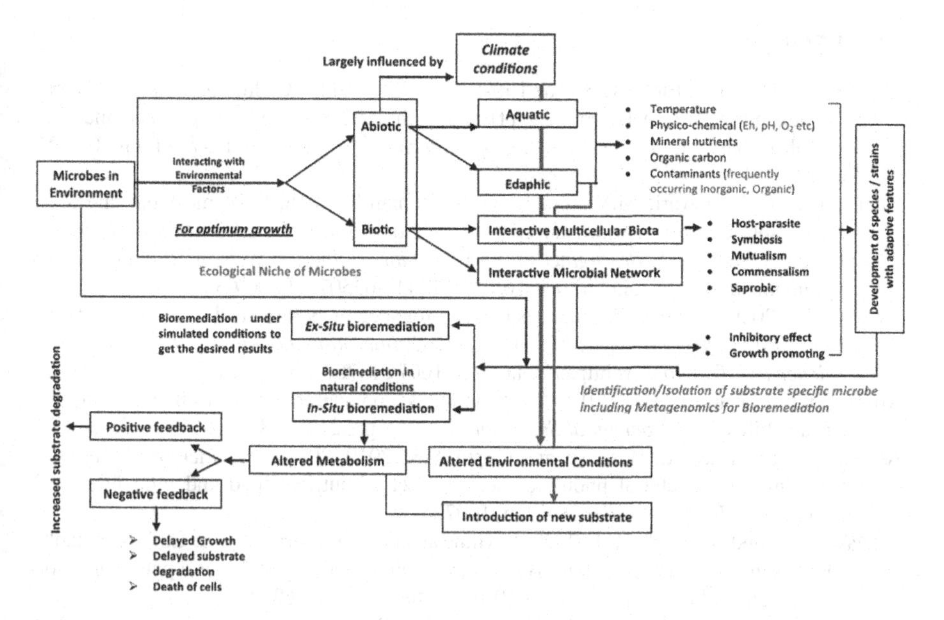

FIGURE 9.6 Simple layout showing factors influencing the microbial ecology, thereby influencing microbial bioremediation processes. (Reproduced with permission from Srivastava et al. (2014) © 2019 Elsevier B.V.)

shuffling to increase protein stability for possible industrial biorefinery applications. These strategies have proven to be quite useful in the production of biotechnological extremozymes. Because unculturable bacteria account for more than 99% of all microorganisms in the environment, present approaches are unable to extract the target enzyme from nature's huge enzyme resource pool. As a result, "omics" technology has given scientists a potent tool for deriving novel enzymes from nature (Juerges and Hansjürgens 2018).

9.11 CONCLUSION

Extremophiles are of great potential in the biodegradation of xenobiotics present in industrial effluents. It is also one of the most effective methods in the sense of cost-effectiveness and easy application. Although the application of extremophiles in bioremediation has been extensively studied, still the application of this procedure on the larger scale is still lacking. Future application of such bioremediation techniques to treat wastewater and in curbing environmental pollution can be a major step in reducing environmental pollution. Hence, developing technologies for exploring for microbial microenvironments and understanding the mechanisms of different microbial activity and metabolic pathways (e.g. redistribution, detoxification, mobilization/immobilization, translocation, transformation, biosorption and bioaccumulation) under diverse climatic and extreme conditions need to be further studied and understood for the future application on a large scale.

REFERENCES

Abd-Elnaby, H., Abou-Elela, G.M., and El-Sersy, N.A. 2011. "Cadmium resisting bacteria in Alexandria eastern harbor (Egypt) and optimization of cadmium bioaccumulation by Vibrio harveyi." *African Journal of Biotechnology* 10(17). https://doi.org/10.5897/ajb10.1933.

Annamalai, N., Rajeswari, M.V., and Balasubramanian T. 2016. "Thermostable and alkaline cellulases from marine sources." In *New and Future Developments in Microbial Biotechnology and Bioengineering: Microbial Cellulase System Properties and Applications*. https://doi.org/10.1016/B978-0-444-63507-5.00009-5.

Aragaw, T.A. 2020. "Functions of various bacteria for specific pollutants degradation and their application in wastewater treatment: a review." *International Journal of Environmental Science and Technology*. https://doi.org/10.1007/s13762-020-03022-2.

Arora, N.K., and Panosyan, H. 2019. "Extremophiles: applications and roles in environmental sustainability." *Environmental Sustainability* 2(3), pp. 217–218.

Ashok, V., Shriwastav, A., Bose, P., and Gupta, S.K. 2019. "Phycoremediation of wastewater using algal-bacterial photobioreactor: effect of nutrient load and light intensity." *Bioresource Technology Reports* 7, p. 100205.

Aragaw, T.A., and Angerasa, F.T. 2020. "Synthesis and characterization of Ethiopian kaolin for the removal of basic yellow (BY 28) dye from aqueous solution as a potential adsorbent." *Heliyon* 6(9). https://doi.org/10.1016/j.heliyon.2020.e04975.

Arulazhagan, P., Al-Shekri, K., Huda, Q., Godon, J.J., Basahi, J.M., and Jeyakumar, D. 2017. "Biodegradation of polycyclic aromatic hydrocarbons by an acidophilic stenotrophomonas maltophilia strain AJH1 isolated from a mineral mining site in Saudi Arabia." *Extremophiles* 21(1). https://doi.org/10.1007/s00792-016-0892-0.

Azubuike, C.C., Chikere, C.B., and Okpokwasili, G.C. 2016. "Bioremediation techniques-classification based on site of application: principles, advantages, limitations and prospects." *World Journal of Microbiology and Biotechnology*. https://doi.org/10.1007/s11274-016-2137-x.

Babu, P., Chandel, A.K., and Singh, O.V. 2015. Survival mechanisms of extremophiles. *Extremophiles and their applications in medical processes*, 9–23. https://doi.org/10.1007/978-3-319-12808-5_2.

Bhandari, S., Poudel, D.K., Marahatha, R., Dawadi, S., Khadayat, K., Phuyal, S., Shrestha, S. et al. 2021. "Microbial enzymes used in bioremediation." *Journal of Chemistry*. https://doi.org/10.1155/2021/8849512.

Białkowska, A., Majewska, E., Olczak, A., and Twarda-clapa, A. 2020. "Ice binding proteins: diverse biological roles and applications in different types of industry." *Biomolecules*. https://doi.org/10.3390/biom10020274.

Blasi, B., Poyntner, C., Rudavsky, T., Prenafeta-Boldú, F. X., De Hoog, S., Tafer, H., and Sterflinger, K. 2016. "Pathogenic yet environmentally friendly? Black fungal candidates for bioremediation of pollutants." *Geomicrobiology Journal* 33 (3–4). https://doi.org/10.1080/01490451.2015.1052118.

Bonch-Osmolovskaya, E., and Atomi, H. 2015. "Editorial overview: extremophiles: from extreme environments to highly stable biocatalysts." *Current Opinion in Microbiology*. https://doi.org/10.1016/j.mib.2015.06.005.

Bonete, M.J., Bautista, V., Esclapez, J., García-Bonete, M.J., Pire, C., Camacho, M., Torregrosa-Crespo, J., and Martínez-Espinosa, R. M. 2015. "New uses of haloarchaeal species in bioremediation processes." In *Advances in Bioremediation of Wastewater and Polluted Soil*. https://doi.org/10.5772/60667.

Chen, H., Teng, Y., Lu, S., Wang, Y., and Wang, J. 2015. "Contamination features and health risk of soil heavy metals in China." *Science of The Total Environment* pp. 512–513. https://doi.org/10.1016/j.scitotenv.2015.01.025.

Chen, G.Q. and Jiang, X.R., 2018. "Next generation industrial biotechnology based on extremophilic bacteria." *Current Opinion in Biotechnology* 50, pp. 94–100.

Collins, T., and Margesin, R. 2019. "Psychrophilic lifestyles: mechanisms of adaptation and biotechnological tools." *Applied Microbiology and Biotechnology*. https://doi.org/10.1007/s00253-019-09659-5.

Das, S. 2014. "Functional analysis of microbial communities." In *Microbial Biodegradation and Bioremediation*.

Daoud, L. and Ali, M.B. 2020. "Halophilic microorganisms: interesting group of extremophiles with important applications in biotechnology and environment." In *Physiological and Biotechnological Aspects of Extremophiles* (pp. 51–64). Academic Press. https://doi.org/10.1016/B978-0-12-818322-9.00005-8

Dennett, G.V. and Blamey, J.M., 2016. "A new thermophilic nitrilase from an Antarctic hyperthermophilic microorganism." *Frontiers in Bioengineering and Biotechnology*, *4*, p. 5.

Deshmukh, R., Khardenavis, A.A., and Purohit, H.J. 2016. "Diverse metabolic capacities of fungi for bioremediation." *Indian Journal of Microbiology*. https://doi.org/10.1007/s12088-016-0584-6.

Donati, E.R., Sani, R.K., Goh, K.M., and Chan, K.G. 2019. "Editorial: recent advances in bioremediation/biodegradation by extreme microorganisms." *Frontiers in Microbiology*. https://doi.org/10.3389/fmicb.2019.01851.

Donato, P.D., Buono, A., Poli, A., Finore, I., Abbamondi, G.R., Nicolaus, B., and Lama, L. 2018. "Exploring marine environments for the identification of extremophiles and their enzymes for sustainable and green bioprocesses." *Sustainability (Switzerland)*. https://doi.org/10.3390/su11010149.

Duleba, M., 2019. "Phycology." *Acta Botanica Hungarica* 61(1–2), pp. 215–217.

Dumorné, K., Córdova, D.C., Astorga-Eló, M. and Renganathan, P. 2017. "Extremozymes: a potential source for industrial applications." *Journal of Microbiology and Biotechnology* *27*(4), pp. 649–659.

Fang, Z. M., Li, T.L., Chang, F., Zhou, P., Fang, W., Hong, Y.Z., Zhang, X.C., Peng, H. and Xiao, Y.Z. 2012. "A new marine bacterial laccase with chloride-enhancing, alkaline-dependent activity and dye decolorization ability." *Bioresource Technology* 111. https://doi.org/10.1016/j.biortech.2012.01.172.

Fonseca, F.S.A.D., Angolini, C.F.F., Zezzi Arruda, M.A., Junior, C.A.L., Santos, C.A., Saraiva, A.M., Pilau, E. et al. 2015. "Identification of oxidoreductases from the petroleum Bacillus safensis strain." *Biotechnology Reports* 8. https://doi.org/10.1016/j.btre.2015.09.001.

Gavrilescu, M., Demnerová, K., Aamand, J., Agathos, J., and Fava, F. 2015. "Emerging pollutants in the environment: present and future challenges in biomonitoring, ecological risks and bioremediation." *New Biotechnology* 32(1). https://doi.org/10.1016/j.nbt.2014.01.001.

Gillespie, I. M.M., and Philp, J.C.. 2013. "Bioremediation, an environmental remediation technology for the bioeconomy." *Trends in Biotechnology*. https://doi.org/10.1016/j.tibtech.2013.01.015.

Giovanella, P., Cabral, L., Costa, A.P., de Oliveira Camargo, F.A., Gianello, C. and Bento, F.M., 2017. "Metal resistance mechanisms in Gram-negative bacteria and their potential to remove Hg in the presence of other metals." *Ecotoxicology and Environmental Safety*, *140*, pp. 162–169. https://doi.org/10.1016/j.jhazmat.2019.121024

Giovanella, P., Vieira, G.A.L., Ramos Otero, I.V., Pellizzer, E.P., de Jesus Fontes, B., and Sette, L.D. 2020. "Metal and organic pollutants bioremediation by extremophile microorganisms." *Journal of Hazardous Materials*. https://doi.org/10.1016/j.jhazmat.2019.121024.

Godheja, J., Shekhar S.K., Siddiqui, S.A., and Modi D.R. 2016. "Xenobiotic compounds present in soil and water: a review on remediation strategies." *Journal of Environmental & Analytical Toxicology* 6(5). https://doi.org/10.4172/2161-0525.1000392.

Gómez-Silván, C., Molina-Muñoz, M., Poyatos, J.M., Ramos, A., Hontoria, E., Rodelas, B. and González-López J. 2010. "Structure of archaeal communities in membrane-bioreactor and submerged-biofilter wastewater treatment plants." *Bioresource Technology* 101(7). https://doi.org/10.1016/j.biortech.2009.10.091.

González, A.G., and Terrón, R.P. 2021. "Importance of extremophilic microorganisms in biogeochemical cycles." *GSC Advanced Research and Reviews* 9(1), pp. 082–093.

Gundala, P., and Chinthala, P. 2017. "Extremophilic pectinases." In *Extremophilic Enzymatic Processing of Lignocellulosic Feedstocks to Bioenergy* (pp. 155-180), Sani R. K. and Krishnaraj R. N., eds. Springer, Cham.

Gunjal, A., Waghmode M., and Patil, N. 2021. "Role of extremozymes in bioremediation." *Research Journal of Biotechnology*, 16, p. 3.

Gupta, S.K., Ansari, F.A., Shriwastav, A., Sahoo, N.K., Rawat, I., and Bux, F. 2016. "Dual role of *Chlorella sorokiniana* and *Scenedesmus obliquus* for comprehensive wastewater treatment and biomass production for bio-fuels." *Journal of Cleaner Production* 115, pp. 255–264.

Gupta, S.K., Kumar, N.M., Guldhe, A., Ansari, F.A., Rawat, I., Nasr, M., and Bux, F. 2018. "Wastewater to biofuels: comprehensive evaluation of various flocculants on biochemical composition and yield of microalgae." *Ecological Engineering*, 117, pp. 62–68.

Gupta, S.K., Singh, B., Mungray, A.K., Bharti, R., Nema, A.K., Pant, K.K., and Mulla, S.I. 2022. "Bioelectrochemical technologies for removal of xenobiotics from wastewater." *Sustainable Energy Technologies and Assessments* 49, p. 101652. https://doi.org/10.1016/j.seta.2021.101652

Haque, S., Singh, R., Pal, D.B., Faidah, H., Ashgar, S.S., Areeshi, M.Y., Almalki, A.H., Verma, B., Srivastava, N., and Gupta, V.K. 2022. "Thermophilic biohydrogen production strategy using agro industrial wastes: current update, challenges, and sustainable solutions." *Chemosphere* 307, p. 136120.

Huber, R. and Stetter, K.O. 2021. The Thermotogales: hyperthermophilic and extremely thermophilic bacteria. In *Thermophilic bacteria* (pp. 185–194). CRC Press.

Jafari, M., Abdollahi, H., Ziaedin Shafaei, S., Gharabaghi, M., Jafari, H., Akcil, A., and Panda, S. 2019. "Acidophilic bioleaching: a review on the process and effect of organic-inorganic reagents and materials on its efficiency." *Mineral Processing and Extractive Metallurgy Review* p. 1481063. https://doi.org/10.1080/08827508.2018.1481063

Juerges, N., and Hansjürgens, B. 2018. Soil governance in the transition towards a sustainable bioeconomy – a review. *Journal of Cleaner Production* 170. https://doi.org/10.1016/j.jclepro.2016.10.143.

Kaushik, S., Alatawi, A., Djiwanti, S.R., Pande, A., Skotti, E., and Soni, V. 2021. "Potential of extremophiles for bioremediation." In *Microbial Rejuvenation of Polluted Environment* (pp. 293–328). Springer, Singapore.

Khemili-Talbi, S., Kebbouche-Gana, S., Akmoussi-Toumi, S., Angar, Y. and Gana, M. L. 2015. "Isolation of an extremely halophilic arhaeon natrialba Sp. C21 able to degrade aromatic compounds and to produce stable biosurfactant at high salinity." *Extremophiles* 19(6). https://doi.org/10.1007/s00792-015-0783-9.

Koga, Y. 2012. "Thermal adaptation of the archaeal and bacterial lipid membranes." *Archaea*. https://doi.org/10.1155/2012/789652.

Lu, X.Y., Zhang, T., and Fang. H.H.P. 2011. "Bacteria-mediated PAH degradation in soil and sediment." *Applied Microbiology and Biotechnology*. https://doi.org/10.1007/s00253-010-3072-7.

Maayer, P.D., Anderson, D., Cary, C., and Cowan, D.A. 2014. "Some like it cold: understanding the survival strategies of psychrophiles." *EMBO Reports*. https://doi.org/10.1002/embr.201338170.

Margot, J., Bennati-Granier, C., Maillard, J., Blánquez, P., Barry, D.A. and Holliger, C., 2013. "Bacterial versus fungal laccase: potential for micropollutant degradation." *AMB Express* 3(1), pp. 1–14. https://doi.org/10.1186/2191-0855-3-63

Meckenstock, R.U., Elsner, M., Griebler, C., Lueders, T., Stumpp, C., Aamand, J., Agathos, S. N. et al. 2015. "Biodegradation: updating the concepts of control for microbial cleanup in contaminated aquifers." *Environmental Science and Technology* 49(12). https://doi.org/10.1021/acs.est.5b00715.

Nájera-Fernández, C., Zafrilla, B., Bonete, M.J. and Martínez-Espinosa, R.M. 2012. "Role of the denitrifying haloarchaea in the treatment of nitrite-brines." *International Microbiology* 15(3). https://doi.org/10.2436/20.1501.01.164.

Navarro, C.A., von Bernath, D., and Jerez, C.A. 2013. "Heavy metal resistance strategies of acidophilic bacteria and their acquisition: importance for biomining and bioremediation." *Biological Research*. https://doi.org/10.4067/S0716-9760201300040 0008.

Pehlivanoglu-Mantas, E., Insel, G., Karahan, O., Ubay Cokgor, E., and Orhon, D. 2008. "Case studies from Turkey: xenobiotic-containing industries, wastewater treatment and modeling." *Water, Air, and Soil Pollution: Focus* 8(5–6). https://doi.org/10.1007/s11267-008-9180-z.

Rathinam, N.K. and Sani, R.K. 2018. "Bioprospecting of extremophiles for biotechnology applications." In *Extremophilic Microbial Processing of Lignocellulosic Feedstocks to Biofuels, Value-Added Products, and Usable Power* (pp. 1–23). Springer, Cham.

Shukla, A.K. and Singh, A.K., 2020. "Exploitation of potential extremophiles for bioremediation of xenobiotics compounds: a biotechnological approach." *Current Genomics* 21(3), pp. 161–167.

Srivastava, J., Naraian, R., Kalra, S.J.S. and Chandra, H. 2014. "Advances in microbial bioremediation and the factors influencing the process." *International Journal of Environmental Science and Technology* 11, pp. 1787–1800. https://doi.org/10.1007/s13762-013-0412-z

Tapadar, S., Tripathi, D., Pandey, S., Goswami, K., Bhattacharjee, A., Das, K., Palwan, E., Rani, M., and Kumar, A. 2021. "Role of extremophiles and extremophilic proteins in industrial waste treatment." In *Removal of Emerging Contaminants through Microbial Processes*. https://doi.org/10.1007/978-981-15-5901-3_11.

Yadav, A.N. 2021. "Biodiversity and bioprospecting of extremophilic microbiomes for agro-environmental sustainability." *Journal of Applied Biology and Biotechnology* 9(3), pp. 1–6.

Yoshimune, K., Galkin, A., Kulakova, L., Yoshimura, T., and Esaki, N. 2005. "Cold-active DnaK of an Antarctic psychrotroph shewanella Sp. Ac10 supporting the growth of DnaK-Null Mutant of escherichia coli at cold temperatures. *Extremophiles* 9(2). https://doi.org/10.1007/s00792-004-0429-9.

Zhu, D., Adebisi, W.A., Ahmad, F., Sethupathy, S., Danso, B., and Sun, J. 2020. "Recent development of extremophilic bacteria and their application in biorefinery." *Frontiers in Bioengineering and Biotechnology*. https://doi.org/10.3389/fbioe.2020.00483.

10 Microbial Removal of Endocrine Disrupting Chemicals from Wastewater

Arindam Rakshit, Arkapriya Nandi, and Priya Banerjee

10.1 INTRODUCTION

With an improvement in the standard of living, people's demand for improved quality of water has also increased. At the same time, the volume of wastewater produced has also increased with a corresponding rise in population density. Continuous release of undesirable chemical substances into water has resulted in poor water quality (Basheer, 2018). These chemicals enter the aquatic environment through effluents discharged by different industries or along with the flow of storm water. Chemicals that hinder with the natural performance of the endocrine systems of exposed individuals are referred to as EDCs (Egger et al., 2017; van der Meer et al., 2021). EDCs can either trigger or prevent a hormonal response. As evident from Figure 10.1, the major sources of EDCs in the aquatic environment are industrial or municipal and are discharged with effluents into natural waterbodies (Haq and Raj, 2019).

Different types of EDCs and their respective chemical structures have been enlisted in Figure 10.2. EDCs include a diverse group of compounds including synthetic lubricants and solvents of their by-products, which are used for industrial purposes. Compounds like detergents, dioxins, food additives, industrial chemicals, pesticides, pharmaceuticals and personal care products (PPCPs), plastics and plasticizers, polybrominated or polychlorinated biphenyls, etc., and even some naturally occurring substances (like estrogen) and their by-products have been widely reported as EDCs (Haq and Raj, 2019). Industrial compounds, steroidal and synthetic estrogens present in wastewaters mimic estrogen. Compounds like estrone (E1), 17β-estradiol (E2), estriol (E3) and 17α-ethinylestradiol (EE2) are widely present in industrial effluents, especially those generated by pharmaceutical industries. Bisphenol A (BPA), nonylphenol (NP) and β-sitosterol are few other EDCs that are widely discharged through industrial effluents. Exposure to EDCs can cause fluctuations of blood hormone levels, result in imposex and intersex, induce alterations in gene expressions, decline in fertility and fecundity, hermaphroditism as well as masculinization/feminization in aquatic organisms (Haq and Raj, 2019).

DOI: 10.1201/9781003517238-10

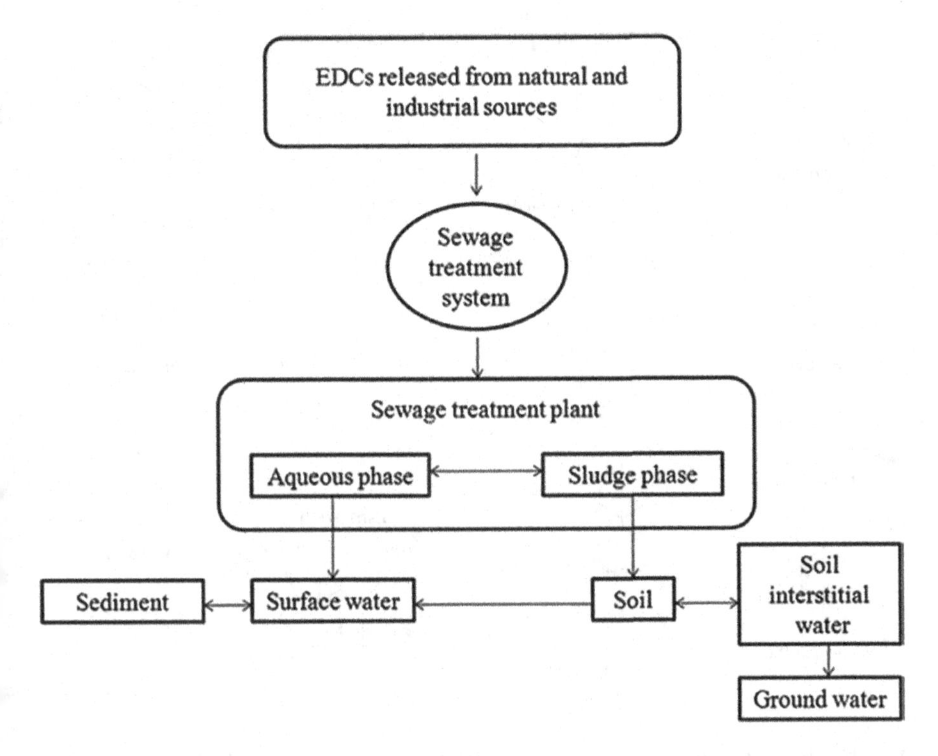

FIGURE 10.1 Distribution of EDCs in environment. (Reproduced with permission from Haq and Raj (2019).)

Both ground and surface water reserves may be contaminated by EDC bearing effluents discharged from the industries and the wastewater treatment plants (WWTPs) (Kumari et al., 2016). Though present in low concentrations, these substances are constantly polluting the natural aquatic ecosystem (Ali et al., 2017; Al-Shaalan et al., 2019; Basheer, 2018). Environmental accumulation of such substances may exert detrimental impacts on both aquatic ecosystems and humans alike (Olasupo and Suah, 2021). Moreover, most EDCs exert genotoxic, mutagenic or estrogenic effects and are biorecalcitrant in nature. Several international organizations as well as different countries have imposed restrictions on the use of such compounds. EDCs have been recognized as emerging micropollutants owing to the risks and hazards posed by the same in aqueous environments even in quantities ranging from ng/L to μg/L (Budeli et al., 2021). As a result, removal of these harmful chemicals from effluents has received major focus in recent research (Al Sharabati et al., 2021).

10.2 EDC REMOVAL BY CONVENTIONAL WWTPS

Conventional WWTPs depend on the microbial communities present in the effluents and the activated sludge for the degradation of EDCs (Roccuzzo et al., 2021). In these plants, the primary treated EDC bearing effluent is retained in an aerated reactor having different microbial species. In this method, microbial biomass is suspended

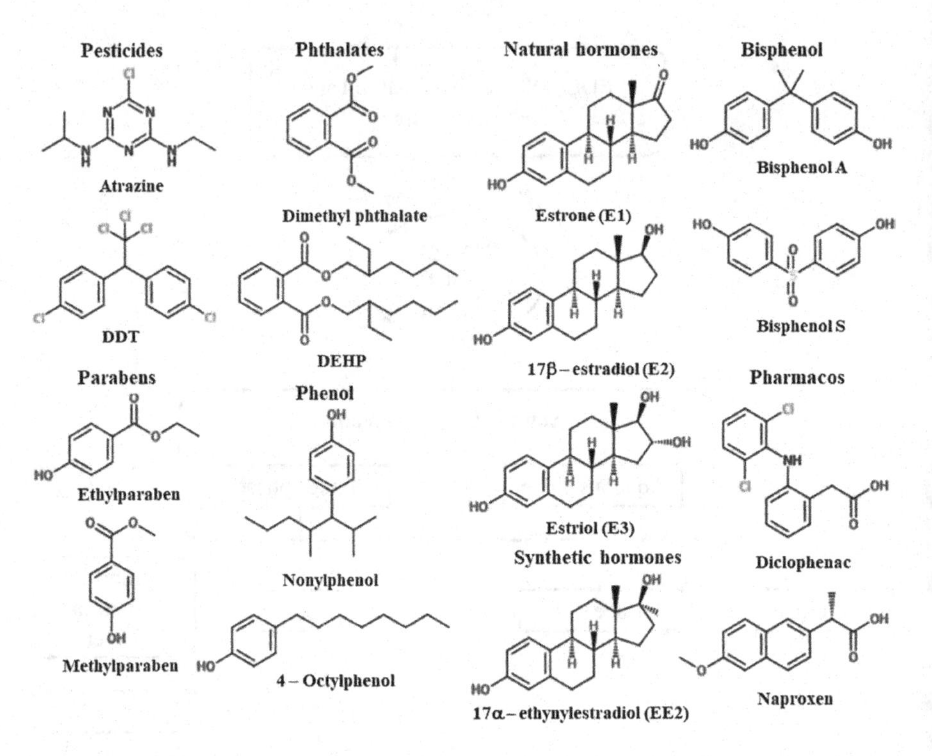

FIGURE 10.2 Chemical structures of the major classes of natural and synthetic endocrine disruptors present in different water matrices. (Reproduced with permission from Vieira et al. (2020).)

as flocs for ensuring maximum contact between the microbial biomass and the effluent under constant agitation. Through this process, the pollutant is adsorbed by the organic matter and is rapidly and efficiently separated from the effluents bearing them (Roccuzzo et al., 2021). Extracellular polymeric substances (EPS) produced by some bacteria also guide the genesis and maintenance of flocs. Organic pollutant adsorption and nutrient accumulation from external environment also depend upon the distribution of nucleic acids, polysaccharides, lipids and proteins in the EPS (Roccuzzo et al., 2021).

However, WWTPs based on activated sludge systems are often unable to achieve complete degradation of EDCs owing to the oscillating levels of EDCs in the influents and variations in sludge parameters like age, hydraulic retention time (HRT) and temperature (Roccuzzo et al., 2021). Integration of different physical and chemical processes like advanced oxidation, nanofiltration, photocatalysis, sorption, etc., reportedly augments the EDC removal efficiency of these WWTPs (Aissani et al., 2018; Ayuba et al., 2019; Bartolomeu et al., 2018; Fernando et al., 2019; Khan et al., 2020; Miralles-Cuevas et al., 2017; Mohammadi et al., 2020; Polloni-Silva et al., 2017). However, these processes are not suitable for commercial scale applications. The costs incurred for reagents used and operation of the setup are significantly higher than other contemporary approaches. Moreover, simultaneous formation of a complex waste sludge renders the concerned process unsuitable for the treatment of high volumes of wastewater (Roccuzzo et al., 2021). Hence, recent research is devoting significant attention toward development of sustainable and environmentally responsive EDC treatment technologies (Schug et al., 2016). In this attempt, researchers are also investigating the improvement of conventional activated sludge (CAS) systems using specialized microorganisms (bacteria/microalgae/fungi) isolated from microbial communities obtained from contaminated sites (Haq and Raj, 2019).

10.3 BIOTREATMENT OF EDCS

Biological treatment methods have been found to be extremely suitable for detoxification of EDCs as they are both cost and energy effective in nature. Eco-friendly EDC treatment can be achieved using bacteria, fungi, algae and plants. Few biological processes for treatment of EDC bearing effluents are discussed as follows.

10.3.1 Sorption

In this method, micropollutant moieties adhere to activated sludge (solid particles). Some pollutants are easily adsorbed by the activated sludge. Pollutants basically adhere to primary and secondary sludge via two methods of sorption – adsorption and absorption. During adsorption process, an electrostatic interaction is formed between the negatively charged surface of microorganisms with the positively charged functional groups of micropollutants (EDCs). On the contrary, during absorption, a hydrophobic interaction occurs between the pollutants and the lipophilic cell membrane of the microbes. The ratio of concentration of pollutant present in the solid to the concentration present in the aqueous phase (at equilibrium) is called the solid–water distribution coefficient (K_d). The ratio of equilibrium concentration of micropollutants adsorbed and in solution can be determined using K_d (Sathe et al., 2018). According to Sathe et al. (2018), pollutants are first absorbed on the activated sludge and then removed by the process of biodegradation. K_d value of the primary sludge is reportedly higher than the K_d value of the activated sludge (Zaman et al., 2021).

10.3.2 Biodegradation

Biodegradation is an integrated and sustainable endeavor that involves the incorporation of various microscopic organisms for the treatment of organic pollutant-rich solid wastes and wastewater generated from different sources. In recent times, bioremediation is considered as a new innovative tool for waste management due to its high efficiency, eco-friendly and cost-effective nature. This technology allows microorganisms to adapt to the toxic environment created by hazardous wastes, facilitating the advent of new toxin-resistant strains. These new strains can convert toxic chemical substances into a less harmful substance through metabolism. Various biological processes like activated sludge, aerated filters, bioaugmentation, biological contactors, bioslurping, sequencing batch reactors, trickling filters, etc., have been reported in previous studies for the treatment of effluents.

Biological treatment of effluents depends upon enzymatic and metabolic efficiency of microorganisms for pollutant degradation. Bioremediation by native microbial populations is considered to be the most environment-friendly, energy-efficient and cost-effective process for effluent treatment (Singh and Borthakur, 2018). The most advantageous fact about biological treatment process is that their final product is non-toxic. Moreover, innocuous gases like nitrogen, carbon and hydrogen synthesized as final products of microbial degradation are easily assimilated within the environment (Singh and Borthakur, 2018). In order to improve the biodegradation potential of native microorganisms, certain parameters like bioavailability, rate of adsorption and mass transfer need to be enhanced. However inclusive research is still needed to determine the impact of bioremediation methods on the environment on a large scale (Zaman et al., 2021).

10.3.2.1 EDC Removal Using Bacteria

EDCs biodegradation in WWTPs has been widely investigated using different species of bacteria. Banerjee et al. (2016) had reported >98% and >99% removal of triclosan and surfactant removal from effluents laden with PPCPs. Al Aani et al. (2020) reported soil isolates of *Rhodococcus* sp. and *Sphingomonas* sp. for degradation of 17β-estradiol present in effluents. Sewwandi et al. (2022) reported *Novosphingobium tardaugens* isolated from activated sludge for degradation of estrogen. Different strains of *Pseudomonas* sp., *Enterobacter asburiae* and *Stenotrophomonas* sp. reportedly degrade NP present in effluents (Haq and Raj, 2019).

In a recent study, Zhang et al. (2022) suggested five possible mechanisms of E2 degradation using *Ochrobactrum* sp. (as shown in Figure 10.3). In pathway A, E2 was hydroxylated to yield an aldehyde. In pathway B, E2 was dehydrogenated to E1 in two possible ways. In pathway C, E2 is converted to E1 which was further converted to a ketone, water and carbon dioxide. In pathway D, E1 so formed is hydroxylated to 4-OH-E1 which is further hydroxylated to product III (via pathway E) or oxidized to product IV (via pathway F). By-products (V, d, VIII) formed in these pathways may have further potential for degradation but are thought to be less toxic in comparison to their precursor molecules (Zhang et al. 2022). In another recent study, authors reported EE2 degradation by *Nitrosomonas europaea* and *Pseudomonas citronellolis* (Prakash and Chaturvedi, 2023). In another study, Zühlke et al. (2020) reported BPA degradation using *Cupriavidus basilensis*. Other bacterial species reported for EDC degradation in recent reports have been listed in Table 10.1.

10.3.2.2 EDC Removal Using Fungi

In their study, Zhuo and Fan (2021) investigated white rot fungi (*Pleurotus ostreatus*) and enzymes derived from the same for degradation of organic pollutants (as shown in Figure 10.4). *P. ostreatus* has been previously reported for >90% degradation of EDCs like BPA, E1, E2, E3, EE2, 4-NP, triclosan, etc., over 12 days (Haq and Raj, 2019). In similar studies, *Trametes versicolor* was found to degrade EDCs like NP polyethoxylates, BPA and Bisphenol S (BPS) from aqueous media (Grelska and Noszczyńska, 2020; Stenholm et al., 2020; Trivedi and Chhaya, 2022). *T. versicolor*, *P. ostreatus*, and *Phanerochaete chrysosporium* are also reportedly capable of degrading different classes of compounds like parabens, phthalates and phenols

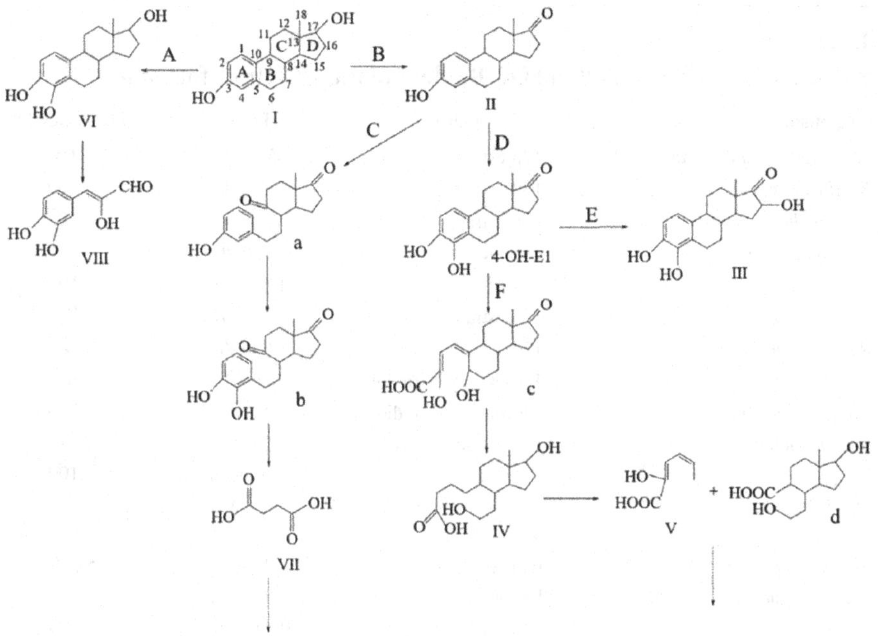

FIGURE 10.3 Possible mechanisms of 17β-estradiol degradation using *Ochrobactrum* sp. (Reproduced with permission from Zhang et al. (2022) © 2021 Elsevier.)

(Pezzella et al., 2017). Carstens et al. (2020) proposed pathways for phthalate, plasticizers and BPA degradation by fungal strains identified as *Ascocoryne* sp., *Phoma* sp., *Stachybotrys chlorohalonata*, *Stropharia rugosoannulata*, and *Trichosporon porosum* (as shown in Figure 10.5).

Fungal enzymes have been widely reported for EDC degradation. However, efficiency of these enzymes decreases in uncontrolled environment, making the same unsuitable for large-scale treatment of real effluents. On the contrary, bacterial enzymes are found to be more stable than those derived from fungi even under adverse environmental and poor physiological conditions (Haq and Raj, 2019). Other fungal species reported for EDC degradation in recent reports have been listed in Table 10.2.

10.4 CONVENTIONAL WASTEWATER TREATMENT

Domestic and industrial effluents are conventionally treated using activated sludge process (ASP). However, ASP requires aeration which in turn increases the energy utilized and cost incurred by this process. Moreover, this process is limited by formation of huge quantities of sludge. As reported previously, anaerobic processes have been found to be more advantageous over aerobic ones in terms of lower energy requirement, higher rates of pathogen removal and significantly reduced sludge formation. Moreover, methane emitted in this process may be used as an alternative source of energy. Anaerobic process-based filters, sequencing batch reactors and

TABLE 10.1
Bacterial Species Engaged in EDC Biodegradation/Biotransformation

Organisms	EDCs	Dose	Efficiency (%)
Virgibacillus halotolerans	17β-estradi	5 mg/L	100
Bacillus flexus	17β-estradiol	5 mg/L	100
Bacillus licheniformis	17β-estradiol	5 mg/L	100
Novosphingobium sp.	Estrone	1.75 μg/L^{-1}	80.43
	Estriol	1.52 μg/L	100
	17β-estradiol	0.71 μg/L	94.76
Rhodococcus sp.	17β-estradiol	30 mg/L	94
Rhodococcus zopfii	17α-ethynylestradiol	2 mg/L	86.5
Pseudomonas putida	17α-ethynylestradiol	0.5 mg/L	73.8
Acinetobacter sp.	17β-estradiol	40 mg/L	90
		0.5 mg/L	100
		1.8 mg/L	100
		5 mg/L	100
Novosphingobium sp.	Estrone	1 mM	No data
Sphingomonas sp.	Estrone	0.05 mg/L	100
	17β-estradiol	0.05 mg/L	100
	Testosterone	0.05 mg/L	100
Thauera sp.	Testosterone	1 mM	No data
Rhodococcus pyridinivorans	Zearalenone	5 mg/L	87.21
Bacillus sp.	17β-estradiol	0.2 mg/L	91.70
Aeromonas punctata	17β-estradiol	0.2 mg/L	94.20
Klebsiella sp.	17β-estradiol	0.2 mg/L	100
Enterobacter sp.	17β-estradiol	0.2 mg/L	77.10
Enterobacter sp.	17β-estradiol	0.2 mg/L	85.40
Aeromonas veronii	17β-estradiol	0.2 mg/L	93.40
Acinetobacter sp.	17β-estradiol	5 mg/L	77.00
Pseudomonas sp.	17β-estradiol	5 mg/L	68.00
Sphingomonas sp.	17β-estradiol	1.8 mg/L	78.70–98.80
Comamonas testosteroni	Testosterone	0.5 mM	60.00
	17β-estradiol	0.5 mM	35.00
	Estrone	0.5 mM	45.00
	Estriol	0.5 mM	25.00
C. testosteroni	17β-estradiol	1 mg/L	76.00
Rhodococcus equi	17β-estradiol	1 mg/L	86.00
Deinococcus actinosclerus	17β-estradiol	10 mg/L	90.00
Stenotrophomonas maltophilia	17β-estradiol	10 mg/L	90.00

Source: Reproduced with permission from Wojcieszyńska et al., (2020).

upflow sludge blankets are reportedly efficient in removal of pollutants present in both high and low concentrations. However, ASPs are not efficient in removal of EDCs from effluents (Haq and Raj, 2019). EDCs are also formed in effluents due to incomplete degradation of pollutants by ASPs. EDCs like alkyl phenols, BPA,

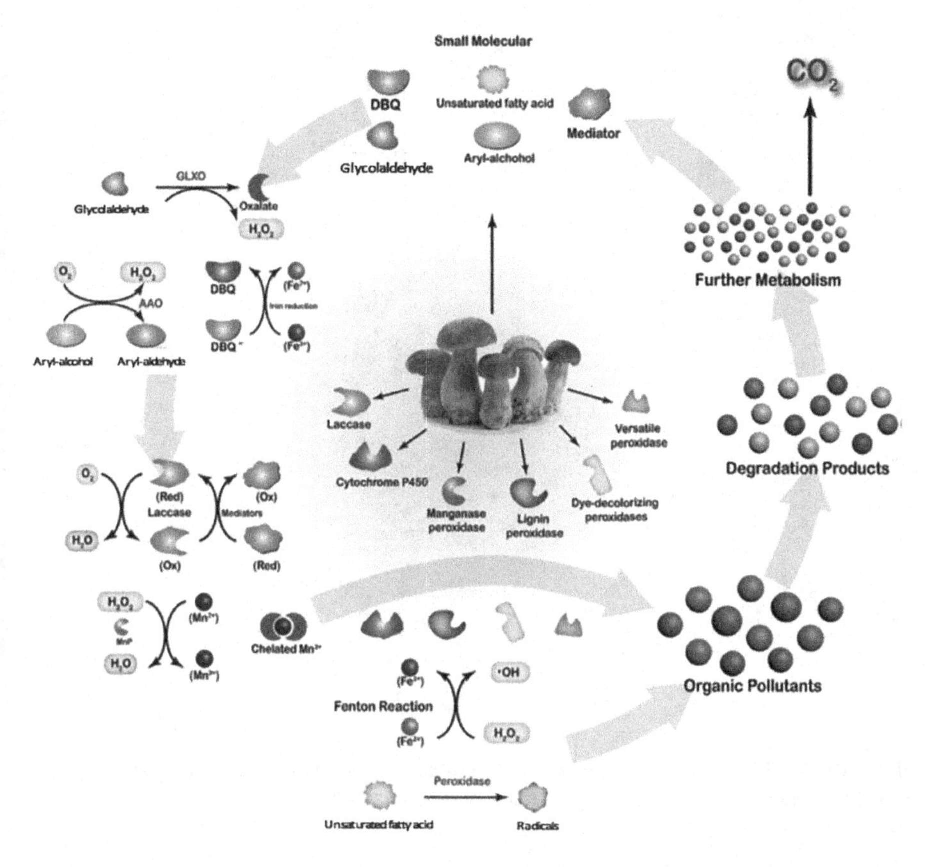

FIGURE 10.4 Degradation of organic pollutant by white rot fungi (WRF) and their oxidative and extracellular ligninolytic systems. (Reproduced with permission from Zhuo and Fan (2021).)

etc., often remain unaffected in WWTPs and are discharged into the environment although in trace quantities. As a result, EDCs are presently being detected in various ground water, surface water as well as drinking water resources.

10.5 EDC DEGRADATION USING IMMOBILIZED MICROBIAL SYSTEMS

In recent times, immobilization of microorganisms (whole cells and enzymes) has gained significant importance as an extremely useful process for the treatment of effluents bearing toxic pollutants. In comparison to ASP, cell and enzyme immobilization has been found to help boost biomass concentration and microbial activities, improve process stability, increase tolerance to adverse and extreme environmental conditions (temperature, pH, salinity, pollutant load, etc.), improve rates of hydraulic loading and enhance the reusability of biocatalysts (Zaman et al., 2021). Moreover, the immobilized cell cultures (ICCs) are efficiently separated in the settling tank. ICCs are also capable of overcoming a larger number of obstacles which have been found to be impossible for suspended cultures. Hence, ICCs have been reported

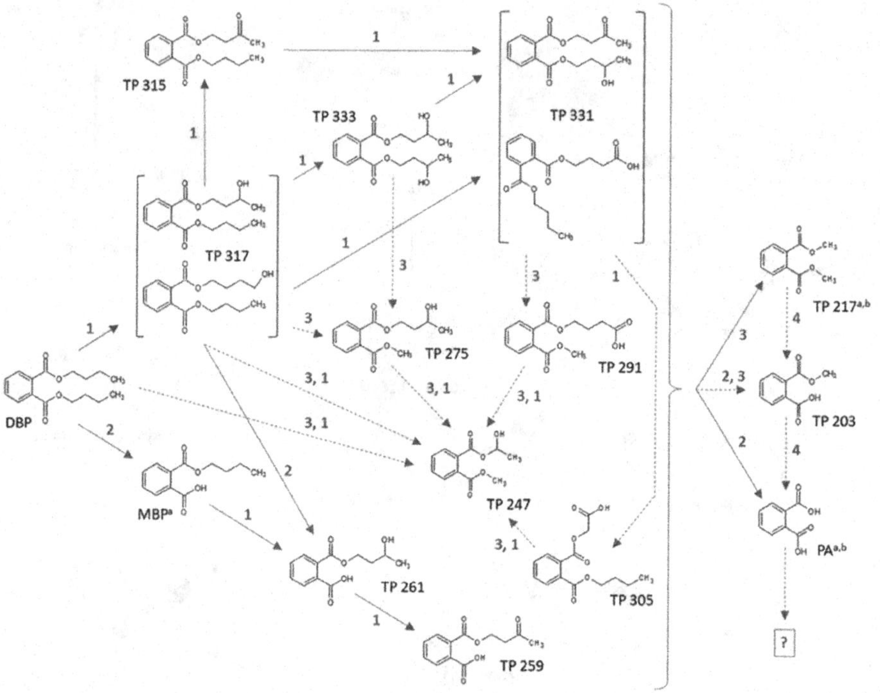

FIGURE 10.5 Proposed pathways for fungi-based degradation of di-*n*-butyl phthalate (DBP). (Reproduced with permission from Carstens et al. (2020).)

to be a more efficient process for pollutant degradation than ASP (Banerjee et al., 2016). A comparative account of ASP- and ICC-based processes has been shown in Table 10.3.

Thus, ICCs are being extensively investigated for removal of EDCs from effluents. Inert substances have been selected for microbe immobilization owing to their non-hazardous nature and ability to facilitate pollutant removal (Liu et al., 2018; Wojcieszyńska et al., 2020). Several natural and artificial carriers have been reported in recent studies for microbial immobilization (Dzionek et al., 2018). *Novosphingobium* sp. immobilized on calcium alginate (CA) reportedly yielded 80.43%, 94.76% and 100% removal of E1, E2 and E3, respectively (Liu et al., 2018; Olajuyigbe et al., 2019). Efficient degradation of E2 was also achieved using *Rhodococcus* sp. immobilized on CA (Wojcieszyńska et al., 2020). Immobilization facilitates lower toxicity for ICCs and rapid exchange of nutrients and metabolites. This in turn reduces cost incurred. However, pores of substrates used for immobilization are often clogged after repeated use. Clogging and low mechanical strength often result in a decrease in pollutant uptake (Liu et al., 2018). Natural biopolymers like CA, chitosan, starch, etc., have been reported for immobilization of microorganisms (Bilal et al., 2019).

TABLE 10.2
Fungal Species Engaged in EDC Biodegradation/Biotransformation

Organisms	EDCs	Dose	Efficiency (%)
Trichoderma citrinoviride	17β-estradiol	200 mg/L	99.60
Trametes versicolor	Dimethyl phthalate	100 μM	60.00
	Methyl paraben	100 μM	100.00
	Butyl paraben	100 μM	100.00
	Nonylphenol	100 μM	85.00
	Bisphenol A	100 μM	100.00
Pleurotus ostreatus	Dimethyl phthalate	100 μM	50.00
	Methyl paraben	100 μM	100.00
	Butyl paraben	100 μM	98.00
	Nonylphenol	100 μM	68.00
	Bisphenol A	100 μM	60.00
Phanerochaete chrysosporium	Dimethyl phthalate	100 μM	65.00
	Methyl paraben	100 μM	69.00
	Butyl paraben	100 μM	58.00
	Nonylphenol	100 μM	69.00
	Bisphenol A	100 μM	66.00
Aspergillus brasiliensis	Progesterone	10 mg/L	No data
Absidia coerulea	Androst-4-ene-3,17-dione	1 g/L	29.00
Beauveria bassiana	Androst-4-ene-3,17-dione	1 g/L	65.00
	Androsta-1,4-diene-3,17-dione	1 g/L	30.00
Drechslera sp.	Androst-4-ene-3,17-dione	1 g/L	4.00
	Androsta-1,4-diene-3,17-dione	1 g/L	8.00
	Testosterone	1 g/L	100.00
Gibberella zeae	Androsta-1,4-diene-3,17-dione	1 g/L	7.00
Cunninghamella echinulata	Androst-4-ene-3,17-dione	1 g/L	6.00
Lentinula edodes	Testosterone	200 mg/L	No data
	17α-ethynylestradiol	0.8 mg/L	No data
Aspergillus nidulans	Progesterone	1 g/L	No data
Circinella muscae	Progesterone	7 g/L	No data
	Testosterone enanthate	7 g/L	No data

Source: Reproduced with permission from Wojcieszyńska et al. (2020).

10.6 APPLICATION OF IMMOBILIZED SYSTEMS FOR EDC REMOVAL FROM EFFLUENTS

Recent studies reporting efficient EDC removal using immobilized microorganisms/enzymes have been listed in Table 10.4. Majority of studies published in recent times have investigated the potential of immobilized fungal laccases for degradation of different EDCs. Enzymes are reportedly more efficient than whole cell

TABLE 10.3
Comparison of Non-immobilized and Immobilized Biocatalytic Processes.

Properties	Non-immobilized Biocatalytic Processes	Immobilized Biocatalytic Processes
Biomass growth (for cell biocatalysts)	Biomass reaches high concentrations in a short time that complicates the control of process	Biomass growth remains same along the process
Contamination	Risk of contamination by reaction mixture	Minimizes or eliminates product contamination
Cost of design	No additional cost is necessary	Additional cost for design of support material and technique
Downstream process	Difficult separation due to biocatalyst/substrate/product mixture	Facilitates separation from the production medium
Industrial application	Can be applied in various industrial production processes	New techniques and support materials need to be improved for application in different industries
Mass transfer and diffusion limitations	Biocatalyst can interact with environment with no limitation	Mass transfer is limited due to the support material
Movement (for cell biocatalysts)	Free movement—high mobility	Limited movement due to physical/chemical interaction with support material
Overall cost-effectiveness	Loss of valuable biocatalysts	Valuable biocatalysts can be reused
Productivity	Low productivity (kg product/kg enzyme)	High catalyst productivity (kg product/kg enzyme)
Recovery and reuse	Minimal or null reuse of biocatalyst	Efficient recovery and reuse
Stability	Low stability	Enhanced operational stability against different operational conditions (temperature, pH)

Source: Reproduced with permission from Zaman et al. (2021).

immobilization. However, application of enzymes is limited to its sensitivity to adverse environmental parameters. More studies are therefore required on process optimization and upscaling.

10.7 EDC REMOVAL USING MEMBRANE BIOREACTORS (MBRS)

EDCs may be removed by membrane-based separation, microbial degradation, air stripping, adsorption and biosorption, and/or photo-transformation. Non-polar pollutants can be removed via membrane retention and sorption of solids (Besha et al., 2017). Nevertheless, sorption has some limitations in case of removal of polar pollutants. Hence, microbial degradation has been considered the most appropriate process in this regard (Besha et al., 2017). Hydrophobic and hydrophilic pollutants are removed by biosorption and biodegradation, respectively. However, despite several

TABLE 10.4
Immobilized Systems for EDC Removal from Effluents

Organism/Enzyme	Support	EDC Treated	Percent (%) Removal	Reference
Bacteria-Mediated Removal				
Halomonas halodurans; *Bacillus halodurans*	Alginate beads	Phenol	83.1	Benit et al. (2022)
Sphingobium estronivorans	Cellulose acetate	E2 and EE2	94.6–100	Qin et al. (2023)
Fungi-Mediated Removal				
Phanerochaete sp.	Polyacrylonitrile nanofibrous membrane	BPA	>85	Conceição et al. (2023)
Trametes versicolor	Polyurethane foam	NP poly-ethoxylates	>90	Stenholm et al. (2020)
Phanerochaete chrysosporium	Ca-alginate	BPA	100	Wang et al. (2022)
Enzyme-Mediated Removal				
Horseradish peroxidase	Amino functionalized metal organic frameworks	Phenol; BPA; and 4-methoxy phenol	48; 51; and 62	Zeyadi and Almulaiky (2023)
Laccase	Zeolitic imidazolate frameworks	Dimethyl phthalate	90.1	Dlamini et al. (2023)
	Chitosan-polyacrylic acid microspheres	Naphthol green B; Indigo carmine	81 (NG); 72 (IC)	Leontieş et al. (2022)
	Porous silica beads	BPA	90	Maryšková et al. (2020)
	PEGA resin	BPA	80	Yamaguchi and Miyazaki (2021)
	Poly(methyl methacrylate)/Fe_3O_4 electrospun material	Tetracycline	94-100	Zdarta et al. (2020)
Lipase	Reduced graphene oxide/ nickel/platin nanowires	Tributyrin	100	Evli et al. (2023)

advantages, these processes are limited by the large quantities of intermediate compounds or secondary metabolites formed as a result of the same (Goswami et al., 2018). The MBR process efficiently addresses the separation of these compounds from the treated waters (Zaman et al., 2021).

In MBRs, several physical and biochemical processes are involved in EDC removal from contaminated waters. This process includes biodegradation of pollutants followed by subsequent separation of mixed liquor from treated effluents using membranes. An MBR offers several advantages including better quality and yield of treated effluents, and reduced formation of sludge. It is also highly flexible in terms of adaptation to a wide range of solid retention time (SRT) and/or HRT. The quality of an MBR is strongly guided by the hydrodynamic conditions of the sedimentation tanks and the characteristics of the sludge formed. The ultimate goal is to ensure maximum efficiency of the membrane. However, this process is often limited by the space required and cost incurred (Haq and Raj, 2019).

10.8 INTEGRATED MBR-BASED PROCESSES FOR EDC REMOVAL

In recent times, multiple processes have been simultaneously applied for achieving efficient treatment of effluents. These processes include active oxidation processes (AOPs), membrane-based separations and other different processes coupled with MBR systems. Integration of other methods with MBR processes enhances permeate quality, reduces fouling and improves pollutant removal efficiency of the latter (Zaman et al. 2021). Different integrated MBR-based processes have been shown in Figure 10.6 and discussed as follows.

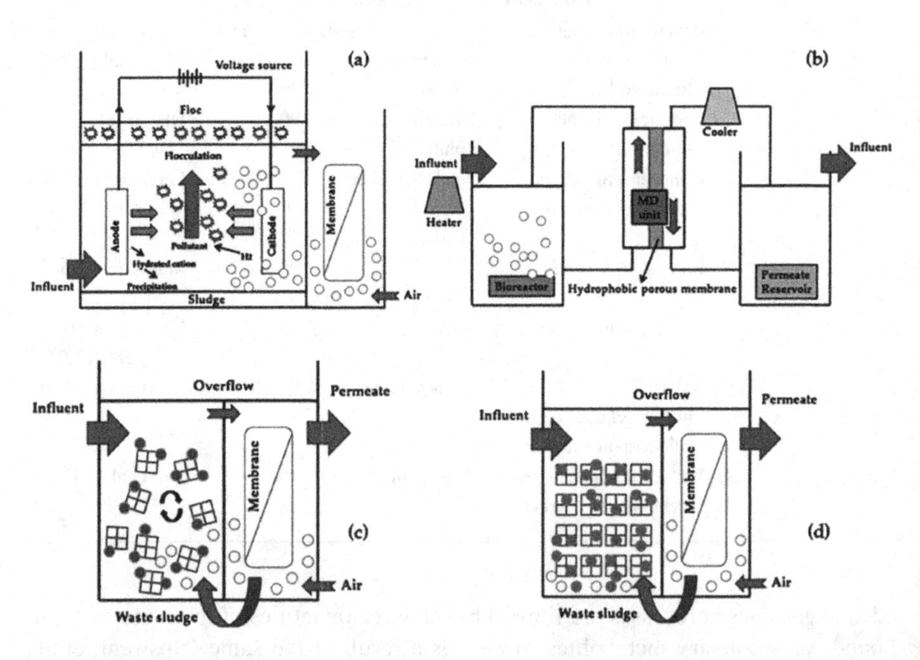

FIGURE 10.6 Schematic diagram of some hybrid MBR systems: (a) electrocoagulation-augmented external side stream membrane bioreactor system, (b) membrane distillation bioreactor (MDBR) with an external side stream membrane, (c) biofilm membrane bioreactor (BF-MBR) and (d) bio-entrapped membrane bioreactor (BE-MBR). (Reproduced with permission from Zaman et al. 2021.)

10.8.1 AOP-MBRs

According to previous studies, AOPs have efficiently removed micropollutants from effluents and converted the same to biodegradable non-toxic compounds. However, efficiency of this process is limited by the presence of suspended solids. An MBR boosts the efficiency of these AOPs by removing the suspended solids of concern and accelerating the process of effluent treatment. Moreover, AOP enhances MBR efficiency by reducing membrane fouling. In an integrated process, AOPs render the recalcitrant pollutants biodegradable while MBRs help convert the same to non-toxic end products. Electrocoagulation processes integrated with an MBR has successfully eliminated different organic and inorganic pollutants from effluents. However, recent studies are attempting to reduce the volume of sludge formed as a by-product of this integrated process (Zaman et al., 2021).

10.8.2 Reverse Osmosis (RO) and Forward Osmosis (FO) Coupled MBR Systems

In a traditional MBR, effluents are placed in a vessel referred to as a bioreactor. Longer SRTs recorded in this setup enhance the chance of membrane fouling. Use of RO and FO membranes in osmotic membrane-coupled bioreactors (OMBRs) for separation of mixed liquor from treated effluents significantly increases pollutant removal efficiency and decreases energy requirement and membrane fouling. However, these processes are limited by low permeate yield and salt accumulation (Zaman et al., 2021). Nevertheless, Ding et al. (2020) suggested that FO membrane-coupled anaerobic membrane bioreactors (AnMBRs) efficiently removed phosphorus and nitrogen from effluents. In a previous study, FO and RO membranes were simultaneously used in a single setup for pumping effluent and filtering as well as recycling the treated water respectively (Goswami et al., 2018). RO membranes have lower cost in comparison to FO membranes. However, the application of RO membrane is limited by osmotic gradients and/or high power consumption. Hence, it is better to use these membranes in an integrated setup, for overcoming limitations posed by each type of membrane when used individually (Zaman et al., 2021).

10.8.3 Granular MBRs

Aerobic granular sludge (AGS)-based processes have attracted significant attention in recent investigations, owing to their efficiency in removal of micropollutants from effluents (Tomar and Chakraborty, 2018a,b). This sludge is reportedly capable of simultaneous execution of nitrification and denitrification within the treatment vessel (Faria et al., 2020). As reported previously by Goswami et al. (2018), AGS has been successfully used in an integrated setup consisting of a membrane airlift bioreactor (MABR) and a sequencing batch airlift reactor (SBAR). While the SBAR promoted nitrification and denitrification process in the presence of high and low aeration respectively, the MABR operated under both aerobic and anoxic conditions. AGS-based processes have reportedly demonstrated reduced sludge formation as well as enhanced removal of PPCPs and efficient removal of ammonia, chemical oxygen demand (COD), nitrogen, and total phosphorous from effluents (Zaman et al., 2021).

10.8.4 Membrane Distillation Bioreactor

In an MDBR, a membrane-based distillation setup is integrated with a thermophilic bioprocess (Zaman et al., 2021). In this process, microporous membrane water vapor resulting in better quality permeates. MDBRs are more efficient than MBRs in terms of treatment of municipal effluents. The quantity of sludge produced by an MDBR is also lower than that by an MBR. Moreover, both membrane fouling and cost incurred are significantly low for an MDBR. A high-capacity MDBR has been shown to reduce greenhouse gas emissions, salt concentrations, total organic carbon as well as eliminate organic pollutants from effluents (Besha et al., 2017).

10.8.5 Biofilm/Bio-Entrapped MBR

In this process, insertion of a bio-support in the MBRs facilitates minimization of suspended solid concentration and membrane fouling. This integrated process greatly enhances the segregation of ammonia and nitrogen (Goswami et al., 2018). This process promotes microbial activity and efficiently removes persistent pollutants with minimum membrane fouling (Goswami et al., 2018; Zaman et al., 2021).

A comparative analysis of EDC removal using different integrated approaches has been shown in Table 10.5.

10.9 COST-EFFECTIVENESS OF EFFLUENT TREATMENT BY MBR-BASED PROCESSES

Microorganisms have been used for decades for effluent treatment. Toward its inception MBRs were not widely considered owing to high costs, improper operational skills and high requirement of trained personnel. With time, this process has evolved significantly to be more user friendly and more cost effective. In recent times, around 2200 MBRs are in use or being installed all across the world (Banerjee et al. 2018). According to recent studies, MBR-based effluent treatments are much more efficient and cost effective in comparison to other contemporary processes for the same. The optimum cost for treatment of effluent was found to be the lowest for bacterial treatment (US$ 1.82 m^{-3} of effluent sans operational and labor costs) in comparison to combined O_3 and H_2O_2 treatments (US$ 5.02 m^{-3}), ozonation (US$ 4.94 m^{-3}) and Fenton's processes (US$ 3.50 m^{-3}). Membrane filtration of the treated effluent may cost an additional US$ 0.478–0.592 m^{-3} for 1 million gallon of effluent per day (Banerjee et al., 2018). Cost of MBR-based treatments often increases due to energy consumed for aeration (required to reduce membrane fouling) (Nie et al. 2017). Biogas and/or methane recovered from an AnMBR can help meet this energy requirement (Goswami et al., 2018; Zaman et al., 2021). Due to automation of these processes, cost incurred for labor has also reduced significantly in recent times.

TABLE 10.5
Comparative Account of Pollutant Removal Efficiency (%) of Different Integrated MBR Systems

Type of Compound	Compound	Pollutant Removal (%)				
		MBR	MBR–NF	MBR–RO	MBR-Powdered Activated Carbon (PAC)	MBR-Granular Activated Carbon (GAC)
Analgesics non-steroidal anti-inflammatory (NSAIDs) and anti-pyretics	Acetaminophen	95.1–99.9	91.0–99.9	99.6–99.9	–	–
	Diclofenac	15.0–87.4	87.5–97.0	88.3–95.9	>98.0	–
	Ibuprofen	73.0–99.8	99.4–99.8	99.4–99.8	–	–
	Ketoprofen	3.7–91.9	–	–	–	>98.0
	Naproxen	40.1–99.3	7.8	–	87.3	>98.0
Antibiotics	Erythromycin	25.2–90.4	–	>99.0	>88.0	–
	Sulfamethoxazole	20.0–91.9	90	>99.0	82	–
Anti-depressant and anti-epileptics	17α-ethynylestradiol	0–93.5	>71.0	>71.0	86.7	–
	17β-estradiol	>99.4	>71.0	–	92.4	–
	Bisphenol A	88.2–97.0	95	96	–	–
	Carbamazepine	42–51.0	81.0–93.0	84.8–99.0	80.0–99.0	>98.0
	Diazepam	67	–	>99	–	–
	Estrone	76.9–99.4	>76.0	>76.0	–	–
Beta-blockers	Atenolol	5–96.9	8.5	>99.0	–	–
	Metoprolol	29.5–58.7	71.2	>99.0	>99.0	–
Lipid regulator and cholesterol lowering drugs	Bezafibrate	88.2–95.8	–	–	–	–
	Clofibric acid	25.0–71.0	–	–	–	–
	Gemfibrozil	32.5–85	–	–	–	–

Source: Reproduced from Besha et al. (2017).

The current cost of membrane replacement accounts for 10–14% of the total operational cost (Besha et al. 2017). Regular cleaning of membranes, better aeration regimes and intermittent air cycling or bubbling are expected to reduce costs even further (Goswami et al. 2018).

10.10 CONCLUSION

EDCs exert detrimental impacts on the health and functioning of ecosystems exposed to the same. This review envisages microbial processes for efficient removal of EDCs from contaminated effluents. MBRs and MBR-based integrated processes reported in recent studies have efficiently removed EDCs from effluents. However, wide-scale application of these processes are limited by space and energy requirements. So far bacterial processes, especially those involving immobilized bacteria, have been found to be most effective for EDC removal from effluents. Enzymes have been found to be more effective. However, application of the same is limited by growing resistance of the specific enzymes for the specific EDCs. So, a thorough review of recent literature highlighting the need for more evolved and cost-effective technologies that could utilize the potential of microorganisms for efficient detoxification of EDCs is needed. Developing the same is expected to be more efficient and sustainable for the environment.

REFERENCES

Aissani T, Yahiaoui I, Boudrahem F, Ait Chikh S, Aissani-Benissad F, Amrane A. The combination of photocatalysis process (UV/TiO2 (P25) and UV/ZnO) with activated sludge culture for the degradation of sulfamethazine. *Separation Science and Technology*. 2018;53(9):1423–33.

Al Aani S, Mustafa TN, Hilal N. Ultrafiltration membranes for wastewater and water process engineering: a comprehensive statistical review over the past decade. *Journal of Water Process Engineering*. 2020;35:101241.

Al Sharabati M, Abokwiek R, Al-Othman A, Tawalbeh M, Karaman C, Orooji Y, Karimi F. Biodegradable polymers and their nano-composites for the removal of endocrine-disrupting chemicals (EDCs) from wastewater: a review. *Environmental Research*. 2021;202:111694.

Ali I, Alothman ZA, Alwarthan A. Supra molecular mechanism of the removal of 17-β-estradiol endocrine disturbing pollutant from water on functionalized iron nano particles. *Journal of Molecular Liquids*. 2017;241:123–129.

Al-Shaalan NH, Ali I, ALOthman ZA, Al-Wahaibi LH, Alabdulmonem H. High performance removal and simulation studies of diuron pesticide in water on MWCNTs. *Journal of Molecular Liquids*. 2019;289:111039.

Ayuba S, Mohammadib AA, Yousefic M, Changanic F. Performance evaluation of agro-based adsorbents for the removal of cadmium from wastewater. *Desalination and Water Treatment*. 2019;142:293–299.

Banerjee P, Barman SR, Swarnakar S, Mukhopadhyay A, Das P. Treatment of textile effluent using bacteria-immobilized graphene oxide nanocomposites: evaluation of effluent detoxification using *Bellamya bengalensis*. *Clean Technologies and Environmental Policy*. 2018;20(10):2287–2298.

Banerjee P, Dey TK, Sarkar S, Swarnakar S, Mukhopadhyay A, Ghosh S. Treatment of cosmetic effluent in different configurations of ceramic UF membrane based bioreactor: toxicity evaluation of the untreated and treated wastewater using catfish (*Heteropneustes fossilis*). *Chemosphere*. 2016;146:133–144.

Bartolomeu M, Neves MG, Faustino MA, Almeida A. Wastewater chemical contaminants: remediation by advanced oxidation processes. *Photochemical & Photobiological Sciences*. 2018;17(11):1573–1598.

Basheer AA. New generation nano-adsorbents for the removal of emerging contaminants in water. *Journal of Molecular Liquids*. 2018;261:583–593.

Benit N, Lourthuraj AA, Barathikannan K, Mostafa AA, Alodaini HA, Yassin MT, Hatamleh AA. Immobilization of *Halomonas halodurans* and *Bacillus halodurans* in packed bed bioreactor for continuous removal of phenolic impurities in waste water. *Environmental Research*. 2022;209:112822.

Besha AT, Gebreyohannes AY, Tufa RA, Bekele DN, Curcio E, Giorno L. Removal of emerging micropollutants by activated sludge process and membrane bioreactors and the effects of micropollutants on membrane fouling: a review. *Journal of Environmental Chemical Engineering*. 2017;5(3):2395–2414.

Bilal M, Jing Z, Zhao Y, Iqbal HM. Immobilization of fungal laccase on glutaraldehyde cross-linked chitosan beads and its bio-catalytic potential to degrade bisphenol A. *Biocatalysis and Agricultural Biotechnology*. 2019;19:101174.

Budeli P, Ekwanzala MD, Unuofin JO, Momba MN. Endocrine disruptive estrogens in wastewater: revisiting bacterial degradation and zymoremediation. *Environmental Technology & Innovation*. 2021;21:101248.

Carstens L, Cowan AR, Seiwert B, Schlosser D. Biotransformation of phthalate plasticizers and bisphenol A by marine-derived, freshwater, and terrestrial fungi. *Frontiers in Microbiology*. 2020;11:317.

Conceição JC, Alvarega AD, Mercante LA, Correa DS, Silva EO. 2023. Endophytic fungus from Handroanthus impetiginosus immobilized on electrospun nanofibrous membrane for bioremoval of bisphenol A.. *World Journal of Microbiology and Biotechnology*, 39(10), 261. https://doi.org/10.1007/s11274-023-03715-z

Ding Y, Guo Z, Hou X, Mei J, Liang Z, Li Z, Zhang C, Jin C. Performance analysis for the anaerobic membrane bioreactor combined with the forward osmosis membrane bioreactor: process conditions optimization, wastewater treatment and sludge characteristics. *Water*. 2020;12(11):2958.

Dlamini ML, Lesaoana M, Kotze I, Richards H. An oxidoreductase enzyme, fungal laccase immobilized on zeolitic imidazolate frameworks for the biocatalytic degradation of an endocrine-disrupting chemical, dimethyl phthalate. *Journal of Environmental Chemical Engineering*. 2023;11(3):109810.

Dzionek A, Wojcieszyńska D, Hupert-Kocurek K, Adamczyk-Habrajska M, Guzik U. Immobilization of Planococcus sp. S5 strain on the loofah sponge and its application in naproxen removal. *Catalysts*. 2018;8(5):176.

Egger G, Binns A, Rössner S. Health and the environment: clinical implications for lifestyle medicine. In: *Lifestyle Medicine* 2017 (pp. 309–315). Academic Press Elsevier B.V. https://doi.org/10.1016/B978-0-12-810401-9.00019-X

Evli S, Öndeş B, Uygun M, Uygun DA. Lipase loaded motion-based multisegmental nanowires for pollutant tributyrin degradation. *International Journal of Environmental Science and Technology*. 2023;20(5):5509–5518.

Faria CV, Ricci BC, Silva AF, Amaral MC, Fonseca FV. Removal of micropollutants in domestic wastewater by expanded granular sludge bed membrane bioreactor. *Process Safety and Environmental Protection*. 2020;136:223–233.

Fernando EY, Keshavarz T, Kyazze G. The use of bioelectrochemical systems in environmental remediation of xenobiotics: a review. *Journal of Chemical Technology & Biotechnology*. 2019;94(7):2070–2080.

Goswami L, Manikandan NA, Dolman B, Pakshirajan K, Pugazhenthi G. Biological treatment of wastewater containing a mixture of polycyclic aromatic hydrocarbons using the oleaginous bacterium Rhodococcus opacus. *Journal of Cleaner Production*. 2018;196:1282–1291.

Grelska A, Noszczyńska M. White rot fungi can be a promising tool for removal of bisphenol A, bisphenol S, and nonylphenol from wastewater. *Environmental Science and Pollution Research*. 2020;27(32):39958–39976.

Haq I, Raj A. Endocrine-disrupting pollutants in industrial wastewater and their degradation and detoxification approaches. In: *Emerging and Eco-Friendly Approaches for Waste Management* 2019 (pp. 121–142). Springer, Singapore.

Khan NA, Khan SU, Ahmed S, Farooqi IH, Yousefi M, Mohammadi AA, Changani F. Recent trends in disposal and treatment technologies of emerging-pollutants-A critical review. *TrAC Trends in Analytical Chemistry*. 2020;122:115744.

Kumari V, Yadav A, Haq I, Kumar S, Bharagava RN, Singh SK, Raj A. Genotoxicity evaluation of tannery effluent treated with newly isolated hexavalent chromium reducing Bacillus cereus. *Journal of Environmental Management*. 2016;183:204–11.

Leontieş AR, Răducan A, Culiţă DC, Alexandrescu E, Moroşan A, Mihaiescu DE, Aricov L. Laccase immobilized on chitosan-polyacrylic acid microspheres as highly efficient biocatalyst for naphthol green B and indigo carmine degradation. *Chemical Engineering Journal*. 2022;439:135654.

Liu J, Li S, Li X, Gao Y, Ling W. Removal of estrone, 17 β-estradiol, and estriol from sewage and cow dung by immobilized Novosphingobium sp. *ARI-1. Environmental Technology*. 2018;39(19):2423–2433.

Maryšková M, Schaabova M, Tomankova H, Novotný V, Rysova M. Wastewater treatment by novel polyamide/polyethylenimine nanofibers with immobilized laccase. *Water*. 2020;12(2):588.

Miralles-Cuevas S, Oller I, Agüera A, Llorca M, Pérez JS, Malato S. Combination of nanofiltration and ozonation for the remediation of real municipal wastewater effluents: acute and chronic toxicity assessment. *Journal of Hazardous Materials*. 2017;323:442–451.

Mohammadi AA, Dehghani MH, Mesdaghinia A, Yaghmaian K, Es' haghi Z. Adsorptive removal of endocrine disrupting compounds from aqueous solutions using magnetic multi-wall carbon nanotubes modified with chitosan biopolymer based on response surface methodology: functionalization, kinetics, and isotherms studies. *International Journal of Biological Macromolecules*. 2020;155:1019–1029.

Nie Y, Kato H, Sugo T, Hojo T, Tian X, Li YY. Effect of anionic surfactant inhibition on sewage treatment by a submerged anaerobic membrane bioreactor: efficiency, sludge activity and methane recovery. *Chemical Engineering Journal*. 2017;315:83–91.

Olajuyigbe FM, Adetuyi OY, Fatokun CO. Characterization of free and immobilized laccase from cyberlindnera fabianii and application in degradation of bisphenol A. *International Journal of Biological Macromolecules*. 2019;125:856–864.

Olasupo A, Suah FB. Recent advances in the removal of pharmaceuticals and endocrine-disrupting compounds in the aquatic system: a case of polymer inclusion membranes. *Journal of hazardous materials*. 2021;406:124317.

Pezzella C, Macellaro G, Sannia G, Raganati F, Olivieri G, Marzocchella A, Schlosser D, Piscitelli A. Exploitation of Trametes versicolor for bioremediation of endocrine disrupting chemicals in bioreactors. *PloS One*. 2017;12(6):e0178758.

Polloni-Silva J, Valdehita A, Fracácio R, Navas JM. Remediation efficiency of three treatments on water polluted with endocrine disruptors: assessment by means of in vitro techniques. *Chemosphere*. 2017;173:267–274.

Prakash C, Chaturvedi V. Degradation of 17 α-ethynyl estradiol by *Pseudomonas citronelolis* BHUWW1 and its degradation pathway. *Biocatalysis and Agricultural Biotechnology*. 2023;54:102937.

Qin D, Kiki C, Ma C, Sun Q, Yu CP. Coupling UV-radiation with immobilized bacteria for the removal of 17β-estradiol and 17 α-ethynylestradiol. *Journal of Water Process Engineering*. 2023;56:104451.

Roccuzzo S, Beckerman AP, Trögl J. New perspectives on the bioremediation of endocrine disrupting compounds from wastewater using algae-, bacteria-and fungi-based technologies. *International Journal of Environmental Science and Technology*. 2021;18(1):89–106.

Sathe SS, Mahanta C, Mishra P. Simultaneous influence of indigenous microorganism along with abiotic factors controlling arsenic mobilization in Brahmaputra floodplain, India. *Journal of Contaminant Hydrology*. 2018;213:1–4.

Schug TT, Johnson AF, Birnbaum LS, Colborn T, Guillette Jr LJ, Crews DP, Collins T, Soto AM, Vom Saal FS, McLachlan JA, Sonnenschein C. Minireview: endocrine disruptors: past lessons and future directions. *Molecular Endocrinology*. 2016;30(8):833–47.

Sewwandi M, Wijesekara H, Soysa S, Gunarathne V, Rajapaksha AU, Vithanage M. Biological treatment of endocrine-disrupting chemicals (EDCs). In: *Biotechnology for Environmental Protection* 2022 (pp. 165–191). Singapore: Springer Nature Singapore.

Singh P, Borthakur A. A review on biodegradation and photocatalytic degradation of organic pollutants: a bibliometric and comparative analysis. *Journal of Cleaner Production*. 2018;196:1669–1680.

Stenholm Å, Hedeland M, Arvidsson T, Pettersson CE. Removal of nonylphenol polyethoxylates by adsorption on polyurethane foam and biodegradation using immobilized *Trametes versicolor*. *Science of The Total Environment*. 2020;724:138159.

Tomar SK, Chakraborty S. Characteristics of aerobic granules treating phenol and ammonium at different cycle time and up flow liquid velocity. *International Biodeterioration & Biodegradation*. 2018b;127:113–123.

Tomar SK, Chakraborty S. Effect of air flow rate on development of aerobic granules, biomass activity and nitrification efficiency for treating phenol, thiocyanate and ammonium. *Journal of Environmental Management*. 2018a;219:178–188.

Trivedi J, Chhaya U. Bioremediation of bisphenol A found in industrial wastewater using *Trametes versicolor* (TV) laccase nanoemulsion-based bead organogel in packed bed reactor. *Water Environment Research*. 2022;94(10):e10786.

van der Meer TP, Chung MK, Van Faassen M, Makris KC, Van Beek AP, Kema IP, Wolffenbuttel BH, van Vliet-Ostaptchouk JV, Patel CJ. Temporal exposure and consistency of endocrine disrupting chemicals in a longitudinal study of individuals with impaired fasting glucose. *Environmental Research*. 2021;197:110901.

Vieira WT, de Farias MB, Spaolonzi MP, da Silva MG, Vieira MG. Removal of endocrine disruptors in waters by adsorption, membrane filtration and biodegradation. A review. *Environmental Chemistry Letters*. 2020;18(4):1113–43.

Wang J, Xie Y, Hou J, Zhou X, Chen J, Yao C, Zhang Y, Li Y. Biodegradation of bisphenol A by alginate immobilized *Phanerochaete chrysosporium* beads: continuous cyclic treatment and degradation pathway analysis. *Biochemical Engineering Journal*. 2022;177:108212.

Wojcieszyńska D, Marchlewicz A, Guzik U. Suitability of immobilized systems for microbiological degradation of endocrine disrupting compounds. *Molecules*. 2020;25(19):4473.

Yamaguchi H, Miyazaki M. Laccase aggregates via poly-lysine-supported immobilization onto PEGA resin, with efficient activity and high operational stability and can be used to degrade endocrine-disrupting chemicals. *Catalysis Science & Technology*. 2021;11(3):934–42.

Zaman A, Banerjee P, Mukhopadhyay A, Das P, Chattopadhyay D. Membrane bioreactors for separation of persistent organic pollutants from industrial effluents. In: *Biological Treatment of Industrial Wastewater* 2021 (pp. 257–293). Royal Society of Chemistry. https://doi.org/10.1039/9781839165399-00257

Zdarta J, Jankowska K, Bachosz K, Kijeńska-Gawrońska E, Zgoła-Grześkowiak A, Kaczorek E, Jesionowski T. A promising laccase immobilization using electrospun materials for biocatalytic degradation of tetracycline: effect of process conditions and catalytic pathways. *Catalysis Today*. 2020;348:127–36.

Zeyadi M, Almulaiky YQ. Amino functionalized metal-organic framework as eco-friendly support for enhancing stability and reusability of horseradish peroxidase for phenol removal. *Biomass Conversion and Biorefinery*. 2023:1–3.

Zhang Q, Xue C, Owens G, Chen Z. Isolation and identification of 17β-estradiol degrading bacteria and its degradation pathway. *Journal of Hazardous Materials*. 2022;423:127185.

Zhuo R, Fan F. A comprehensive insight into the application of white rot fungi and their lignocellulolytic enzymes in the removal of organic pollutants. *Science of the Total Environment*. 2021;778:146132.

Zühlke MK, Schlüter R, Mikolasch A, Henning AK, Giersberg M, Lalk M, Kunze G, Schweder T, Urich T, Schauer F. Biotransformation of bisphenol A analogues by the biphenyl-degrading bacterium *Cupriavidus basilensis* - a structure-biotransformation relationship. *Applied Microbiology and Biotechnology*. 2020;104:3569–3583.

Index

Note: **Bold** page numbers refer to tables.

Printed in the United States
by Baker & Taylor Publisher Services